COGNITIVE NETWORKS

COGNITIVE NETWORKS

Towards Self-Aware Networks

Edited by

Qusay H. Mahmoud

University of Guelph, Canada

John Wiley & Sons, Ltd

Telephone (+44) 1243 779777

Email (for orders and customer service enquiries): cs-books@wiley.co.uk
Visit our Home Page on www.wileyeurope.com or www.wiley.com

Other Wiley Editorial Offices

John Wiley & Sons Inc., 111 River Street, Hoboken, NJ 07030, USA

Jossey-Bass, 989 Market Street, San Francisco, CA 94103-1741, USA

Wiley-VCH Verlag GmbH, Boschstr. 12, D-69469 Weinheim, Germany

John Wiley & Sons Australia Ltd, 42 McDougall Street, Milton, Queensland 4064, Australia

John Wiley & Sons (Asia) Pte Ltd, 2 Clementi Loop #02-01, Jin Xing Distripark, Singapore 129809

John Wiley & Sons Canada Ltd, 6045 Freemont Blvd, Mississauga, Ontario, L5R 4J3, Canada

Wiley also publishes its books in a variety of electronic formats. Some content that appears in print may not be available in electronic books.

Anniversary Logo Design: Richard J. Pacifico

Library of Congress Cataloging-in-Publication Data:

Cognitive networks : towards self-aware networks / edited by Qusay H. Mahmoud.
p. cm.
ISBN 978-0-470-06196-1 (cloth)
1. Software radio. 2. Wireless communication systems. 3. Autonomic computing.
I. Mahmoud, Qusay H., 1971-
TK5103.4875.C62 2007
621.384 – dc22

2007011302

British Library Cataloguing in Publication Data

A catalogue record for this book is available from the British Library

ISBN 978-0-470-06196-1 (HB)

Typeset in 10/12pt Times by Laserwords Private Limited, Chennai, India
Printed and bound in Great Britain by Antony Rowe Ltd, Chippenham, Wiltshire
This book is printed on acid-free paper responsibly manufactured from sustainable forestry in which at least two trees are planted for each one used for paper production.

To the authors of the individual chapters, without whom this book would not exist. And to readers of this book, who will be the future innovators in the field

Contents

Contributors

Raouf Boutaba
David R. Cheriton School of Computer Science
University of Waterloo
200 University Ave. W., Waterloo, Ontario, Canada N2L 3G1
rboutaba@cs.uwaterloo.ca

R. Chandramouli
Department of Electrical and Computer Engineering
Stevens Institute of Technology
Hoboken, NJ 07030, USA
mouli@stevens.edu

Luiz A. DaSilva
Virginia Tech
4300 Wilson Blvd. Suite 750
Arlington, VA 22203, USA
ldasilva@vt.edu

Hervé Debar
France Télécom R&D
42 rue des Coutures
BP 6243
F-14066 Caen Cedex 4, France
herve.debar@orange-ftgroup.com

Thomas G. Dietterich
School of Electrical Engineering and Computer Science
1148 Kelley Engineering Center
Oregon State University
Corvallis, OR 97331-5501 USA
tgd@eecs.oregonstate.edu

Linda Doyle
CTVR, Trinity College
University of Dublin, Ireland
ledoyle@tcd.ie

Royta Egashira
Information and Computer Science
University of California, Irvine
Irvine, CA 92697-3435, USA
egashira@ics.uci.edu

Tim K. Forde
CTVR, Trinity College
University of Dublin, Ireland
timforde@mee.tcd.ie

Daniel H. Friend
Department of Electrical and Computer Engineering (mail code 0111)
Virginia Tech
302 Whittemore Hall
Blacksburg, VA 24061-0111 USA
dhfriend@vt.edu

Keita Fujii
Information and Computer Science
University of California, Irvine
Irvine, CA 92697-3435, USA
kfujii@ics.uci.edu

Allen Ginsberg
The MITRE Corporation
7515 Colshire Dr.
McLean, VA 22102-7142, USA
aginsberg@mitre.org

William D. Horne
The MITRE Corporation
7515 Colshire Dr.
McLean, VA 22102-7142, USA
whorne@mitre.org

Harikeshwar Kushwaha
Department of Electrical and Computer Engineering
Stevens Institute of Technology
Hoboken, NJ 07030, USA
harikeshwar@gmail.com

Pat Langley
Institute for the Study of Learning and Expertise
2164 Staunton Court
Palo Alto, CA 94306, USA
langley@isle.org

Kenji Leibnitz
Graduate School of Information Science and Technology
Osaka University
1–5 Yamadaoka, Suita, Osaka 565–0871, Japan
leibnitz@ist.osaka-u.ac.jp

Jun Lu
Information and Computer Science
University of California, Irvine
Irvine, CA 92697-3435, USA
lujun@ics.uci.edu

Allen B. MacKenzie
Department of Electrical and Computer Engineering
Virginia Tech
302 Whittemore Hall
Blacksburg, VA 24061-0111, USA
mackenab@vt.edu

Qusay H. Mahmoud
Department of Computing and Information Science
University of Guelph
Guelph, Ontario, N1G 2 W1 Canada
qmahmoud@cis.uoguelph.ca

Chetan Mathur
Department of Electrical and Computer Engineering
Stevens Institute of Technology
Hoboken, NJ, 07030, USA
cnanjund@gmail.com

Joseph Mitola III
The MITRE Corporation
Tampa, Fl, USA
jmitola@mitre.org

Mehul Motani
Department of Electrical and Computer Engineering
National University of Singapore
10 Kent Ridge Crescent
Singapore 119260
motani@nus.edu.sg

Masayuki Murata
Graduate School of Information Science and Technology
Osaka University
1–5 Yamadaoka, Suita, Osaka 565–0871, Japan
murata@ist.osaka-u.ac.jp

Yi Pan
Information and Computer Science
University of California, Irvine
Irvine, CA 92697-3435, USA
ypan@ics.uci.edu

Jeffrey D. Poston
The MITRE Corporation
7515 Colshire Dr.
McLean, VA 22102-7142, USA
jdposton@mitre.org

John Strassner
Motorola Labs
1301 East Algonquin Road
Mail Stop IL02-2240
Schaumburg, IL 60010, USA
john.strassner@motorola.com

Vineet Srivastava
HelloSoft, Inc.
8-2-703, Road No.12,
Banjara Hills,
Hyderabad - 500 034,
Andhra Pradesh, India
vineet.personal@gmail.com

K.P. Subbalakshmi
Department of Electrical and Computer Engineering
Stevens Institute of Technology
Hoboken, NJ, 07030, USA
ksubbala@stevens.edu

Tatsuya Suda
Information and Computer Science
University of California, Irvine
Irvine, CA 92697-3435, USA
suda@ics.uci.edu

Ryan W. Thomas
Department of Electrical and Computer Engineering
Virginia Tech
302 Whittemore Hall
Blacksburg, VA 24061-0111, USA
rwthomas@vt.edu

Naoki Wakamiya
Graduate School of Information Science and Technology
Osaka University
1–5 Yamadaoka, Suita, Osaka 565–0871, Japan
wakamiya@ist.osaka-u.ac.jp

Jin Xiao
David R. Cheriton School of Computer Science
University of Waterloo
200 University Ave. W., Waterloo, Ontario, Canada N2L 3G1
j2xiao@cs.uwaterloo.ca

Yiping Xing
Department of Electrical and Computer Engineering
Stevens Institute of Technology
Hoboken, NJ 07030, USA
yipingxing@gmail.com

Ariffin Yahaya
Information and Computer Science
University of California, Irvine
Irvine, CA 92697-3435, USA
ariffin@ics.uci.edu

Foreword 1

Professor Qusay H. Mahmoud ('Q') has contributed a landmark to the evolution of cognitive radio technologies with the publication of his text, *Cognitive Networks*. Specifically, this text clearly marks the beginning of the transition of cognitive radio from enabling technologies in relatively abstract, academic and pre-competitive settings to more clearly address what might be called the critical business-case enablers for broad market relevance. From its introduction at KTH, The Royal Institute of Technology, Stockholm, in 1998, cognitive radio has been about market relevance, overcoming the superficial lack of radio spectrum by pooling, prioritization and space–time adaptation to not just the radio frequency (RF) situation, but also to the needs of the users in the scene. Textbooks to date, however, have contributed more significantly to the foundations of the technology or to relatively small market segments such as secondary spectrum or ad hoc networks than to the broader Internet-scale wireless markets of increasingly heterogeneous cellular services and commodity consumer electronics. Communications networks for these broader real-world markets typically must be (or become) secure, scalable, reliable, controllable, interoperable and, of course, billable. It is difficult to state the importance of Q's book in a few words. Although a few chapters (like my own) are more foundational than transitional, most of the chapters move cognitive networks solidly along a transition from theory to practice.

Let me offer a few highlights that qualify as paradigm shifts. First, Hervé Debar's (France Télécom) chapter (Chapter 12) on intrusion detection in cognitive networks draws on the IETF's intrusion detection working group to characterize the problem of knowledge attack and to structure management and policy plans with operations that define a paradigm shift in network security. Since cognitive networks are inherently flexible in the extreme, it would be difficult or impossible for cognitive wireless networks (CWNs) to transition from theory and niche applications to broadly applicable practice without a solid handle on security. This chapter addresses in a systematic way, the intrusion detection aspect of information security framed comprehensively by Mathur and Subbalakshmi in their companion chapter on security issues, which characterizes security issues from availability and access to privacy and non-repudiation, along with classes of attack that would be problematic for CWNs in the literature. Although Debar's chapter addresses only intrusion detection in detail, this pair of chapters set a high standard indeed for the articulation of a critical transition issue – security – and the related contribution that shows the way towards transition.

In another important chapter (Chapter 2), Motorola calls the foundational layers of reconfigurable wireless networks the 'autonomic' layers, adapting IBM's term, an allusion to the autonomic nervous system of mammals that controls involuntary actions such as heart beat and respiration. Although I prefer my own original cognition cycle – observe, orient, plan, decide, act and learn – to the simpler Motorola–IBM cycle – monitor, analyze, plan and execute (with knowledge in the middle, but without learning) – their FOCALE architecture that realizes the cycle in cognitive networks ties learning to business goals. That's a big improvement on my own OOPDAL loop, which wasn't explicitly tied to the inevitable business logic from which the revenue comes. This chapter, like several others, also introduces 'layers' and related objects and relationships for services, management and the control of the foundational reconfigurable networks. I found the differences in device interface language for controlling different kinds of routers compelling evidence for what I call computational semantics interoperability: they proposed an ontology-mapping construct as essential for the transition towards practical cognitive networks.

The MITRE chapter by Ginsberg, Horne and Poston (Chapter 10) develops this notion of semantic interoperability among heterogeneous networks further, drawing on the web ontology language (OWL) from the semantic web community, in some sense a technology looking for a problem that may in fact have found a home in the creation of machine-readable specifications. When I teach radio engineering courses in the US, almost no radio engineers have used Z.100, the International Telecommunications Union's standard specification and description language (SDL) for digital systems, particularly for the state machines and message sequence charts (MSC) of radio. Not so in Europe where Z.100 was invented so that engineers from across the EC could collaborate in the mathematical language of SDL. For me, it was a watershed event when a few years ago, the GSM MoU committee determined that the machine-readable SDL would be normative while the human-readable text would be explanatory, reversing decades-long practice that the natural language text of a specification be normative while the figures and computer-readable code in the specification be illustrative.

In this context, Boutaba and Xiao's chapter (Chapter 4) on self-managing networks shows how the technical ideas of semantic interoperability have substantial cost leverage: 80% of information technology is expended on operations and maintenance, with nearly half of service outages caused by human error. Although the telecommunications industry continues to automate, it is on the edges of heterogeneous networks – the wired and the wireless, the multi-standard to the core network – that the contribution of cognitive networks becomes most evident. Lu *et al.* survey a broad scope of ideas, approaches and architectures for self-management in Chapter 3 on adaptive networks. One of the more intriguing is the ant colony idea developed in greater detail by Liebnitz *et al.* in Chapter 1 on biologically inspired networking. Behavior, after all, is crucial. Their first figure shows how the price of scaling up to Internet-sized networks is a loss of determinism because the overhead of centralized control becomes prohibitive as network size increases into the millions and billions of nodes. Reminiscent of foundational work in artificial life by Stuart Kaufman (*At Home in the Universe* and the Artificial Life proceedings of the Santa Fe Institute), Leibnitz *et al.* remind us that biologically inspired architectures may fall short of optimal in some respects, but also may be more robust in dealing with catastrophe. Their mathematical treatment and simulations each relate back to important properties of cognitive networks that enable transition to Internet scale.

Not to condemn with faint praise, all of the chapters, including those not yet mentioned in this brief note, each contribute importantly to critical issues in cross-layer optimization, coding, distributed learning and overall cognitive network robustness. I'm sure the growing community of cognitive radio will benefit greatly from this important work.

Dr. Joseph Mitola III
Consulting Scientist
The MITRE Corporation
Tampa, FL, USA

Foreword 2

Cognitive networking is being hailed as the next holy grail of (and a potentially disruptive technology in) wireless communication. Although it is a very broad term, it simply means an intelligent communication system (consisting of both the wireline and/or the wireless connections) that is aware of its environment, both internal and external, and acts adaptively and autonomously to attain its intended goal(s). This implies that all the network nodes and the end devices are self-aware and context-aware all of the time. This is a grand long-term vision; still, for a good period of time, policy-based approaches will continue to rule in large parts of the solution while cognitive approaches slowly make their inroads, which is already happening today. The interest in cognitive networking is mainly driven from the need to manage complexity and the efficient utilization of available resources to deliver applications and services as cheaply as possible.

Increasing complexity of the communication networks is a growing challenge for network designers, network operators and network users. It would seem that the recent advances in cognitive radio and networks might offer the perfect solutions to these problems, even though the Internet certainly continues to enable the integration of a wide-range of transport mediums, connecting billions of mobile and fixed terminals, devices and sensors at homes and businesses. (The estimates are that by the year 2017, there will be roughly 7 trillion wireless devices serving 7 billion people around the globe – *Technologies for the Wireless Future, WWRF,* John Wiley & Sons, Ltd, 2006). With the applications and use cases growing unabatedly across body, personal, home, business, vehicle and wide area networks, the spatio-temporal complexity and traffic dynamics continue to increase. These are increasing the burden on the available network resources and their operational management, making the lives of the network administrators and users cumbersome. This begs the answer to the question of how this increased complexity can be reasonably managed without adding more complexity while also distancing the users and the network owners from spending their time on network operations and management.

Until now, cognitive networking has been mostly studied from the spectrum and radio perspective. Recently much interest has arisen in applying the concepts from other fields, such as machine learning, semantics, complex network theories from the natural systems, statistical physics, chemistry, evolution, ecology, mathematics and economics, that mimic the behavior of the communication networks of today. The latter constitutes a more revolutionary approach to cognitive networks. After all, the natural systems embody complex communication networks that are self-healing and robust. This attempt to cross-fertilize the ideas is clearly in an embryonic stage and most of the actual work still lies ahead of us. Nevertheless, I believe that such an effort is well warranted to bring an order of magnitude of improvement in the performance, improvement in the reliability and robustness, improvement in the utilization of resources, reduction in the cost of operations, and lowering of the cost of services to the user.

This is the first book of its kind that brings together the ideas from very different fields in one place to show the interplay between them and how can they be used to build the cognitive networks of the future that we envision. I am glad to see chapters on biologically inspired networking, autonomic networking, adaptive networking, self-management, machine learning, cross-layer design, distributed learning and reasoning, semantics and security, all of which open up new vistas to research in cognitive networks. I congratulate Dr. Mahmoud for successfully assembling the well-known experts in those fields to provide a cohesive and step-by-step approach to cognitive networks. I have enjoyed reading the manuscript, and I am sure you will too!

Dr. Sudhir Dixit
Nokia Siemens Network
Palo Alto, California

Preface

Networks touch every part of our lives, but managing such networks is problematic and costly. Networks need to be self-aware to govern themselves and provide resilient applications and services. Self-awareness means that learning is a crucial component to reduce human intervention, and hence the need for biologically inspired, or non-deterministic, approaches. While some of the existing breakthroughs in machine learning, reasoning techniques and biologically inspired systems can be applied to build cognitive behavior into networks, more innovations are needed.

In the wireless space, the Industrial/Scientific/Medical (ISM) band has inspired impressive technologies, such as wireless local area networks, but interference is becoming increasingly problematic due to the overcrowding in this popular band. In addition, the cellular wireless market is in transition to data-centric services including high-speed Internet access, video, audio and gaming. While communications technology can meet the need for very high data link speeds, more spectrum is needed because the demand for additional bandwidth is continuously increasing due to existing and new services as well as users' population density. This calls for intelligent ways for managing the scarce spectrum resources.

The cognitive radio terminology was coined by Joseph Mitola III and refers to a smart radio that has the ability to sense the external environment, learn from the history and make intelligent decisions to adjust its transmission parameters according to the current state of the environment. Cognitive radio leverages the software defined radio (SDR), which offers a flexible configurable platform needed for cognitive radio implementation. Cognitive radio offers the potential to dramatically change the way spectrum is used in systems and increase the amount of spectrum available for wireless communications.

Cognitive (wireless) networks are the future, and they are needed simply because they enable users to focus on things other than configuring and managing networks. Without cognitive networks, the pervasive computing vision calls for every consumer to be a network technician. The applications of cognitive networks enable the vision of pervasive computing, seamless mobility, ad hoc networks and dynamic spectrum allocation, among others.

This is the first book on cognitive networks which clearly indicates that cognitive network design can be applied to any type of network, being wired or wireless. It provides a state-of-the-art guide to cognitive networks and discusses challenges for research directions. The book covers all important aspects of cognitive networks including chapters on concepts and fundamentals for beginners to get started, advanced topics and research-oriented chapters.

This book offers an unfolding perspective on current trends in cognitive networks. The lessons learned and issues raised in this book pave the way toward the exciting developments of next-generation networks.

Audience

This book is aimed at students, researchers and practitioners. It may be used in senior undergraduate and graduate courses on cognitive networks, adaptive networks, wireless networks and future-generation networks. Students and instructors will find the book useful as it provides an introduction to cognitive networks, their applications and research challenges. It provides researchers with a state-of-the-art guide to cognitive networks, as well as pointers to who else is doing what in the field. Practitioners will find the book useful as general reading and as a means of updating their knowledge on particular topics such as adaptive networks, self-managing networks, autonomic networking, cross-layer-design, cognitive radio technology, machine learning and security issues in cognitive networks.

Acknowledgements

I am deeply grateful to my editors Rowan January and Birgit Gruber for providing me with the opportunity to edit this book, but more importantly for providing me with comments, suggestions and guidelines over the course of the project.

This book would not exist without the authors of the individual invited chapters that make up this book. I would like to thank all authors for allowing their contributions to appear in this book. I also want to thank those who helped in reviewing the chapters for their thoughtful comments and suggestions.

I have benefited greatly from the professionalism and support of Sarah Hinton, the project editor, and Brett Wells, the content editor. They deserve special thanks for their support and guidance throughout the publication process. I also want to thank the copy-editor, Tessa Hanford, for her attention to detail.

My family has provided me so much throughout this project – and many others. Your patience and love sustain me and keep me going.

Qusay H. Mahmoud
Toronto, Canada

Introduction

Motivation

As the Internet moves well beyond the classical services of email, file transfer and remote login, and as the explosion of Internet portable devices continues, the applications and services needed to support mobility will have different network requirements. This has motivated research on active and programmable networks, on a large scale in the 1990s, which represents a novel approach to network architecture in which the switches perform customized computation on the messages flowing through them [11]. The objective was to make the network adaptive and programmable, but despite the many research projects, networks still have to be managed by human administrators. This is simply because the network is still not aware of its state or needs, and doesn't have knowledge of its goals and how to achieve them, and it is not able to reason for its actions [10]. Such properties would render the network adaptive and self-governed. Current data networking technology perform reactive adaptation by responding to changes in the environment after a problem has occurred [12]. Instead, adaptation should be proactive by actively affecting the network when the environment changes.

Mobile computing is becoming pervasive but in order to reach its full potential, significant improvements are needed in usability of end users and manageability of administrators [8], which can be realized by incorporating intelligence into the network environment, and autonomic networking is the key into this area. Autonomic networking connotes the self-regulating capability of the human nervous system. An autonomic computing system needs to know and understand itself. The system will need detailed knowledge of its components, operating environments and connections to other systems. To function properly, it will need to know the resources it owns, the ones it can borrow, lend, buy or simply share with others.

Networks need to be self-aware in order to provide resilient applications and services. Such networks should exhibit cognitive properties where actions are based on reasoning, autonomic operations, adaptive functionality and self-manageability. However, it is important to note that the concept of cognitive networks is different from intelligent networks that have been considered by the telecommunications community in the past, in the sense that in a cognitive network actions are taken with respect to the end-to-end goals of a data flow. In such a network, the collection of elements that make up the network observes network conditions and based on prior knowledge gained from previous interactions, it plans, decides and acts on this information [12]. However, as discussed in this book, there is a performance issue with cognitive techniques since such systems would not reach their performance level instantly; there is an adaptation step. Such systems, however, have a higher resilience towards critical errors.

Cognitive Networks

The word 'cognitive' has recently been used in several different contexts related to computing, communications and networking technologies, including cognitive radio [4] and discussed later, and cognitive networks [2, 8, 12]. Several definitions of cognitive networks have been proposed.

For example, Ramming [8] presents a vision for cognitive networks in which networks should be self-aware in the sense that they can make configuration decisions in the context of a mission and a specific environment. He argues that networks that manage themselves require a new kind of technology known as cognitive technology. He also argues that the network should understand what the application is trying to accomplish, and an application should be able to understand what the network is capable of doing at any given moment. This would allow a network to make use of new capabilities by learning application requirements and dynamically choosing the network protocol that will meet these requirements. Ramming argues for domain-specific languages that could enable users and operators to describe their goals and requirements, and the statements in such languages would be used by the cognitive network to determine the proper balance of resources.

Sifalakis *et al.* [10] define the term 'cognitive' when applied to networks to refer to the intended capability of the network to adapt itself in response to conditions or events, based on reasoning and the prior knowledge it has acquired. Their proposal of adding a cognitive layer on top of an active network, however, doesn't make that network a cognitive network. It needs to provide an end-to-end performance, security, quality of service and resource management, and satisfy other network goals.

Boscovic [2] defines a cognitive network as a network that can dynamically alter its topology and/or operational parameters to respond to the needs of a particular user while enforcing operating and regulatory policies and optimizing overall network performance. He suggests that cognitive networks would use self-configuration capability to respond and dynamically adapt to the operational and context change, and feature a distributed management functionality that can be implemented in accordance with the autonomic computing paradigm.

The chapters in this book clearly indicate that cognitive network design is not limited to wireless networks, but can be applied to any network. Hence, the definition proposed by Thomas *et al.* [12] is applicable, which states that a cognitive network has a cognitive process that can perceive current network conditions, and then plan, decide and act on those conditions. The network can learn from these adaptations and use them to make future decisions, all while taking into account end-to-end goals. The end-to-end scope is important because without it we only have a cognitive layer but not a cognitive network. There are two levels of cognitive processes, at the node level and at the network level, and the end-to-end aspect refers to the collective decision taken by the whole network to achieve the stated goals.

All of the above definitions have one thing in common: the ability to think, learn and remember. Hence a cognitive network has the capability to adapt in response to conditions or events based on the reasoning and prior knowledge it has acquired. Furthermore, cognitive networks can be characterized by their ability to perform their tasks in an autonomous fashion by using their self- attributes such as self-managing, self-optimizing, self-monitoring, self-repair, self-protection, self-adaptation, self-healing [1] to adapt dynamically to changing requirements or component failures while taking into account the end-to-end goals.

Cognitive networks promise to provide better protection against security attacks and intruders by analyzing feedback from the various layers of the network. This is rightly so because of the end-to-end goals of cognitive networks – collaborative ability for threat detection and response; in other words network nodes do not cope with threats individually. However, the wealth of information provided by cognitive networks might provide a greater opportunity for attackers as discussed in the book.

Cross-layer Design

Computer networks have been designed following the principle of protocol layering, so network functionalities are designed in isolation of each other (separate layers) and interfaces between layers. Each layer uses the services provided by the layer below it and provides services to the layer above it. Interlayer communication happens only between adjacent layers and is limited to procedure calls and responses, examples of this are the seven-layer Open Systems Interconnection (OSI) and the four-layer TCP/IP model.

In cognitive networks, however, there is a need for greater interaction between the different layers of the protocol stack in order to achieve the end-to-end goals and performance in terms of resource management, security, QoS or other network goals. Cross-layer design [5] refers to protocol design done by actively exploiting the dependence between the protocol layers to obtain performance gains. Cognitive networks will employ cross-layer design and optimization techniques in order to adapt, simply because a great level of coordination is needed between the traditional protocol layers.

Feedback Control Loops

Biological systems operate with feedback-based control loops to adapt to changes in the environment. As a consequence, the fundamental management element of an autonomic computing architecture is a control loop. This control loop starts with gathering sensor information, which is then analyzed to determine if any correlation to the managed resource is needed. If so, then that correlation is planned, and appropriate actions are executed using effectors that translate commands back to a form that the managed resource can understand [4].

As discussed in [12], the Observe–Orient–Decide–Act (OODA) loop approach provides an interesting support for modeling systems where states change according to information provided by external resources. It is important to note that the OODA loop doesn't include learning, which will play a central role in cognitive networks. The cognition cycle proposed in [7] is the basis for many of the architectures used in cognitive radio research. Several chapters in the book discuss feedback control loops in more detail.

Cognitive Radio Technology

Wireless services have moved well beyond the classical voice-centric cellular systems, and demand for wireless multimedia applications is continuously increasing. Access to radio spectrum is regulated either as licensed or unlicensed. In licensed spectrum, the rights to use specific spectrum bands are granted exclusively to an individual operator, and in unlicensed spectrum, certain bands are declared open for free use by any operator or individual following specific rules. This impressive success of unlicensed services has motivated the development of novel approaches to utilize unused spectrum in an intelligent, coordinated and opportunistic basis, without causing harm to existing services. For example, the US Federal Communications Commission (FCC) has already started working on dynamic spectrum access – whereby unlicensed users borrow spectrum from spectrum licensees. This is due to the fact that measurement studies have found that licensed spectrum is relatively unused across time and frequency [9], which may suggest that spectrum scarcity is artificial [13]. Cognitive radio is being considered as the enabling technology for dynamic spectrum access due to its ability to adapt operating parameters to changing requirements and conditions.

The cognitive radio terminology was coined by Mitola [7] and refers to a smart radio which has the ability to sense the external environment, learn from the history and make intelligent decisions to adjust its transmission parameters according to the current state of the environment. Cognitive radios have recently received much attention for two reasons: flexibility and potential gains in spectral efficiency. They can rapidly upgrade, change their transmission protocols and schemes, listen to the spectrum as well as quickly adapt to different spectrum policies.

Organizations have proposed methods that exploit cognitive radios to obtain higher spectral efficiency. They involve the concept of spectrum sharing, or secondary spectrum licensing. This is in contrast to current network operation where one licensee has exclusive access to a designated portion of the frequency spectrum. Under this model, much of the licensed spectrum remains unused. To alleviate this, proposals which involve cognitive radios sensing these gaps in the spectrum and opportunistically employing unused spectral holes have recently emerged [3].

Overview of the Book

One of the major driving application areas of cognitive networks in the wireless world is dynamic spectrum access and seamless mobility. The adoption of cognitive radio will lead to wide social and business consequences as smart devices exploit the wireless web to displace traditional cellular phones. Applications of cognitive networks in the wired space include automatic configuration and reconfiguration, self-governing networks and provisioning resilient applications and services.

The remainder of this book contains a collection of invited chapters, both introductory and advanced, that are authored by leading experts in the field. The chapters cover topics that are related to wired and wireless networking and the future of networking. The first few chapters cover biologically inspired networking and autonomic networking, followed by chapters that discuss machine learning and cross-layer design. In addition, several chapters address emerging research issues related to cognitive radio. Finally, there are a couple of chapters covering security issues and intrusion detection in cognitive networks.

The topics covered in this book are the holy grails where future innovations in cognitive networks will come from. The individual chapters provide state-of-the-art material as well as directions for future research. Here is an overview of the individual chapters:

Chapter 1: Biologically Inspired Networking

Biologically inspired networking is becoming a very active area of research simply because researchers are realizing that methods from biology have attractive features such as scalability and resilience to changes in the environment. In this chapter, Kenji Leibnitz, Naoki Wakamiya and Masayuki Murata introduce general concepts from biological systems and show their possible application in computer networking. Biological systems can inspire and help us in designing fully distributed, self-organizing and self-governing networks. However, biologically inspired methods are often slower in reaction than conventional control algorithms.

Chapter 2: The Role of Autonomic Networking in Cognitive Networks

In this chapter, John Strassner discusses the role of autonomic networking in cognitive networks. The chapter starts by presenting the concepts of autonomic computing by examining IBM's autonomic computing architecture, followed by a discussion of why existing autonomic approaches were not completely suitable for network management. The chapter proposes enhancements to IBM's autonomic computing architecture, and then presents Motorola's FOCALE, which is an autonomic networking architecture specifically designed to support network management. This architecture features novel features such as the use of multiple control loops, information and data models, ontologies, context awareness and policy management. Finally, the chapter discusses how to apply FOCALE to cognitive wireless network management, and uses Motorola's Seamless Mobility as a use-case.

Chapter 3: Adaptive Networks

Adaptive networks, or networks that employ adaptive schemes, can provide better service continuity in dynamic environments. This chapter by Jun Lu, Yi Pan, Ryota Egashira, Keita Fujii, Ariffin Yahaya and Tatsuya Suda provides a comprehensive examination and analysis of network schemes that adapt to changes in various dynamic factors including resources, application data and user behaviors.

Chapter 4: Self-Managing Networks

In this chapter, Raouf Boutaba and Jin Xiao present state of the art of the concepts, theories and advances in self-managing networks. They argue that self-management is in essence another aspect of the cognitive network concept in the sense that it envisions a system that can understand and analyze its own environment, and can plan and execute appropriate actions with little or no human inputs. In addition, they discuss the strong tie between self-awareness and self-management where self-awareness must be present to enable self-management. They brought forth two important issues in self-management: benchmarking and validation, and the concept of self-stabilization. They also argue that a much stronger emphasis must be placed on system reliance and trustworthiness if the vision of self-management is to become a reality, and hence their design must follow a rigorous and formal approach that emphasizes on verifiable theories and formal design methodologies.

Chapter 5: Machine Learning for Cognitive Networks: Technology Assessment and Research Challenges

Cognitive networks pose enormous challenges for machine learning research. In this chapter, Thomas G. Dietterich and Pat Langley examine various aspects of machine learning that touch on cognitive approaches to networking. They present the state of the art in machine learning with emphasis on those aspects that are relevant to the emerging vision of the Knowledge Plane, which is a pervasive system within the network that builds and maintains high-level models of what the network is supposed to do, in order to provide services and advise to other elements of the network.

Chapter 6: Cross-Layer Design and Optimization in Wireless Networks

Cognitive networks will employ cross-layer design and optimization techniques. In this chapter, Vineet Srivastava and Mehul Motani offer a detailed treatment of cross-layer design and optimization as applied to wireless networks. They taxonomize the different cross-layer design proposals according to the kind of architecture violations they represent, and describe some cross-layer design ideas that have made their way into commercial products and industry standards. There is no doubt that cross-layer design holds tremendous potential in wireless communications, but they argue that in order for that potential to be fulfilled, it is imperative to view cross-layer design holistically by considering both performance and architectural viewpoints.

Chapter 7: Cognitive Radio Architecture

The cognitive radio terminology was coined by Joseph Mitola III, and in this chapter he discusses the important aspects of the architecture of the ideal cognitive radio (iCR) with respect to the critical machine learning technologies. The iCR vision includes isolated radio devices and cognitive wireless networks with machine perception, such as vision, speech and other language skills. He argues that while many information-processing technologies from e-business solutions to the semantic web are relevant to iCR, the integration of audio and visual sensory perception into software-defined radio (SDR) with suitable cognition architectures remains both a research challenge and a series of increasingly interesting radio systems designs.

Chapter 8: The Wisdom of Crowds: Cognitive Ad Hoc Networks

In this chapter, Linda Doyle and Tim Forde refocus the definition of ad hoc around its true meaning in order to explore the real potential of ad hoc networks, and to illustrate the natural endpoint for the evolution of ad hoc networks is a fully cognitive ad hoc network. They suggest that the cognition cycle be adopted by the network-wide system, and present discussion of some observed social conventions which enable loosely associated groups of people to pool knowledge so as to collectively learn and to make decisions which would otherwise require the intervention of a centralized expert.

Chapter 9: Distributed Learning and Reasoning in Cognitive Networks: Methods and Design Decisions

In this chapter, Daniel H. Friend, Ryan W. Thomas, Allen B. MacKenzie and Luiz A. DaSilva discuss distributed learning and reasoning for cognitive networks. They demonstrate how a cognitive network fits into the multi-agent system and cooperative distributed problem-solving contexts, and describe the application of three classes of distributed reasoning methods to a cognitive network and provide justification for their use, as well as potential drawbacks. They offer their perspective on how existing research in sensor networks can be used to benefit cognitive networks, and discuss high-level design decisions and their interrelation with distributed learning and reasoning in a cognitive network.

Chapter 10: The Semantic Side of Cognitive Radio

There is a growing need to standardize the data and knowledge structures related to the spectrum environment so as to enable mechanism and automated methods for spectrum access. In this chapter, Allen Ginsberg, William D. Horne and Jeffrey D. Poston provide the foundation for understanding the role of formal semantics, as exemplified in ontology languages such as the web ontology language (OWL) and associated rule languages such as semantic web rule language (SWRL), in the field of cognitive radio. They argue that it is necessary for radios to share a common language in which they could communicate special circumstances that would cause operational or policy rules to be altered or suspended.

Chapter 11: Security Issues in Cognitive Radio Networks

In this chapter, Chetan N. Mathur and K.P. Subbalakshmi present the security issues that are unique to cognitive radio communications. They discuss several novel security attacks on different layers of the protocol stack in cognitive networks that make use of inherent vulnerabilities such as sensitivity to weak primary signals, unknown primary receiver location, lack of common control channel and tight synchronization requirement in centralized cognitive networks. They argue that more research is needed in order to fulfill the vision of cognitive networks in offering resilient services and keeping intruders out, and offer directions for future research.

Chapter 12: Intrusion Detection in Cognitive Networks

The wealth of information provided by cognitive networks might provide a greater opportunity for attackers. In this chapter, Hervé Debar discusses intrusion detection or the process of detecting unauthorized use, or an attack upon a computer or network, and then presents a model for adapting intrusion detection systems to cognitive networks. The model, which is based on the Observe–Orient–Decide–Act (OODA) loop, describes the information that needs to flow between the nodes of the network and the operations they must support to maintain the properties specified by the security policy.

Chapter 13: Erasure Tolerant Coding for Cognitive Radios

In this chapter, Harikeshwar Kushwaha, Yiping Xing, R. Chandramouli and K.P. Subbalakshmi provide an overview of some coding aspects of dynamic spectrum access with cognitive radios, followed by the spectrum pooling concept for secondary usage of spectrum. Then the loss on a secondary user link is modeled as an erasure, and an erasure tolerant coding is proposed to compensate for it. They describe multiple description codes as a source coding approach for robust communication with applications in secondary spectrum access with cognitive radios.

Finally, I hope readers will find the selection of chapters for this book useful. What I have tried to provide is a state-of-the-art guide to cognitive networks offering a convenient access to exemplars illustrating the diversity of cognitive networks, the techniques being used, and the applications of this emerging field. I believe that the topics covered in this book pave the way toward the exciting developments of next generation networks.

References

[1] Autonomic Network Architecture. ANA Project, http://www.ana-project.org. Accessed December 8, 2006.

[2] Boscovic, D. (2005) Cognitive networks. Motorola Technology Position Paper, http://www.motorola.com/mot/doc/6/6005_MotDoc.pdf. Accessed June 1, 2006.

[3] Devroye, N., Mitran, P. and Tarokh, V. (2006). Cognitive decomposition of wireless networks. *Proceedings of the 1st International Conference on Cognitive Radio Oriented Wireless Networks and Communications (CROWNCOM'06)*, June, Greece.

[4] IBM and autonomic computing (2006) An architectural blueprint for autonomic computing, http://www-03.ibm.com/autonomic/pdfs/ACwpFinal.pdf. Accessed December 8, 2006.

[5] Kawadia, V. and Kumar, P.R. (2005) A cautionary perspective on cross-layer design. *IEEE Wireless Communications*, **12**(1), 3–11.

[6] Kephart, J.O. and Chess, D.M. (2003) The vision of autonomic computing. *IEEE Computer*, **36**(1), 41–50.

[7] Mitola III, J. (2000) *Cognitive Radio: An Integrated Agent Architecture for Software Defined Radio*, PhD thesis, Royal Institute of Technology, Sweden.

[8] Ramming, C. (2004) Cognitive networks. *Proceedings of DARPATech Symposium*, March 9–11, Anaheim, CA, USA.

[9] Shared Spectrum Company. Spectrum Occupancy Measurements, http://www.sharedspectrum.com/?section=measurements. Accessed December 11, 2006.

[10] Sifalakis, M., Mavrikis, M. and Maistros, G. (2004) Adding reasoning and cognition to the Internet. *Proceedings of the 3rd Hellenic Conference on Artificial Intelligence*, May, Samos, Greece.

[11] Tennenhouse, D.L., Smith, J.M., Sincoskie, W.D., Wetherall, D.J., and Minden, G.J. (1997) A survey of active network research. *IEEE Communications Magazine*, **35**(1), 80–6.

[12] Thomas, R.W., DaSilva, L.A. and MacKenzie, A.B. (2005) Cognitive networks. *Proceedings of the IEEE Symposium on New Frontiers in Dynamic Spectrum Access Networks (DySPAN)*, November, Baltimore, MD, USA.

[13] United Stated Frequency Allocations, The Radio Spectrum, http://www.ntia.doc.gov/osmhome/allochrt.pdf. Accessed December 11, 2006.

1

Biologically Inspired Networking

Kenji Leibnitz, Naoki Wakamiya and Masayuki Murata
Osaka University, Japan

1.1 Introduction

The development of computer networks has seen a paradigm shift from static, hierarchical network structures to highly distributed, autonomous systems without any form of centralized control. For networking nodes, the ability to self-adapt and self-organize in a changing environment has become a key issue. In conventional network structures, e.g., the Internet, there is usually a hierarchical order with centralized and static control. For example, hosts are aggregated to local area networks (LANs), which are connected via gateways to wide area networks (WANs) and network domains, etc., all using static connections and addressing. Recently, however, the trend leads more and more to networks that dynamically set up connections in an ad-hoc manner. Mobile ad-hoc networks (MANETs) are a prominent example, but also overlay structures such as peer-to-peer (P2P) networks require a scalable, robust and fully distributed operation with self-adaptive and self-organizing control mechanisms. The main control functions are no longer performed at intermediate nodes like routers, but shifted to the end-user nodes. Additionally, the location of these nodes may now be no longer static but can be mobile, imposing new challenges on the search for shared information in P2P networks or the location of a node in an ad hoc network.

For these types of new dynamic networks, the following three requirements for network control are considered mandatory:

- *Expandability (or scalability)*: facing a growing number of nodes and end users, as well as an increasing variety of devices attached to the network, the network must be able to continue its normal operation.
- *Mobility*: in addition to the end users' mobility, we should also consider the mobility and churn, i.e., the process of users entering and leaving the system, of intermediate

Cognitive Networks: Towards Self-Aware Networks Edited by Qusay H. Mahmoud

network nodes. This implies that deterministic packet forwarding cannot be expected, but must be performed in a probabilistic way.

- *Diversity*: the hardware and software of new types of network devices may generate entirely new traffic patterns imposing further challenges on the performance of the network. Additionally, networks must be able to cooperate with each other, both horizontally and vertically.

The only solution to meet the above characteristics seems to be that end hosts must be equipped with mechanisms permitting them to adapt to the current network status, for instance when finding peers or controlling congestion. For this reason, biologically inspired approaches seem promising since they are highly capable of self-adaptation, although they can be rather slow to adapt to environmental changes. Of course, the application of biologically inspired approaches in information technologies is not a new issue, but most of the previous attempts have been concentrated on optimization problems in network control. However, we can learn further important lessons from nature and the focus should lie on the scalability, adaptability, robustness and self-organization properties of biological systems. Especially, exploring the symbiotic nature of biological systems can result in valuable knowledge for computer networks.

The purpose of this chapter is to introduce general concepts from biological systems and to show their possible application in the field of computer networking. While nature in itself is of such high diversity and the topic of biologically inspired networking comprises many different aspects, only a small selection can be presented here. We will first discuss in Section 1.2 common principles found in biological networks and focus in particular on self-organization and the role of fluctuations. Then, we will show some analogies between biological processes found in the human body and computer networks. We will present some case studies of methods from swarm intelligence (Section 1.3), which describes the way individuals cooperatively interact with each other like swarms of bees or ants. Then, in Section 1.4 we will summarize some approaches which are based on evolutionary and adaptive systems, for instance featuring predator–prey-like behavior.

1.2 Principles of Biologically Inspired Networking

There are several key factors which can be observed in biological systems. Especially, features such as self-organization and robustness are of great importance when biological methods are applied to computer networks. However, there is also a trade-off to make when it comes to considering self-organized, distributed systems over those which are centrally controlled. Although scalability is improved, the approach towards fully distributed topologies comes at a cost of performance. Since there is no global view of the entire network, global optimization of network parameters is no longer feasible. Methods searching for local ad hoc solutions may yield only inferior results, so a trade-off must be found which balances scalability with controllability and performance.

Dressler [9] illustrates this trade-off as shown in Figure 1.1. The distinction is made between systems with centralized control and those that are fully self-organized. Distributed systems could be considered as intermediate, hybrid networks which allow the management of large numbers of nodes in a scalable way while preserving the benefits from a centralized control.

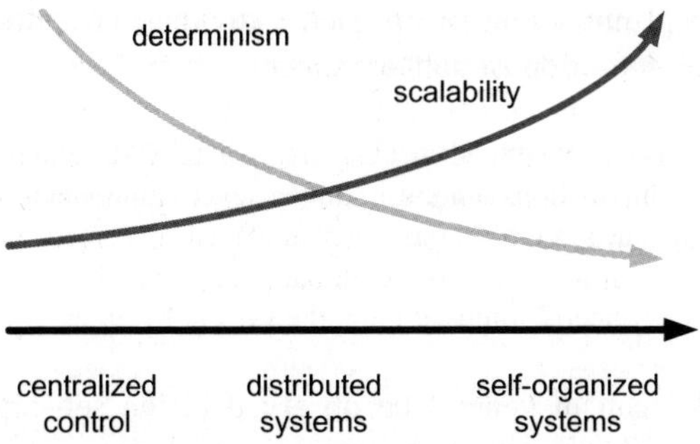

Figure 1.1 Trade-off between controllability and scalability in system control

Furthermore, while most computer networks are designed to optimally function under controlled conditions, biologically inspired systems rely on fluctuations and randomness. The benefits of the latter strategy can be seen when we consider the reaction to critical errors. Due to their probabilistic nature, they are less reliable on critical entities (e.g., servers, bottleneck links) with a single point of failure. Instead, several possible solutions may be considered as reaction to the failure, which although perhaps inferior in performance permits the operation at a sufficient quality. This leads to the behavior which can compensate critical errors, thus making the system more resilient and reliable, see Figure 1.2. However, biological systems usually do not instantly reach their operational performance level, but slowly approach this level by small fluctuations. This may require several (perhaps hundreds or thousands) of adaptation steps until reaching an acceptable level of performance.

1.2.1 Self-Organization

Perhaps the most important property of biological systems is their ability of self-organization, which describes systems consisting of autonomous individuals (cells, swarms of insects, etc.) grouping together into certain structures without any explicit rules. Self-organization is closely related to the emergence property of some biological systems, in which the outcome of the system depends on the collection of the individual behaviors

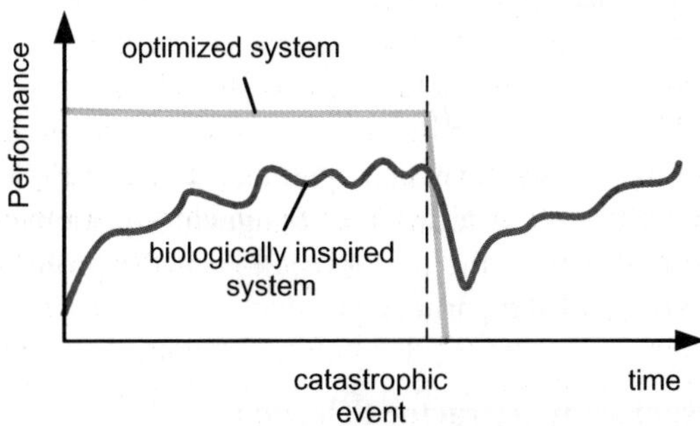

Figure 1.2 Biologically inspired systems may show inferior performance compared to optimized systems, but have a higher resilience towards critical errors

and their interaction and not so much on each individual task they perform. Bonabeau *et al.* [2] give a formal definition of self-organization as follows.

> Self-organization is a set of dynamical mechanisms whereby structures appear at the global level of a system from interactions among its lower-level components. The rules specifying the interactions among the system's constituent units are executed on the basis of purely logical information, without reference to the global pattern which is an emergent property of the system rather than a property imposed upon the system by an external ordering influence.

Four basic principles can in general be observed in the self-organization of biological systems. *Positive feedback*, e.g., recruitment or reinforcement, permits the system to evolve, and promotes the creation of structure. Positive feedback acts as an amplifier for a desired outcome, whereas *negative feedback* regulates the influence from previous bad adaptations. Negative feedback also prevents the system from getting stuck in local solutions and may take the form of saturation, exhaustion or competition. Another important feature is that nature-inspired systems usually do not rely on any global control unit, but operate in an entirely distributed and autonomous manner. This means that each individual acquires its information, processes it and stores it locally. However, in order that a self-organized structure is generated, individuals need to exchange information with each other. This is done by either *direct or indirect interactions* among each other. Finally, a characteristic found in self-organized structures is that they often rely on *randomness and fluctuation* to enable the discovery of new solutions and to enhance the stability and resilience of the system.

1.2.2 Noise and Fluctuations

Biological systems are also characterized by the existence of an inherent stochastic component. In nature, nothing is perfect and there is no determinism. For example, if we consider cells of the same type, the values of the quantities describing them will vary from cell to cell and for a single cell, these values will also fluctuate over time [18]. When randomness is referred to in physics, we often consider the term of *Brownian motion*, which is the random motion of particles immersed in a fluid. The mathematical process used to describe it is called the *Wiener process*. Brownian motion is a continuous-time stochastic process and the position of the Brownian particle is described approximatively by the *Langevin equation*, see Equation (1.1).

$$\frac{dv}{dt} = -\gamma v + \eta \tag{1.1}$$

Here v is the velocity of the Brownian particle, γ is a friction coefficient of the underlying fluid, and η is a noise term. The Langevin equation can be solved by the Fokker–Planck equation to obtain the time evolution of the probability density function of the position and velocity of the particle.

1.2.2.1 Adaptive Response by Attractor Selection

Fluctuation can also be found as the driving force for information processing through the nonlinear behavior inherent to gene networks. These networks do not necessarily require a preprogrammed connectivity for signal transduction from environmental inputs to DNA. For example, Kashiwagi *et al.* [19] introduce the concept of *adaptive response by attractor selection* (ARAS), which shows that a gene network composed of mutually inhibitory operons allows its host *Escherichia coli* cells to adapt to changes in the availability of a nutrient for which no molecular machinery is available for signal transduction. Since ARAS is driven by noise, it is noise-tolerant and can even be stimulated by noise. A cell activity or vigor leads to an alternative expression of the operon that produces the enzyme adaptively to the designated nutrient availability.

In ARAS, the concept of attractors is used to describe the multiple states of gene expression. An *attractor* is the region to which the orbit of a dynamical system recurrently returns regardless of the initial conditions [18]. Even if a state is perturbed by fluctuations, the system state will be drawn over time to an attractor. By formulating the desired outcome as a stochastic differential equation system and an appropriate mapping of the activity of the system, ARAS can be considered as an adaptive control mechanism with many possible areas of application. Leibnitz *et al.* [22] use ARAS to self-adaptively select the best paths in a multipath routing scenario in overlay networks. The method can be basically described as follows. The discovery of paths from source to destination is done in the *route setup* phase by broadcasting route request packets as in conventional routing methods, e.g., ad hoc on-demand distance vector (AODV) routing [25]. Then, in the *route maintenance* phase ARAS is used to determine the transmission probabilities for this set of paths. The addition or removal of paths is easily compensated.

Consider that there are M paths from source to destination. Then, for each packet sent from the source node, the path i is chosen with a probability or rate m_i which is determined by the stochastic differential equation system in (1.2).

$$\frac{dm_i}{dt} = f(m_1, \ldots, m_M) \times \alpha + \eta_i \qquad i = 1, \ldots, M \tag{1.2}$$

Note that the basic structure of Equation (1.2) is similar to that in Equation (1.1). However, there is one important difference. Beside the function f which determines the attractor locations, the dynamics of each m_i depends also on the term α, which represents the *activity* of the system. Activity is directly influenced by the environment, which is in the case of the routing scenario expressed by the path metrics. So, when the activity approaches zero, the dynamic behavior of Equation (1.2) is dominated by the random noise term η_i, resulting in a random walk in the phase space. On the other hand, if α is large, the random component introduced by η_i becomes negligible and the system converges to an attractor solution in a deterministic way.

The transmission probabilities obtained by ARAS are illustrated in Figure 1.3 for an example scenario with $M = 6$ paths between a source node and destination node. The system automatically adapts to changes in the environment by searching for better solutions. In the case of the multipath scenario, this means that due to changes in the path metrics, the currently selected primary path is no longer appropriate, as can be seen at time steps 2000 and 6000 in Figure 1.3. The metric change causes activity to drop and the transmission probabilities approach each other until a better solution is found by random walk. In Figure 1.3, we can also recognize that the duration of the random walk phase varies for each search due to the memoryless search for a new solution. This causes a slight delay until the adaptation to the new solution is completed. The advantage, however, is that

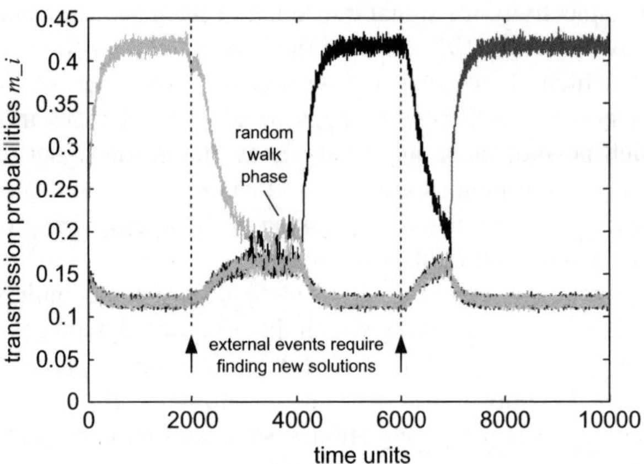

Figure 1.3 ARAS uses random walk to find new solutions when the environment conditions change

no explicit rule is necessary to trigger reactions to the environment since the behavior is implicitly encoded in the ARAS equations.

1.2.3 From Cell Biology to Computer Networks

In the field of cell and molecular biology, the behavior of cells is studied at a microscopic level. Cells are the smallest structural and functional units found in all living organisms and store their genetic information in the DNA. They take in nutrients and process these into energy (*metabolism*) and have the ability to reproduce. Basically, similar features are also found in autonomous systems in computer science [8].

1.2.3.1 Molecular Processes

Comparing organisms to computer networks, Dressler [8] states that there is a highly hierarchical structure found in both systems. Processes (e.g., movements) are organized by interactions of several organs, which are composed of tissues and these in turn are built by many different cells. One cell type consists of identical cells, which are associated and communicate with each other to fulfill a common function within the tissue.

Information exchange in a cellular environment is performed through signaling pathways from cell to cell, comparable to multihop transmission in communication networks. Cells communicate with each other by sending signal molecules which are bound by specific receptors and which are then forwarded to the cell nucleus causing a specific response, see Figure 1.4. The resulting cellular processes may mean that not only a single cell but several cells are activated, resulting in a coordinated reaction of the organ.

Several types of intercellular signaling exist depending on the distance over which the signal is transmitted. Over short range, there is *contact-dependent* signaling in which cells must have direct membrane-to-membrane contact and *paracrine* signaling where a local signal is diffused to react with multiple neighboring cells. On the other hand, when a signal needs to be transmitted over a long distance, signaling is performed either *synaptic* at the end of the axon or *endocrine*, by which hormones are distributed in the bloodstream

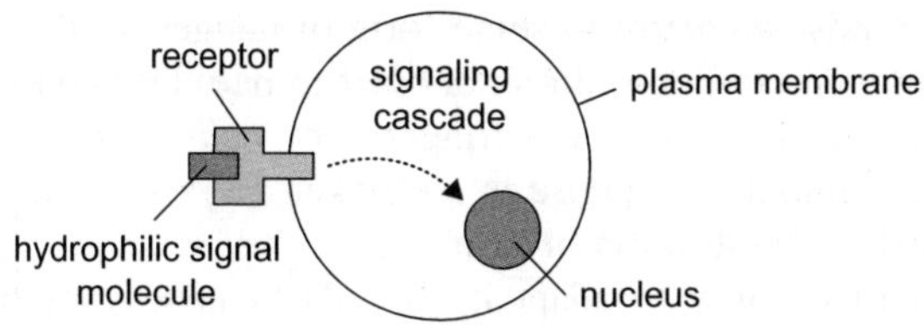

Figure 1.4 Local intercell communication is performed by binding signal molecules

of the body [1]. Mechanisms found in cell communication can be applied to computer networks in order to achieve a self-maintaining and self-healing approach to security in autonomous networking [20].

1.2.3.2 Artificial Immune Systems

Another similar approach for replicating biological behavior found in the body can be seen in the development of *artificial immune systems* (AIS). The main purpose of the biological immune system that has evolved in vertebrates is to protect the body from invading pathogens [4]. Its key function is to distinguish between *self* and *non-self* cells. Depending on the type and method of intrusion, the immune system uses different *immune response* mechanisms to destroy the invader or neutralize its effects. Biological immune systems consist of *innate immunity* and *adaptive immunity*. The innate immunity reacts to any pathogen that is recognized as being an intruder. If this fails, the body activates the adaptive immune system which has the ability to recognize and remember specific pathogens, and to mount stronger attacks each time the pathogen is encountered. Adaptive immunity uses two types of lymphocyte cells, B-cells and T-cells, which are created in the bone marrow and the thymus, respectively. Both types of cells are covered with antibodies which bind themselves to non-self antigens, see Figure 1.5.

The most important types of AIS are the *negative selection, danger signal, clonal selection* and *immune network models*. Negative selection is a mechanism to detect unknown antigens while not reacting to self-cells. Detectors are generated randomly and the thymus performs the negative selection by destroying those detectors that react against self-proteins. Those that do not bind self-proteins leave the thymus and act as memory cells circulating the body and protecting it from foreign antigens.

However, this self/non-self view does not fit experimental observations well, so the danger signal model [13] was introduced as an alternative approach. In the danger model, it is assumed that cells do not react directly to non-self elements, but the immune system

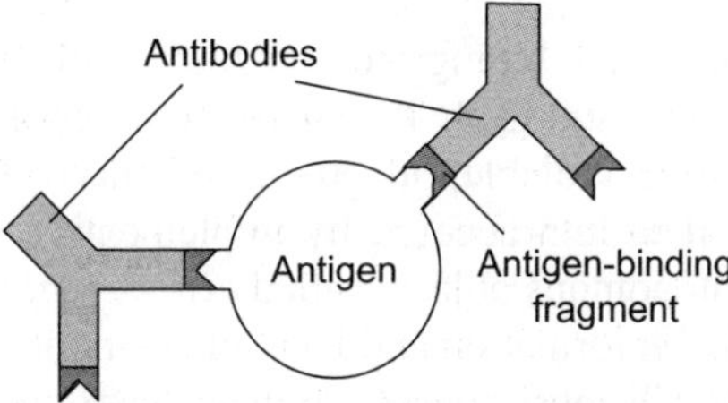

Figure 1.5 Immune response results in an intruding antigen being bound by different types of antibodies

responds to cells that raise an alarm to some form of danger. In this way, it can also be explained why the immune response does not react to harmless non-self cells, e.g., food, fetus. The danger signals are particularly triggered by *cell necrosis*, i.e., the unexpected death of cells, and the immune response is no longer a system-wide response, but must be seen in the context of the location of necrosis.

The immune system's ability to adapt its B-cells to new types of antigen is driven by clonal selection. Depending on the match strength between a B-cell and an antigen, B-cells are cloned and mutated with rates inversely proportional to the match strength (*affinity maturation*). This permits a more focused response to further attacks and the adaptation to changing non-self cells.

A network of interconnected B-cells for antigen recognition is the basis of the artificial immune network model. By stimulating and suppressing each other, a stable network between antibodies is formed which is capable of storing its state as immunological memory, comparable to a Hopfield neural network [17].

Artificial immune systems have been applied to intrusion detection and network security [12]. The term *computer virus* already indicates the analogy that can be found in computer networks and biological processes. Therefore, AIS are often used as mechanisms to detect virus infection of individual hosts by identifying abnormal behavior in terms of system calls (*host-based intrusion detection*). In the same way, TCP network traffic can be monitored for irregularities from the normal behavior (network-based intrusion detection). In [26] an AIS is constructed to detect misbehavior in a mobile ad hoc network, for example caused by faulty hardware or intentional disruption of the network. The model presented there uses a virtual thymus which causes the AIS to become tolerant to previously unseen normal behavior of the nodes, but automatically learns and detects new misbehavior. Further applications of AIS to computing and networking as well as to other fields can be found in [28].

In summary, artificial immune systems can provide a flexible and efficient method to monitor the integrity of a single host or a network and to react to potential dangers. Since the detector cells are small and lightweight, they can be efficiently used in a distributed environment, thus showing good scalability. This distributed operation also results in a very robust system, where the failure of a single detector cell does not result in the failure of the complete detection mechanism. Another useful characteristic is the adaptability found in AIS. The information obtained from distinguishing between self, harmless non-self, and harmful non-self remains in memory permitting a quick reaction to new unknown threats.

1.2.3.3 Bio-Networking Architectures

The application of principles from biology to computer network architectures is reflected in the bio-networking architecture [32]. It can be considered as both a paradigm and a middleware for constructing scalable, adaptive, and survivable applications [27]. The concept of emergence is taken into account by implementing network applications as a group of distributed and autonomous objects called *cyber-entities* (CE). The CE agents act autonomously based on local information and local interactions. Natural selection and evolution is performed and the CE must housekeep their energy in order to avoid starvation.

The bio-net platforms provide the execution environment for the CE, usually a Java virtual machine. Resources on a bio-net platform must be purchased by the CE with energy.

Furthermore, the bio-net is responsible for system services and information services to the CE as well as their scheduling. The self-managing features of the bio-networking architecture are also briefly discussed in Chapter 4.

BIONETS is a similar framework that was proposed for pervasive communication environments [3]. Its main differences from the architecture in [32] are that it is not limited to agent-based services and that the network connections and services are considered as evolutionary processes. BIONETS focuses on scalability, heterogeneity and complexity challenges found in mobile networks. On the lowest plane of a hierarchical structure, tiny nodes (*T-nodes*) gather data from the environment and pass them on to the user nodes (*U-nodes*) on the next plane. U-nodes form islands of connected devices and may mutually exchange information if they are within communication range. Services reside on the next higher plane and may also interact to reflect the social networks of the users.

1.3 Swarm Intelligence

In biology, the intelligent and well-organized behavior of a group of social insects can often be observed. For example, ants or bees solve complex tasks like building nests or searching for food by distributing simple tasks among each other. In such emergent systems, it is not so much the individual work which determines the outcome, but rather the collection of all single activities. This behavior is generally referred to as *swarm intelligence* [2]. The collaboration of insect societies is based on the principle of *division of labor* where specialized workers perform specialized tasks in parallel, leading to a better efficiency. Division of labor can be seen for all insect types in the distinction of worker and reproductive castes. In some cases, several different types of workers coexist which have distinctive tasks as well, e.g., protection of the nest or nurturing the brood.

Each individual insect makes its own decisions, solely based on locally available information. In other words, the insect is not aware of the global situation but only gains its input by interaction with other members of its species. This interaction may be either by direct contact (visual or chemical) or indirect interaction. In the latter case, one individual influences the environment and another reacts to this change at a later time. Ants, for example, interact with each other by laying a trail of pheromones which influences other ants to follow this trail, thus, reinforcing it. Such indirect interaction through the environment is called *stigmergy*.

1.3.1 Ant Colony Optimization

The *ant colony optimization* (ACO) meta-heuristic was introduced by Dorigo [7] as a method for probabilistic optimization in finding the shortest path in a graph. It is an extension of the *ant system* (AS) method based on the foraging behavior of ants that try to find their way between a food source and their nest. ACO can produce nearly optimal solutions to the *traveling salesman problem* (TSP) and its strength lies in the ability to dynamically adapt to changes in the graph topology. For this reason it is well suited for the application to routing in computer networks.

Basically, the method can be described as follows. Agents (artificial ants) randomly wander around searching for food and return to the nest when they have found a source, see Figure 1.6. They lay trails of pheromones which are also followed by other ants should they encounter them. As pheromones evaporate over time, unsuccessful paths are

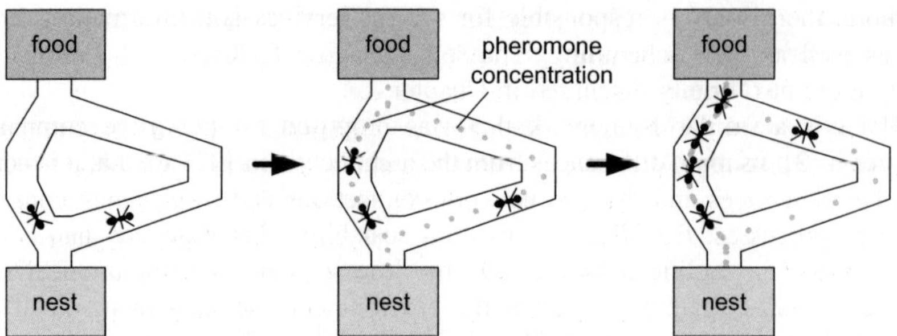

Figure 1.6 In ACO new paths are taken by ants randomly. As the pheromones evaporate over time, the probability for choosing unsuitable paths decreases

soon discarded which constitutes a negative feedback. On the other hand, there is positive feedback for successful paths leading to the food source as they are reinforced by other ants following the trail and increasing the pheromone density.

1.3.1.1 AntNet

AntNet by Di Caro and Dorigo [5] is an application of ACO to routing in packet-switched networks. The operation of AntNet is performed by two different types of ants. At regular intervals, each router in the network sends *forward ants* to probe the network for a minimal cost route towards randomly selected destinations. Whenever forward ants encounter a router on their way, they randomly choose the next hop depending on the probabilities in the routing table of the router. In the case that the destination is unknown at a router, the next hop is selected uniformly among all possible candidates. Once the forward ants reach their destination, they transform into *backward ants* following exactly the same path in the reverse direction to the source. During their return trip, they update the entries of the routing tables at each intermediate router according to the goodness of the path which they encountered. As soon as a backward ant reaches its originating node, it dies and is removed from the system.

1.3.1.2 AntHocNet

Due to the capabilities of operating in a changing environment, ACO also provides an effective principle for routing in mobile ad hoc networks. *AntHocNet* [6] is built on ACO routing and uses a hybrid approach to multipath routing. While AntNet uses a proactive scheme by periodically generating ants for all possible destinations, AntHocNet consists of both reactive and proactive components. Paths are only set up to destinations when they are needed by reactive forward ants that are launched by the source in order to find multiple paths to the destination. Backward ants return to the source to set up the paths. Data is routed stochastically over different paths stored in the pheromone tables. These tables are constantly updated by proactive forward ants. The algorithm reacts to link failures with either a local route repair or by warning preceding nodes on the paths. The authors show that AntHocNet can outperform AODV, which is a well-known routing protocol for ad hoc networks. Due to its emergent operation, AntHocNet also shows much better scalability than AODV.

AntNet and AntHocNet provide flexible and resilient methods for routing in packet-switched networks. The key point in both approaches is the use of probabilities for determining the next hop. These probabilities are constantly updated with the information from probing forward ants which register any changes in the network topology and thus make ACO-based routing very adaptive. However, the probabilistic search for the destination node may require a long time and a large number of forward ants if the network topology is large.

1.3.2 Synchronization with Pulse-Coupled Oscillators

Self-organized and fully distributed synchronization can also be found in nature where it is attained only by mutual interactions among individuals without the existence of a controlling unit. For example, groups of fireflies in Southeast Asia flash in synchrony. Each individual firefly flashes independently of each other at its own interval defined by an internal timer. However, when a firefly meets a group and observes flashes from other fireflies, it reacts to the perceived stimuli and adjusts its internal timer so as to flash at the same rate as the others.

Mutual synchronization of firefly flashing is modeled as a set of *pulse-coupled oscillators* [23]. Each oscillator has a state between zero and one, that monotonically increases with the phase of a timer, see Figure 1.7. The phase cyclically shifts from zero to one as time passes and when the state reaches one, the oscillator fires a pulse, after which it resets the phase back to zero. This happens at time t^+ immediately after the firing instant at time t. This notation is required, since oscillators react after they register the flashing of coupled oscillators, so a slight processing delay is necessary. The fired pulse stimulates other oscillators which are coupled to the firing oscillator and they react by updating their own state. After a series of such mutual interactions, the system achieves synchronization.

Wakamiya and Murata [31] apply a model of pulse-coupled oscillators to data gathering in wireless sensor networks. The purpose of using pulse-coupled oscillators is that each sensor node independently determines the cycle and timing at which it emits a message. For efficient data gathering, it has the effect that sensor information is propagated in concentric circles from the edge of the network to the base station to which all sensed data is transmitted. Thus, sensor nodes at a similar distance from the base station must transmit their information simultaneously, slightly before their inner neighboring nodes are about to transmit. The information and states of a sensor node i consist of its own

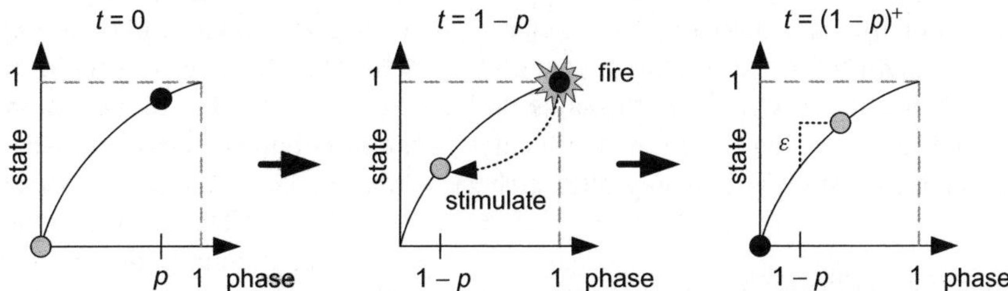

Figure 1.7 Pulse-coupled oscillator model with two oscillators. When one of them fires at time $t = 1 - p$, the other is stimulated causing a change in state

sensor data and that received from its neighbors, a phase φ_i of its timer, the state x_i defined by a monotonically increasing function $x_i = f(\varphi_i)$, a level l_i which corresponds to the number of hops from the base station, and an offset value δ.

Initially, none of the sensor nodes knows its distance to the base station. They broadcast sensor data at their own timing and frequency when the regulated state $x_i' = f(\varphi_i + \delta)$ reaches one. When a sensor node receives a radio signal with a level smaller than itself, it is stimulated and raises its state as shown in Equation (1.3).

$$x_i(t^+) = B(x_i(t) + \varepsilon) \quad \text{where} \quad B(x) = \begin{cases} 0 & \text{if } x < 0 \\ 1 & \text{if } x > 1 \\ x & \text{otherwise} \end{cases} \tag{1.3}$$

At the desired frequency of data gathering, the base station periodically broadcasts beacon signals within its RF transmission range. Sensor nodes which receive this signal recognize that they are within the innermost of the concentric circles around the base station and set their level to 1. In addition, by repeatedly being stimulated by the beacon signals, they become synchronized with these. A synchronized sensor node begins to transmit messages including sensor data and their level at the same frequency as the beacon signals, but at a time instant earlier than the beacon given by the regulated state x_i'. Sensor nodes receiving signals from a level-one node recognize themselves as being two hops away from the base station. In general, each node sets its own level to one level above the minimum from all received messages. Their timers get synchronized with the message transmissions of level-one nodes and they begin their own transmissions at 2δ earlier than the beacon signals. Their signals further stimulate outer nodes. Consequently, all sensor nodes identify their correct levels and behave in synchrony so that the waveform data gathering is accomplished.

Each sensor node can save battery power by turning on the transceiver only from the time it receives a message from its outer neighbors until it completes its own transmission to an inner neighbor. When a sensor node is moved to another location, it is stimulated by its new neighbors and joins the appropriate level of concentric circles. When a sensor node notices the disappearance of an inner node, it rejoins the network by synchronizing with another sensor node in its vicinity. The advantages of using such a synchronization scheme over other conventional methods are manifold. In this scheme, global synchronization is established and maintained only through mutual interactions among neighboring nodes. Each sensor node only transmits its sensor data at its own timing and the signals for data gathering are used for synchronization. Therefore, there is no need for any additional communication that may consume battery power and no time stamp is required for the messages. Finally, no additional routing protocol is needed which would also require additional signaling for path establishment and maintenance. However, it should be noted that this scheme also suffers from the same problems usually found in biological systems: global synchronization and reaction to changes in the environment due to failed sensors is never attained instantly, but only after a short delay.

1.3.3 Reaction–Diffusion

Pattern formation on the coat of mammals (e.g., spots of a leopard) and fish (stripes of an emperor angelfish) can be explained by reaction–diffusion (RD) equations. According to Turing [30], these patterns can arise as a result of instabilities in the diffusion of morphogenetic chemicals in the animals' skins during the embryonic stage of development.

The two parts involved in this process are the long-ranged, slowly propagating diffusion effect of a chemical substance over space and the local reaction to it. Patterns are formed by chemical kinetics among two *morphogens*, the *activator* A and *inhibitor* B. Their concentrations, $A(\mathbf{z}, t)$ and $B(\mathbf{z}, t)$, at a specific location $\mathbf{z}$ on the surface determine the state (or color) at time t. The mechanism is expressed by a pair of second order partial differential equations, where the concentration of one substance is defined by the reaction to the concentration of substances at the location and the respective diffusion components, D_A and D_B.

$$\frac{\partial A}{\partial t} = F(A, B) + D_A \nabla^2 A$$

$$\frac{\partial B}{\partial t} = G(A, B) + D_B \nabla^2 B$$

The constants D_A and D_B with $D_B > D_A$ describe the speed of diffusion, the functions F and G are the nonlinear reaction kinetics, and the Laplacian $\nabla^2 A$ indicates the difference in concentration of morphogen A at a certain location with respect to surrounding concentrations. Depending on the sign of $\nabla^2 A$, morphogen A will either diffuse toward that nearby location or away from it.

Beginning from a homogeneous and even distribution of chemical substances over a surface and by regulating initial random influences and parameters in the RD equations, a variety of patterns (stripes, spots, rings) can be formed by evaluating the equations at each point independently. Figure 1.8 shows some sample patterns generated from the reaction–diffusion equations with different parameter settings.

Reaction–diffusion has been applied to several different areas in computer science and communication networks. In computer graphics, reaction–diffusion is used to generate realistic textures on surfaces. Other cases include using RD for solving optimization problems. Yoshida *et al.* [33] propose using reaction–diffusion as a cooperative control mechanism in a surveillance system of cameras. Each camera adjusts its observation area in a decentralized way based on reaction–diffusion equations to reduce blind spots. The

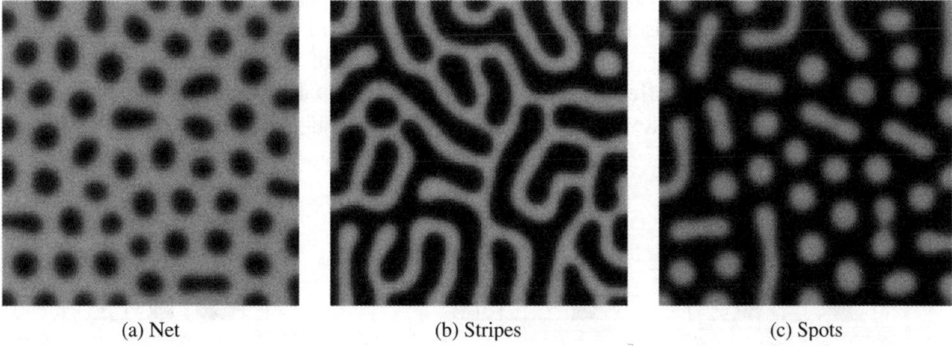

(a) Net (b) Stripes (c) Spots

Figure 1.8 Patterns generated by reaction-diffusion equations resemble those found in nature on the fur of animals

method is able to reconfigure itself to maintain total coverage even when cameras are moved or rearranged.

Henderson *et al.* [16] use reaction–diffusion to form paths in a sensor network along which messages or agents move. A surface point corresponds to a node in the sensor network and the diffusion of chemical substances among neighboring points is accomplished by radio transmission. Without any centralized control the desired pattern is formed in the network where each sensor node only acts according to simple RD equations. A medium access control (MAC) for ad hoc networks based on reaction–diffusion is presented by Durvy and Thiran [10]. Global transmission patterns are set up only by local interactions among ad hoc nodes and a collision-free transmission pattern is set up. Furthermore, an efficient scheduling of the transmissions in multihop wireless networks with high traffic loads is possible.

One drawback in the applicability of reaction–diffusion is that it suffers from the same problems that were previously mentioned for all biological systems. In order to reach a stable pattern, reaction–diffusion requires time until convergence is reached. Therefore, it is not well suited for application in networks which have dynamically changing environments with fast-moving nodes.

1.4 Evolutionary and Adaptive Systems

So far we discussed that biological systems operate with feedback-based control loops to adapt to changes in the environment. This phenomenon can be seen in the natural selection process as well. Species evolve over many generations to be able to better survive in a dynamically changing habitat by adopting new features. Influences from mutation introduce diversity and adaptability to new challenges.

Genetic algorithms (GA) are a popular class of evolutionary algorithms as a heuristic method for the solution of optimization problems. While they are not necessarily related to networking problems, they are mentioned here because they nicely illustrate the operation of genetic evolution, which is a common principle in biologically inspired networks. In GA, a problem is encoded as (binary) individuals (*chromosomes*) and its evolution is performed over many generations applying the typical operations of natural selection, mutation and reproduction. Selection is the process where the fitness of individuals is evaluated upon which the new generation is selected. For the pool of selected individuals, parents are chosen which then produce a child by crossover and mutation, see Figure 1.9. In crossover, parts of the chromosomes of the parents are broken and recombined to form new different chromosomes. Mutation further adds to the diversity by modifying the genetic material. This process repeats itself until a termination condition

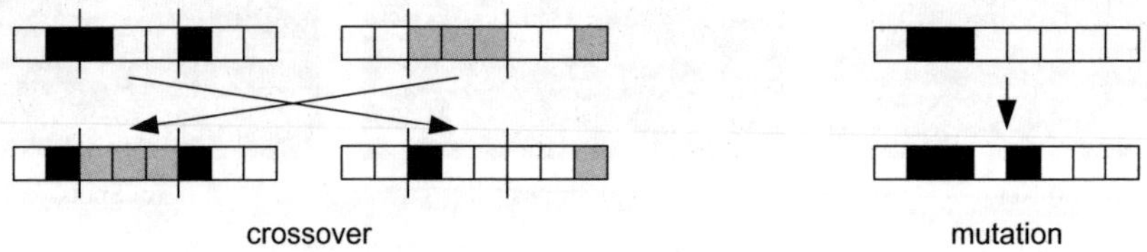

Figure 1.9 Reproduction by crossover of parental chromosomes and mutation

is reached. The resulting solution corresponds to the evolutionary adaptation of a species to its environment.

1.4.1 Models with Interacting Populations

While in genetic evolution the adaptation process is performed over the course of many generations, adaptive feedback models can also be used on a much smaller time scale to model the behavior of interacting populations. Interaction among species causes that the involved populations are affected by each other. In this section we discuss some approaches where models from mathematical biology are used in control algorithms for communication networks.

Basically, population models are used to investigate the dynamic interaction between different kinds of animal species. Murray [24] distinguishes three main types of interaction among populations depending on their growth rates:

- The populations are in *predator–prey* situation if the growth rate of one population is increased at the cost of that of the other.
- The populations are in *competition* if the growth rate for both populations is decreased.
- The populations are in *symbiosis* (or *mutualism*) if the growth rate for both populations is increased.

Although we speak of species, the models can be directly applied to other fields like communication network protocols. In this case, the population of a species would correspond to, for example, transmission rates.

1.4.1.1 Predator–Prey Models

Perhaps the most well-known example is the predator–prey model, which is also known as the *Lotka–Volterra model*. In general, the model is based on the *logistic population model*, which describes the evolution of the population over time and is bounded by a theoretical *carrying capacity*. The carrying capacity is the maximum population limit which can be supported by the environment. For a single population $N(t)$, growth rate $\varepsilon > 0$, and carrying capacity K, the logistic equation takes the following form.

$$\frac{dN}{dt} = \varepsilon \left(1 - \frac{N}{K}\right) N \tag{1.4}$$

The Lotka–Volterra model can be easily extended to n species, all influencing each other with a competition coefficient $\gamma < 1$, which we will assume to be equal for all species.

$$\frac{dN_i}{dt} = \varepsilon \left(1 - \frac{N_i + \gamma \sum_{j=1, j \neq i}^{n} N_j}{K}\right) N_i \tag{1.5}$$

It can be noted that when new species sequentially enter the system, the populations adapt in such a way that they converge to equal values among competing species. This

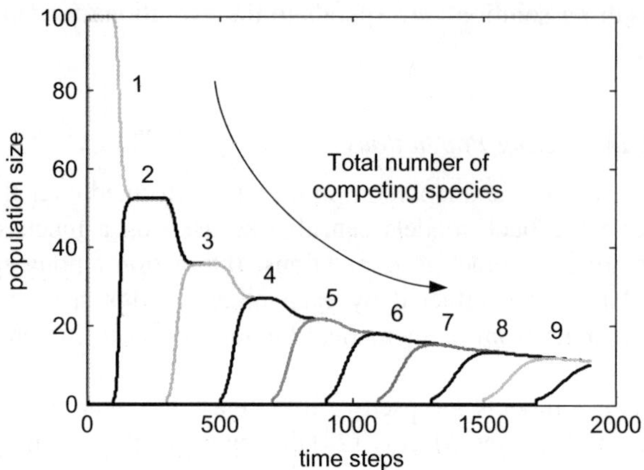

Figure 1.10 Competition model with 10 species that are successively added

feature can be used for example to control the adaptation rate of TCP's congestion control mechanism in a faster, fairer and more efficient way than by the *additive increase multiplicative decrease* (AIMD) approach currently found in TCP [15].

Figure 1.10 illustrates the changes in populations of 10 species when using Equation (1.5), where $K = 100$, $\varepsilon = 1.95$, and $\gamma = 0.90$. Each of the 10 species successively joins the system and their survival and convergence conditions are identical by using the same mutual influence factor γ for all of them. Even when two or more species exist, each independently utilizes Equation (1.5) to obtain N_i, and the population of the species converges to the value equally shared among all competing species. The changing population trends of the species depicted in Figure 1.10 are ideal for controlling the transmission speed of TCP. That is, by using Equation (1.5) for the congestion control algorithm of TCP, rapid and stable link utilization can be realized, whereas each TCP connection can behave independently as an autonomous distributed system. The population size of a species N_i corresponds to the transmission rate of a TCP connection i and K can be seen as the physical bandwidth of the bottleneck link. The method requires, however, knowledge of the transmission rates of all other connections sharing the bottleneck link represented by the sum over all other N_j. This is estimated by obtaining the available bandwidth on the bottleneck link through inline measurements. Unlike other TCP variants, the method using a multispecies Lotka–Volterra model for TCP transmission rate control does not show the typical sawtooth-shape when packet loss is encountered, but rather converges smoothly to the optimal solution. Simulation studies also indicate that the method shows higher scalability than other TCP variants and is therefore well suited to be applied as transport layer protocol for future high-speed networks.

1.4.2 Epidemic Diffusion

Another idea which originates from interacting population models in biology is the concept of using epidemic protocols for information diffusion [11]. Especially, in P2P networks, epidemic models can easily be applied due to their scalability, robustness, resilience to failure and ease of deployment. Epidemic algorithms are based on the spread of a

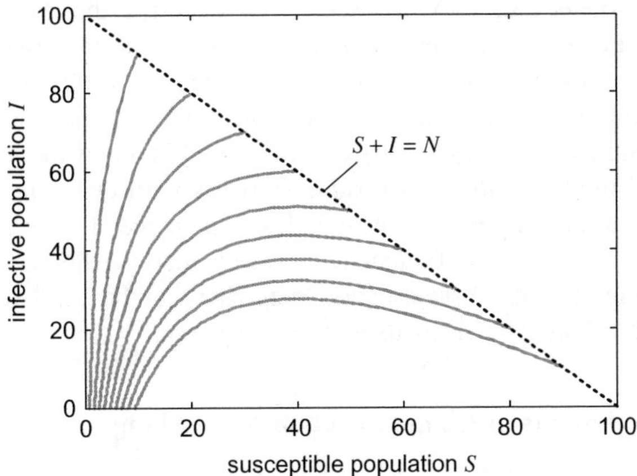

Figure 1.11 Trajectory plot of the SIR model in the S–I phase plane

contagious disease where individuals of an initially healthy population of susceptibles become infected by those bearing the disease. After infection, they may themselves infect other susceptibles. The key aspect of this behavior is that the selection of a susceptible is entirely random and each individual has the same probability of becoming infected.

1.4.2.1 The SIR Model

Mathematical models are used to predict if a disease will turn into an epidemic or can be contained by vaccination [24]. Consider the outbreak of a disease in a closed system with a total population of N which remains constant, i.e., no individual enters or leaves the system. Then the population can be divided into three classes: the *susceptibles* S, who can be infected by the disease; the *infectives* I, who already have been infected and can transmit it to members of the susceptible class; and finally, there is the *removed class* R, who have either recovered from the disease, or become immune to it. This model is usually referred to as the *SIR model* and variations of it exist in the form of SI or SEIR, where a further class of individuals in which the disease is latent is added before being infectious.

Let $r > 0$ and $a > 0$ be constant parameters of the model corresponding to the *infection rate* and *removal rate* of infectives, respectively. Then, the transitions between the populations can be described by the following equation system.

$$\frac{dS}{dt} = -r\,S\,I \qquad \frac{dI}{dt} = r\,S\,I - a\,I \qquad \frac{dR}{dt} = a\,I \tag{1.6}$$

The mathematical formulation requires a set of initial conditions, given by the initial populations $S(0) > 0$, $I(0) > 0$ and $R(0) = 0$. If $S(0) < \rho = a/r$, which is called the *relative removal rate*, then $I(t)$ decreases and approaches 0 for $t \to \infty$, so the infection dies out. On the other hand, if $S(0) > \rho$, there is an epidemic since $I(t)$ initially increases. Due to this threshold behavior of $S(0)$ at ρ, we also speak of a *threshold problem*. The phase plot in the $S-I$ phase plane is shown in Figure 1.11 for $N = 100$, $a = 2$, and $r = 0.05$. Since $R(0) = 0$, all trajectories start on the line with $S + I = N$ and do not exceed this line.

Such types of models can also be used to characterize the diffusion properties of network protocols. For example, Leibnitz *et al.* [21] use a model based on SIR to predict the influence of pollution, i.e., maliciously injected contents, in P2P file-sharing networks. In a finite population of peers connected to a P2P network, several peers sharing a corrupt version of the file are assumed. The mathematical model includes the popularity of certain files and a detailed model of the downloading process with error checks and bandwidth sharing, as well as a user model which includes impatience. It was shown that a small number of malicious peers is sufficient to severely disrupt the file diffusion process. Studies investigating the file dissemination behavior can be used for designing more efficient content distribution systems using P2P networks.

1.4.2.2 Epidemic Methods in Communication Networking

A lot of theoretical research has been conducted on the mathematics of epidemics and many variants of epidemic information dissemination methods for communication networks have been studied. While mathematical models like the SIR model are used to predict how to contain a disease, the goal in epidemic diffusion is rather to 'infect' as many nodes as possible by sending information to them.

Basically, every node buffers the message it receives and forwards it a number of times to another node from a set of limited size, denoted as the *fan-out*. The outcome of the dissemination process is usually bimodal as indicated by the threshold behavior described in the previous section. Either almost all nodes receive the information at the end of the process or almost none. Eugster *et al.* [11] identify four key requirements on epidemic algorithms as:

- *Membership* – how nodes perceive neighboring nodes.
- *Network awareness* – how to make the connections between nodes reflect the actual network topology.
- *Buffer management* – which information to drop when the storage buffer of a node is full.
- *Message filtering* – how to make the nodes only filter information that is relevant to them.

The main advantage of epidemic algorithms is that they do not need to detect or reconfigure in the case of failures as they proactively choose the next hops when they are needed. Furthermore, the decision of forwarding a packet is made autonomously by each node, so if a single node fails, the system will nevertheless continue in operation.

While epidemic diffusion is used for distributing information to a large group of nodes, the basic principle can also be observed in *gossip*-based routing [14] in ad hoc networks. Due to the high degree of mobility and nondeterministic topology in ad hoc networks, the destination node must first be queried and a path set up to it. Usually, this is done by flooding, i.e., duplicating packets at each node and passing them on to each neighbor which has not yet processed this request. However, this causes a lot of overhead packets that can be reduced when the next hop node is chosen only with a certain probability.

1.5 Conclusion

In this chapter we discussed some basic principles and examples of biologically inspired networking. Methods from biology find increasing interest among the research community due to their attractive features like scalability and resilience to changes in the environment. By getting inspiration from biological systems, we can establish fully distributed and self-organizing networks. Especially networks which operate under no coordinating unit with highly autonomous nodes can profit from such methods. Furthermore, by utilizing the inherent fluctuations and noise in biologically inspired networks, they become more tolerant to perturbations, resulting in a greater stability and resilience.

However, we should also keep the following limitations in mind. Biologically inspired methods are often slower in reaction than conventional control algorithms. The reason for this lies in the way the adaptation is performed in nature. Often many thousand generations are needed for a species to evolve and adapt to a changing environment. Furthermore, adaptation is often done without knowing the target function, but simply by negative feedback. Individuals which have mutated to bear unfavorable features with disadvantages compared to other individuals will die out (*survival of the fittest*). This trial-and-error development is often driven by fluctuations and is thus not a directed development to a certain genetic feature. When biological methods are applied, it should be considered that they are in many cases useful heuristics which perform well for certain problem types, especially in the presence of incomplete or fuzzy input data, but may not be as good in performance as real optimization methods.

For this reason, we should be careful not to simply mimic biology, but to also keep a close eye on the problem for which a solution is sought. It is of great importance to thoroughly understand the specific features of the biological system before exploiting it in an interdisciplinary manner [29]. Therefore, after designing a new algorithm based on a biological method and its desirable features, it is necessary to move toward the application of the model and establish a more concrete mechanism with the detailed networking problem in mind. Not all self-adaptive approaches are equally applicable to all network types. However, we should certainly take inspiration from biology since nothing is more adaptive and resilient than nature.

References

[1] Alberts, B., Johnson, A., Lewis, J. *et al.* (2002) *Molecular Biology of the Cell*, Garland Science, New York.

[2] Bonabeau, E., Dorigo, M. and Theraulaz, G. (1999) *Swarm Intelligence: From Nature to Artificial Systems*, Oxford University Press, New York.

[3] Carreras, I., Chlamtac, I., De Pellegrini, F., *et al.* (2007) BIONETS: bio-inspired networking for pervasive communication environments, *IEEE Transactions on Vehicular Technology*, **56**(1), 218–29.

[4] Dasgupta, D. (ed.) (1999) *Artificial Immune Systems and Their Applications*, Springer, Berlin.

[5] Di Caro, G. and Dorigo, M. (1998) Distributed stigmergetic control for communications networks. *Journal of Artificial Intelligence Research*, **9**, 317–65.

[6] Di Caro, G., Ducatelle, F. and Gambardella, L.M. (2005) AntHocNet: an adaptive nature-inspired algorithm for routing in mobile ad hoc networks. *European Transactions on Telecommunications (Special Issue on Self-organization in Mobile Networking)*, **16**(5), 443–55.

[7] Dorigo, M. and Stützle, T. (2004) *Ant Colony Optimization*, MIT Press, Cambridge.

[8] Dressler, F. (2005) Efficient and scalable communication in autonomous networking using bio-inspired mechanisms – an overview. *Informatica*, **29**(2), 183–8.

[9] Dressler, F. (2006) Benefits of bio-inspired technologies for networked embedded systems: an overview. *Proceedings of Dagstuhl Seminar 06031 on Organic Computing – Controlled Emergence*, January, Schloss Dagstuhl, Wadern, Germany.

[10] Durvy, M. and Thiran, P. (2005) Reaction-diffusion based transmission patterns for ad hoc networks. *Proceedings of the IEEE Conference on Computer Communications (INFOCOM)*, March, Miami, FL.

[11] Eugster, P.T., Guerraoui, R., Kermarrec, A.-M. *et al.* (2004) Epidemic information dissemination in distributed systems. *IEEE Computer*, **37**(5), 60–7.

[12] Forrest, S., Hofmeyr, S. and Somayaji, A. (1997) Computer immunology. *Communications of the ACM*, **40**(10), 88–96.

[13] Garrett, S. (2005) How do we evaluate artificial immune systems? *Evolutionary Computation*, **13**(2), 145–78.

[14] Haas, Z., Halpern, J. and Li, L. (2006) Gossip-based ad hoc routing. *IEEE/ACM Transactions on Networking*, **14**(3), 479–91.

[15] Hasegawa, G. and Murata, M. (2006) TCP symbiosis: congestion control mechanisms of TCP based on Lotka–Volterra competition model. *Proceedings of the Workshop on Interdisciplinary Systems Approach in Performance Evaluation and Design of Computer & Communication Systems (Inter-Perf)*, October, Pisa, Italy.

[16] Henderson, T.C., Venkataraman, R. and Choikim, G. (2004) Reaction–diffusion patterns in smart sensor networks. *Proceedings of the IEEE International Conference on Robotics and Automation (ICRA)*, April, New Orleans, LA.

[17] Hopfield, J.J. (1982) Neural networks and physical systems with emergent collective computational abilities. *Proceedings of the National Academy of Sciences of the USA*, **79**(8), 2554–8.

[18] Kaneko, K. (2006) *Life: An Introduction to Complex Systems Biology*, Springer, Berlin.

[19] Kashiwagi, A., Urabe, I., Kaneko, K. *et al.* (2006) Adaptive response of a gene network to environmental changes by fitness-induced attractor selection. *PLoS ONE*, **1**(1), e49.

[20] Krüger, B. and Dressler, F. (2005) Molecular processes as a basis for autonomous networking. *IPSI Transactions on Advanced Research: Issues in Computer Science and Engineering*, **1**(1), 43–50.

[21] Leibnitz, K., Hoßfeld, T., Wakamiya, N. *et al.* (2006) On pollution in eDonkey-like peer-to-peer file-sharing networks. *Proceedings of the 13th GI/ITG Conference on Measurement, Modeling, and Evaluation of Computer and Communication Systems (MMB)*, March, Nuremberg, Germany.

[22] Leibnitz, K., Wakamiya, N. and Murata, M. (2006) Biologically inspired self-adaptive multi-path routing in overlay networks. *Communications of the ACM*, **49**(3), 62–7.

[23] Mirollo, R.E. and Strogatz, S.H. (1990) Synchronization of pulse-coupled biological oscillators. *Society for Industrial and Applied Mathematics Journal on Applied Mathematics*, **25**(6), 1645–62.

[24] Murray, J.D. (2002) *Mathematical Biology I: An Introduction*, Springer, Berlin.

[25] Perkins, C. and Royer, E. (1999) Ad hoc on-demand distance vector routing. *Proceedings of the 2nd IEEE Workshop on Mobile Computing System and Applications*, February, New Orleans, LA.

[26] Sarafijanovic, S. and Le Boudec, J.-Y. (2005) An artificial immune system for misbehavior detection in mobile ad-hoc networks with virtual thymus, clustering, danger signal and memory detectors. *International Journal of Unconventional Computing*, **1**(3), 221–54.

[27] Suzuki, J. and Suda, T. (2005) A middleware platform for a biologically inspired network architecture supporting autonomous and adaptive applications. *IEEE Journal on Selected Areas in Communications*, **23**(2), 249–60.

[28] Timmis, J., Knight, T., De Castro, L.N., *et al.* (2004) An overview of artificial immune systems. In R. Paton, H. Bolouri, M. Holcombe *et al.* (eds.), *Computation in Cells and Tissues: Perspectives and Tools for Thought*, Natural Computation Series, pp.51–86, Springer, Berlin.

[29] Timmis, J., Amos, M., Banzhaf, W. *et al.* (2006) Going back to our roots: second generation biocomputing. *International Journal of Unconventional Computing*, **2**(4), 349–78.

[30] Turing, A.M. (1952) The chemical basis of morphogenesis. *Philosophical Transaction of the Royal Society (London)*, B(237), 37–72.

[31] Wakamiya, N. and Murata, M. (2005) Synchronization-based data gathering scheme for sensor networks. *IEICE Transactions on Communications (Special Issue on Ubiquitous Networks)*, E88-B(3), 873–1.

[32] Wang, M. and Suda, T. (2001) The bio-networking architecture: a biologically inspired approach to the design of scalable, adaptive, and survivable/available network applications. *Proceeding of the IEEE 2001 Symposium on Applications and the Internet (SAINT)*, January, San Diego, CA.

[33] Yoshida, A., Aoki, K. and Araki, S. (2005) Cooperative control based on reaction-diffusion equation for surveillance system. *Proceedings of the International Conference on Knowledge-Based & Intelligent Information & Engineering Systems (KES)*, September, Melbourne, Australia.

2

The Role of Autonomic Networking in Cognitive Networks

John Strassner

Waterford Institute of Technology, Waterford, Ireland; Autonomic Computing, Motorola Labs

2.1 Introduction and Background

Consumers always want more for less. Network equipment vendors, service providers and network operators have responded by building smarter devices that are full of features; concurrently, network operators are offering new types of access methods based on technological advances. This is exacerbated by the desire of service providers to converge different types of networks, both as a means to offer new applications as well as to support new service models. These and other factors are changing the way people communicate and access entertainment, news, music and other types of media, whether using fixed or mobile devices. This gives rise to a very complex environment. At the same time, additional complexity exacerbating this can be found in many areas. [17] lists five principal sources of complexity:

- Separation of business- and technology-specific information that relate to the same subject.
- Inability to harmonize network management data that is inherently different.
- Inability to cope with new functionality and new technologies due to lack of a common design philosophy used by all components.
- Isolation of common management data into separate repositories.
- Inability to respond to user and environmental changes over the course of the system lifecycle.

Cognitive Networks: Towards Self-Aware Networks Edited by Qusay H. Mahmoud

Furthermore, the above problems don't take into account challenges in the radio space, which include:

- Optimization of spectrum usage (including unlicensed spectrum).
- Inability of mobile terminals to implement efficient network selection without consuming significant battery power and taking a relatively long time.
- Coexistence of different radio access technologies to meet user demands while providing fairness and revenue balancing to participating operators.

The so-called 'Beyond the 3rd Generation (B3G)' [5] trend predicts that the set of discrete and often competitive radio access technologies will eventually be transformed into one seamless global communication infrastructure. This requires advanced networking topologies, such as cognitive networks, which use self-configuration to dynamically adapt the functionality offered to meet changing operational and context changes. Unfortunately, this increases the overall complexity of the network and devices using network resources and services even more.

Autonomics is first and foremost, a way to *manage complexity*. The name 'autonomic' was deliberately chosen by IBM to invite comparisons to biological mechanisms [3]. To see this, consider the following analogy: the autonomic nervous system is the part of the body that governs involuntary functions, such as pumping blood and regulating the heart rate, which frees the human mind to perform cognitive functions, such as planning which tasks to perform during the course of a day. Similarly, if the autonomic system can perform manual, time-consuming tasks (such as device configuration changes in response to simple problems) on behalf of the network administrator, then these actions will save the system administrator valuable time, enabling him or her to perform higher level cognitive network functions, such as network planning and optimization. Please see Chapter 1 for more information on biologically inspired approaches to networking.

The remainder of this chapter is organized as follows. A brief introduction to autonomic computing is described in Section 2.2. The differences between autonomic *computing* and autonomic *networking* are examined in Section 2.3. An autonomic networking architecture specifically designed for network management, FOCALE, is presented in Section 2.4. Section 2.5 gives an overview of its application in the context of cognitive networks – both wired and wireless. Section 2.6 discusses challenges and future developments for autonomic networking, while Section 2.7 presents conclusions for this chapter. The final sections provide a glossary of terms and references.

2.2 Foundations of Autonomic Computing

This section will describe the IBM autonomic computing architecture in order to illustrate the fundamental points of autonomic computing.

The fundamental management element of an autonomic computing architecture is a control loop, as defined in [3] [4] and [6]. This control loop starts with gathering sensor information, which is then analyzed to determine if any correction to the managed resource(s) being monitored is needed. If so, then those corrections are planned, and

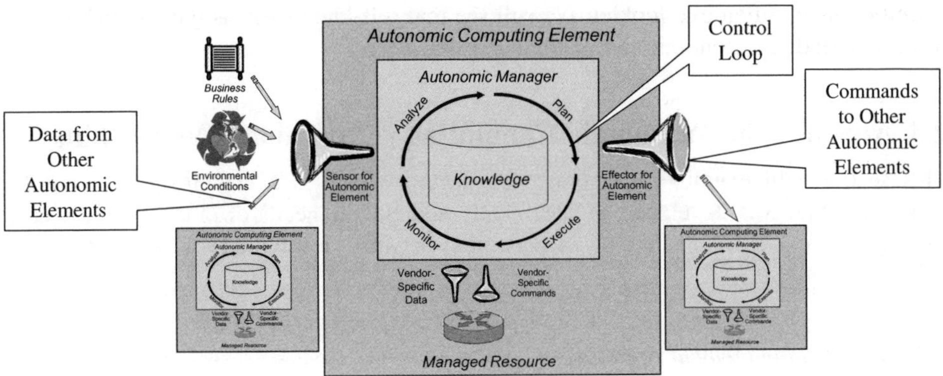

Figure 2.1 IBM's autonomic control loop

appropriate actions are executed using effectors that translate commands back to a form that the managed resource(s) can understand. This is conceptually illustrated in Figure 2.1.

Autonomic computing uses a control loop to monitor the *state* of the resource being managed, and take corrective action if the current state of the resource is not equal to its desired state. This usually results in the reconfiguration of that managed resource, though it can also result in the reconfiguration of other managed resources that are affecting the state of the managed resource that is being monitored. More specifically, the elements of the autonomic control loop are responsible for monitoring the managed resource and other relevant data about the environment in which it is operating, analyze those data and take action if the state of the managed resource and/or system is changed to an undesirable state. Here, undesirable means 'non-optimal' as well as 'failed' or 'error'. The autonomic element is a building block, in which the autonomic manager communicates with other types of autonomic and non-autonomic managers using the sensors and effectors of the autonomic manager.

The autonomic manager provides the overall guidance for the autonomic element in collecting, analyzing and acting on data collected from the managed resource via its sensors. It consists of four parts that govern the functionality of the control loop. The monitor portion gathers data as directed by the autonomic manager, filters and collates it if required, and then presents it to the analysis portion. This portion seeks to understand the data, and to determine if the managed resource is acting as desired. The planning part takes the conclusions of the analysis part and determines if action should be taken to reconfigure the managed resource being monitored (and/or other managed resources as appropriate) using predefined policies that establish the goals and objectives that the autonomic manager enforces. The execute portion translates the plan into a set of commands that direct any reconfiguration required.

This simple model of autonomic computing leads to its definition in terms of a set of self-functions, the usual being self-configuration, self-healing, self-optimization and self-protection. However, in order to implement any of these four self-functions, the system must first be able to *know* itself and its environment. This is called self-knowledge, and is unfortunately often overlooked. We will see that self-knowledge is the key to cognition in general and to autonomic networking in particular.

2.3 Advances in Autonomic Computing – Autonomic Networking

This section will examine autonomic *networking*, which is a new form of autonomic computing that was developed to specifically address the needs of embedding increased intelligence and decision making in networks, network devices and networked applications.

2.3.1 A New Management Approach

The name 'autonomic' was deliberately chosen by IBM to invite comparisons to biological mechanisms. The vast majority of past and current work on autonomic computing focuses on IT issues, such as host systems, storage and database query performance. However, networks and networked devices and applications have different needs and operate under different assumptions than IT issues. Autonomic networking focuses on the application of autonomic principles to make networks and networked devices and applications more intelligent, primarily by enabling them to make decisions without having to consult with a human administrator or user.

The motivation behind autonomic networking is simple: identify those functions that can be done without human intervention to reduce the dependence on skilled resources. This not only lowers operational costs, but also enables these valuable resources to be used for more complex tasks. It also enables the network to respond faster and more accurately to the changing needs of users and businesses. The left of Figure 2.2 shows a human body, while the picture on the right is a UML model of a simple (honest! ☺) application. The autonomic nervous system and the brain in the human body are the 'management plane' and 'inference plane', respectively, of autonomic networking, which are explained below.

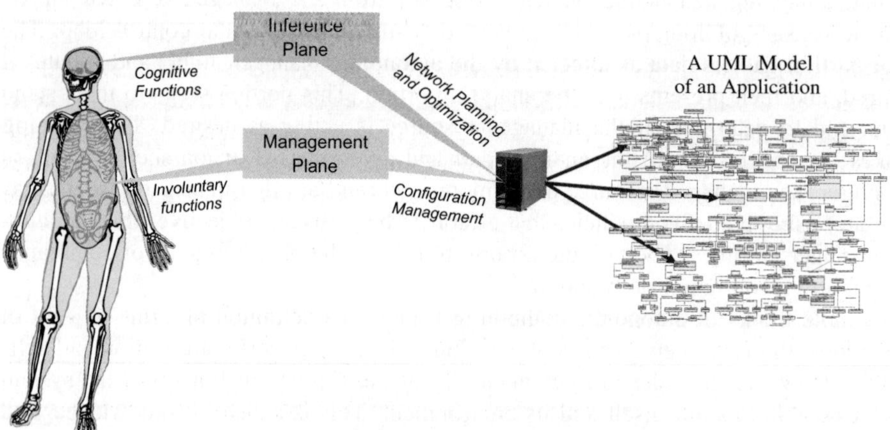

Figure 2.2 Analogies for Motorola's autonomic research

The first, and most important analogy, is that complexity is everywhere. Hence, autonomics is, first and foremost, a *way to manage complexity*. This is conceptualized by the complexity of the UML application shown in figure 2 above – the number of components and their complicated interconnection approaches that of the complexity of the interconnection of the autonomic nervous system with components of the surrounding human body. The networking analogy to this is as follows: the *involuntary* functions of the body equate to manual, time-consuming tasks (such as configuration management) performed by the network administrator, while the cognitive functions of the brain compare to higher level, more complex system-wide functions, such as planning and optimization of the network. In other words, autonomics does not necessarily *replace* humans; rather, it performs time-consuming tasks so that skilled human resources can be better leveraged.

This idea – automating simple, *safe* decisions and corrective actions – can be extrapolated into a new management approach, one that will be particularly useful for next generation networks as well as cognitive networks. This new management approach is founded on the realization that multiple networks as well as networks with different functionalities will use multiple control mechanisms that are not compatible with each other. Figure 2.3 shows a conceptual view of how networks are traditionally managed. In this approach, the data plane is a contiguous path used for data communications, while the control plane is used to set up, manage and terminate connections in the data plane.

Clark *et al.* [2] proposed a new concept, called the Knowledge Plane. The Knowledge Plane, as shown in Figure 2.4, is a fundamentally new construct that breaks the traditional approach used in Figure 2.3.

The objective of this research is to '...to build a fundamentally different sort of network that can assemble itself given high level instructions, reassemble itself as requirements change, automatically discover when something goes wrong, and automatically fix a detected problem or explain why it cannot do so.' Specifically, *cognitive* techniques are relied upon (instead of traditional algorithmic approaches) in order to *deal with the complexity*.

Motorola has defined a different approach, based on maintaining compatibility with legacy devices and applications, thereby preserving existing investments. Its approach is to define two new planes to augment the functions of the data and control planes: the *management* plane and the *inference* plane. The reasoning behind formalizing these two planes is as follows. While the data plane is contiguous, the control plane is not. In fact, if there are multiple networks and/or network technologies in use, then the current state-of-the-art requires multiple control planes. For example, think of the differences in control functions between wired and wireless networks. Unfortunately, these diverse control planes can use completely different mechanisms, and hence, managing an end-to-end service

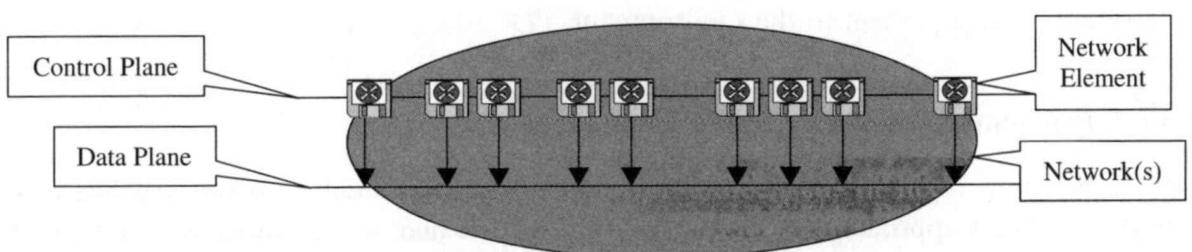

Figure 2.3 Traditional network management approach

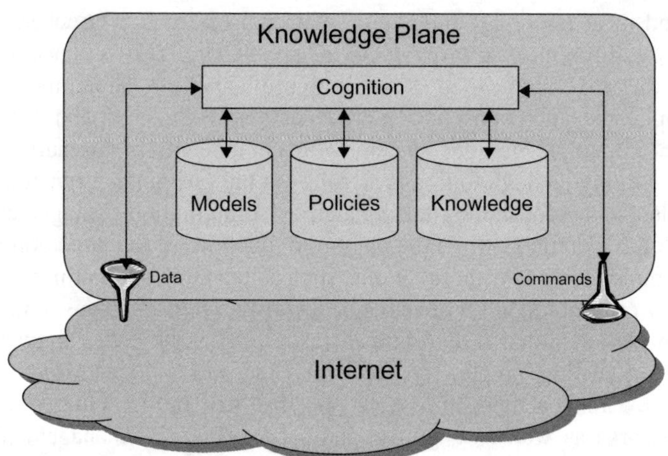

Figure 2.4 The Knowledge Plane

is difficult since different management mechanisms must be coordinated. Therefore, the management plane is used to ensure that the different control planes are *coordinated* and *working together*.

The reasoning behind building a separate *inferencing* plane is simple. In current environments, user needs and environmental conditions can change without warning. This will be exacerbated by next-generation networked applications such as seamless mobility (see Section 2.5.1 later in this chapter). Therefore, the inference plane is used to analyze the system, the environment and the needs of the users with respect to business objectives, instructing the management plane to reconfigure one or more components in order to protect the current business objectives. The FOCALE architecture, defined in Section 2.4 of this chapter, embodies this approach.

2.3.2 Managing Complexity

The technical aspects of controlling system complexity are well described in the literature. However, the business aspects of controlling complexity have received significantly less attention. The TeleManagement Forum's NGOSS program ([15] [16] [26] [27] [28]) has a well-articulated methodology to address the needs of different stakeholders (e.g., business analysts vs. system architects vs. programmers and technicians) throughout the lifecycle of a solution. Previous and current efforts [25] have addressed the potential link between NGOSS and autonomic computing, but have not addressed how to manage and integrate diverse knowledge present in the environment. FOCALE addresses these issues.

2.3.2.1 Problems with Management Languages

One of the difficulties in dealing with complexity is accommodating legacy hardware and software. Current approaches to network configuration and management are technology-specific and network-centric, and do not take business needs into account. For example, monitoring and (re)configuration are performed on most network devices using one or more management protocols and languages. The most pervasive of these are the Simple Network Management Protocol (SNMP), and proprietary vendor-specific Command Line Interfaces (CLIs). SNMP and CLI have three important problems. First, the proliferation of important management information that is only found in different, vendor- and technology-specific forms, such as private MIBs (management information bases, a tree-structured set of data for network management) and the many different incompatible dialects of CLIs, exacerbate this problem. Second, common management information, defined in a standard representation, is not available. While there will most likely never be a single unified information model (just as there will never be one single programming language), there must be an extensible, common modeling basis for integrating these diverse data, or else common management operations and deductions cannot be performed across heterogeneous devices. For this, the DEN-ng model ([1] [14] [15]) is used as a unified information model; diverse knowledge is added to existing knowledge using a novel ontology-based approach, described later in this chapter. Third, neither SNMP nor CLI is able to express business rules, policies and processes, which make it impossible to use these technologies to directly change the configuration of network elements in response to new or altered business requirements [15]. This disconnects the main stakeholders in the system (e.g., business analysts, who determine how the business is run, from network technicians, who implement network services on behalf of the business).

In addition, don't underestimate the problems in different device operating system versions! Script-based configuration solutions, such as PERL or EXPECT, cannot cope with syntax changes. Command signature changes, such as additional or reordered parameters, exacerbate the problem and commonly occur as network device operating systems are versioned. Finally, note that most network equipment vendors structure their device software to address different management scenarios (e.g., the needs of an enterprise customer vs. those of a service provider customer, as well as special customer-specific releases). These releases make management even harder.

2.3.2.2 Different Programming Models

Another problem is the use of different programming models. Figure 2.5 shows an example of the complexity caused by different programming models of network devices – a system administrator defining two different routers from two different manufacturers as BGP (the Border Gateway Protocol, http://www.ietf.org/rfc/rfc1772.txt) peers. BGP is a standard, so this should be easy, right? *Wrong!*

As can be seen from Figure 2.5, syntactical differences are the least of the problems encountered. Semantics are only sometimes evident from the command structure, and then only if the user has intimate knowledge with the particular version of the command language being used. Behavior, such as resource usage and side effects, is completely hidden from the user. Worse, the two devices shown have completely different programming models! The Cisco router on the left uses command configuration *modes*, which the Juniper router on the right does not use. Furthermore, the Juniper router supports *virtual routing*, which enables the resources of the router to be assigned to different processes. Hence, the Juniper router actually supports *multiple BGP instances*. In contrast, the Cisco router on the left has no such capability, and hence only supports a single, global BGP instance.

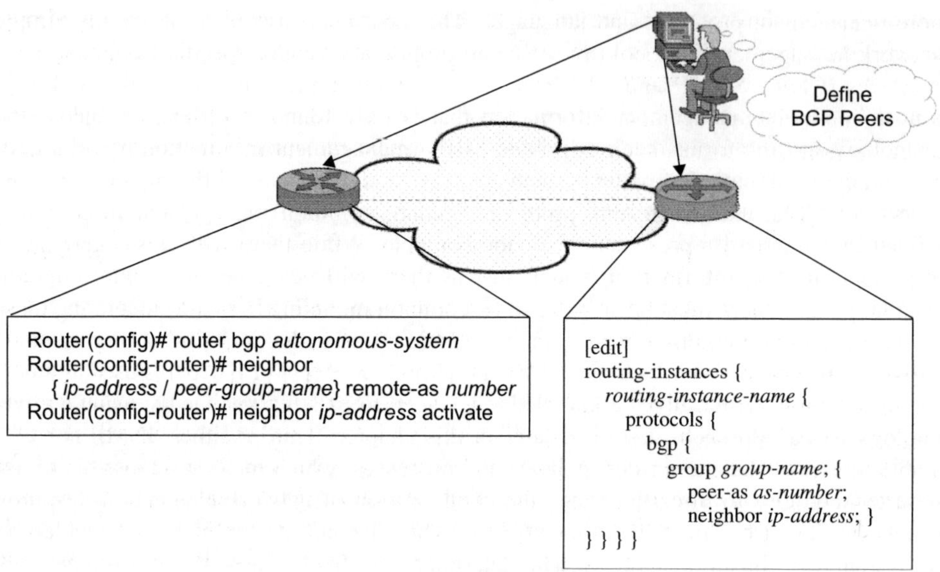

Figure 2.5 Different programming models of a router

2.3.2.3 Lack of a Common Network Lingua Franca

Standards are an important means to define common data, protocols, and approaches for multiple vendors to interoperate. However, most standards are either not detailed enough or deliberately avoid trying to standardize management information at too specific a level. This is typified in the design of SNMP MIBs, where high-level common data is standardized, while vendor-specific data is defined only in vendor-specific proprietary MIBs. CLIs are, if anything, worse – these are pseudo-languages that are specifically designed to accomplish a set of functions for a targeted vendor-specific platform. As stated earlier in Section 2.3.2.1, there must be an extensible basis for integrating these diverse data in order to synthesize related information from diverse sources. Our approach is based on using information and data modeling as defined in the DEN-ng information model as *facts*; diverse knowledge, along with the ability to *reason* about facts, is added to existing knowledge using a novel ontology-based approach, described in Section 2.3.4 later in this chapter.

Figure 2.5 illustrates the following problems:

- The syntax of the language used by each vendor is different.
- The semantics of the language used by each vendor is different.
- The router on the left uses *modes*, which aren't present in the router on the right, which implies their programming models are different.
- The router on the right supports virtual routing (via its routing-instances command) which the router on the left does not support.

There are other differences as well in their functionality that aren't shown. In effect, by using these two different routers, the operator has built two different stovepipes in its network, since administrators are forced to learn the vendor-specific differences of each router. This is how the industry builds Operational Support Systems (OSSs) and Business Support Systems (BSSs) today.

Therefore, if this cannot be automated (due to its complexity), how can it be solved? Certainly not manually, since besides being the source of errors, humans may not even be present, which means that problems can perpetuate and become worse in the system. To better understand this problem, consider the following analogy: the language spoken by a device corresponds to a language spoken by a person, and different dialects of a language correspond to different capabilities in different versions of the device operating system. Most vendors change syntax and semantics of their languages with major and sometimes minor releases of their device operating systems, thereby producing more and more language dialects. How can a single administrator, or even a small number of administrators, learn such a large number of different languages for managing the different types of network devices in a network? More importantly, if there is an end-to-end path through the network consisting of multiple devices, how can this path be configured and managed without a single 'über-language'? Remember that most systems use vendor-specific CLIs for configuration, but typically rely on SNMP or other standards for monitoring. This is because most vendors do not support SNMP configuration, and even if they do, that functionality is limited. CLIs are almost always more powerful because they are typically used in the development of the product. In contrast, network management vendors typically use SNMP because they can (in theory) avoid the problems of parsing different CLIs. However, this is only partly true, since SNMP supports vendor-specific management information. Hence, the problem of translating multiple vendor-specific languages to and from SNMP exists.

2.3.3 The Role of Information and Data Models in Autonomic Computing

Recently, interest in using a single information model ([13] [15] [17] [21]) has increased. The idea is to use a single information model to define the management information definitions and representations that *all* OSS and BSS components will use. This is shown in Figure 2.6

The top layer represents a single, enterprise-wide information model, from which multiple standards-based data models are derived (e.g., one for relational databases, and one for directories). Since most vendors add their own extensions, the lowest layer provides a

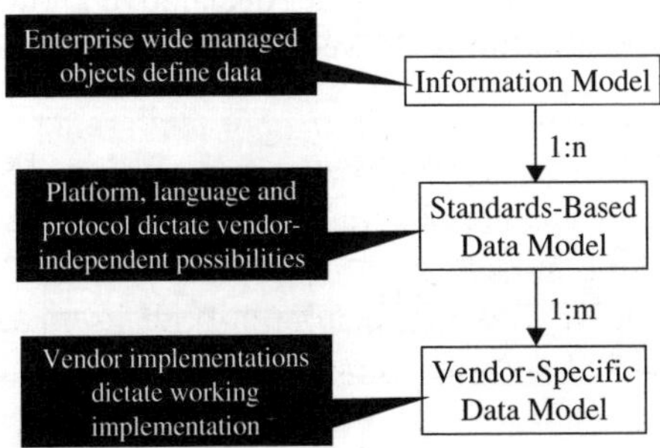

Figure 2.6 The relationship between information and data models

second model mapping to build a high-performance vendor-specific implementation (e.g., translate SQL92 standard commands to a vendor-proprietary version of SQL). Note that this is especially important when an object containing multiple attributes is split up, causing some of the object's attributes to be put in one data store and the rest of the object's attributes to be put in a different data store. Without this set of hierarchical relationships, data consistency and coherency is lost.

This works with legacy applications by forming a single mediation layer. In other words, each existing vendor-specific application, with its existing stovepipe data model, is mapped to the above data model, enabling applications to share and reuse data. In the BGP peering example shown in Figure 2.5, a single information model representing common functionality (peering using BGP) is defined; software can then be built that enables this single common information model to be translated to vendor-specific implementations that support the different languages of each router manufacturer. Hence, we reduce the complexity of representing information from n^2 (where each vendor's language is mapped to every other language) to n mappings (each vendor's language to a common data model).

2.3.4 *The Role of Ontologies in Autonomic Computing*

Knowledge exists in many different forms. In network management, humans and applications are used to using knowledge in a particular form. This is why private knowledge bases have been and continue to be developed – they represent information in a way that is easy for users of a particular application to understand and use.

The obvious problem with this approach is that there is no easy way to share data in private knowledge bases with other applications. This is a serious problem, one that most likely will not be solved by restructuring existing knowledge bases (since they exist because users like them!). The solution requires two advances. First, a common lexicon is needed – this enables mapping multiple existing data models to a single common model. Second, a means for representing the semantics associated with management data is required. This is where ontologies can help.

Facts are provided by information and data models. However, models cannot be used to resolve differences in these facts. Ontologies can be used to provide a set of formal mechanisms for defining a set of definitions for facts, as well as a rich set of relationships between those facts. This can be used to define an interoperable set of definitions and meanings (not just synonyms and antonyms, but more complicated relationships as well, such as holonyms and meronyms). This is shown in Figure 2.7.

More formally, ontologies are used to model declarative knowledge. By this, we mean knowledge in which the relationships between the data are declared, or stated. We use

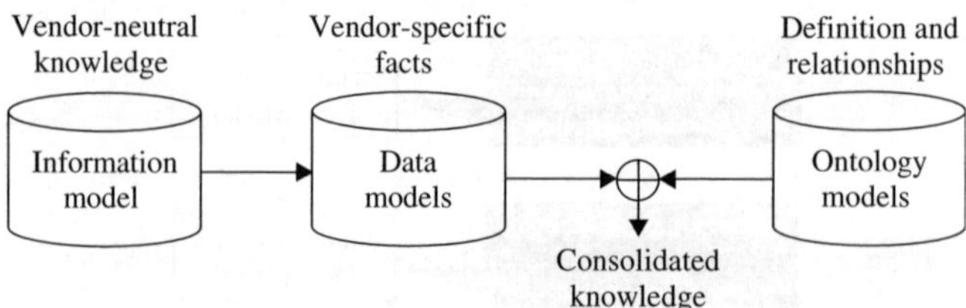

Figure 2.7 Using ontologies and models to build a common lexicon

ontologies to build a lexicon by providing a *set of definitions* for each term (e.g., a DEN-ng modeled object) as well as a *set of relationships* between each term. The set of relationships enables us to choose which definition best fits our current context by examining the different management data and *inferring* the correct definition to use through using the appropriate relationships [19].

Ontologies use *inference engines* to answer queries about the data. Representing knowledge in ontologies enables a formal, extensible set of mechanisms to be used to define mappings between the terms of knowledge bases and their intended meanings. In this way, we achieve knowledge interoperability by using a set of ontologies to precisely and unambiguously identify syntactic and semantic areas of interoperability between each vendor-specific language and programming model used. The former refers to reusability in parsing data, while the latter refers to mappings between terms, which require content analysis. Conceptually, we have the construction of knowledge from facts, augmented by the appropriate definition and relationships, to fit a particular context. The role of context in our architecture is described in the next section.

2.3.5 The Role of Context

A simplified form of the DEN-ng context model is shown in Figure 2.8. This context model is unique, in that it relates Context to Management Information to Policy. At a high level, this model works as follows: Context determines the working set of Policies that can be invoked; this working set of Policies defines the set of Roles and Profiles that can be assumed by the set of ManagedEntities involved in defining Context. Significantly, this model also defines the set of management information that is used to determine how the Managed Element is operating.

Note that in Figure 2.8, the combination of Context and Policy is used to determine not just which roles are allowed to be assigned, but also the type of management information that should be monitored. The use of roles provides two advantages: (1) it avoids explicitly identifying any particular entity and instead indirectly identifies entities through

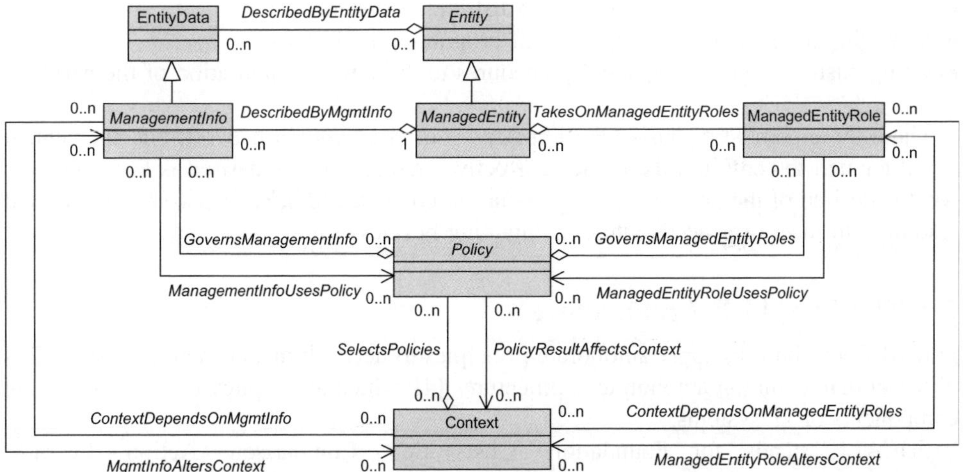

Figure 2.8 Simplified form of the DEN-ng context model

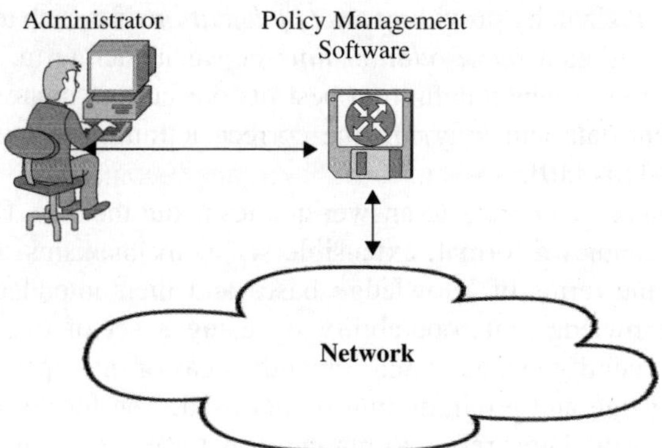

Figure 2.9 The promise of policy management

enabling and disabling roles; and (2) it enables access control schemes, such as role-based access control (RBAC; http://csrc.nist.gov/rbac/). Similarly, the identification of roles not only defines which functionality a managed entity can take on; it also defines which management information should be monitored as a function of context.

2.3.6 A Better Policy Management Approach

The promises of policy management are varied and powerful. These promises are often conceptualized as a single, simple means to control the network, as illustrated in Figure 2.9.

In particular, the ability to hide vendor-specific interfaces behind a uniform policy model is very important. Without this ability, a common interface to managing the programming of the same function in different network devices cannot be accomplished.

While there are many value propositions for policy-based network management [14], three in particular are applicable to cognitive networking:

- Defining the behavior of a network or distributed system.
- Managing the increasing complexity of programming devices.
- Using business requirements and procedures to drive the configuration of the network.

These three points emphasize the importance of *cognition* in network management. It is not enough to configure the device correctly – rather, what is desired is to *understand* what is desired of the network under the current context, and *adapt* network services and resources to meet user needs while maintaining business rules.

2.4 The FOCALE Architecture

FOCALE was built to apply autonomic principles to network management. As such, it is different than common autonomic architectures [4], which are applied to non-networking components of IT systems.

FOCALE stands for Foundation – Observation – Comparison – Action – Learn – rEason. It was named because these six elements describe the six key principles required to support autonomic networking.

The idea behind FOCALE is simple enough, and is summarized as follows. As discussed in Sections 2.2 and 2.3, autonomic principles manage complexity while enabling the system to adjust to the changing demands of its users and environmental conditions. In order for the network to dynamically adjust the services and resources that it provides, its components must first be appropriately configured and reconfigured. Assume that behavior can be defined using a set of finite state machines, and that the configuration of each device is determined from this information. FOCALE is a closed loop system, in which the current state of the managed element is calculated, compared to the desired state (defined in the finite state machines), and then action taken if the two states aren't equal. This can be expanded as follows: define a closed control loop, in which the autonomic system senses changes in itself and its environment; those changes are then analyzed to ensure that business goals and objectives are still being met. If they are, keep monitoring; if they aren't, then plan changes to be made if business goals and objectives are threatened, execute those changes, and observe the result to ensure that the correct action was taken.

However, since networks are complex, highly interconnected systems, the above approach is modified in several important ways. First, FOCALE uses multiple control loops, as will be explained below. Second, FOCALE uses a combination of information models, data models and ontologies for three things: (1) develop its state machines; (2) determine the actual state of the managed element; and (3) understand the meaning of sensor data so that the correct set of actions can be taken. Third, FOCALE provides the ability to *change* the functions of the control loop based on context, policy and the semantics of the data as well as the current management operation being processed. Fourth, FOCALE uses reasoning mechanisms to generate hypotheses as to why the actual state of the managed element is not equal to its desired state, and develops theories and axioms about the system. Finally, FOCALE uses learning mechanisms to update its knowledge base. These differences will be explained in the following sections.

2.4.1 FOCALE's Control Loops

Figure 2.10 shows that in reality, there are two control loops in FOCALE (as opposed to one, which is pictured in most other autonomic systems). The desired state of the managed resource is predefined in the appropriate state machine(s), and is based on business

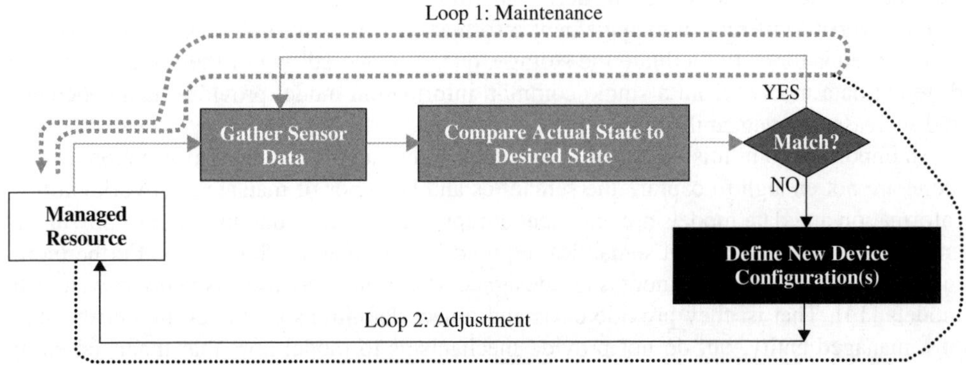

Figure 2.10 Control loops of FOCALE

goals [16] [20] [22]. The top (maintenance) control loop is used when no anomalies are found (i.e., when either the current state is equal to the actual state, or when the state of the managed element is moving towards its intended goal). The bottom (adjustment) control loop is used when one or more reconfiguration actions must be performed. Note that most alternative architectures propose a single loop; FOCALE uses multiple control loops to provide better and more flexible management. Remember that in network management, correcting a problem may involve more actions affecting more entities than the original managed entity in which the problem was noticed. Put another way, why should a control loop designed for *monitoring* be useful for performing a completely different action?

The use of two different control loops, one for maintenance operations and one for reconfiguration operations, is fundamental to overcoming the limitations of using a single static control loop having fixed functionality. Since FOCALE is designed to *adapt* its functionality as a function of *context*, the control loop controlling the reconfiguration process must be able to have its functionality adapted to suit the vendor-specific needs of the different devices being adapted.

Another important reason to use multiple control loops is to protect the set of business goals and objectives of the users as well as the network operators. The implementation of these objectives is different and sometimes in conflict – having a single control loop to protect these objectives is simply not feasible.

The reconfiguration process uses dynamic code generation based on models and ontologies ([18] [23] [20] [22]). The models are used to populate the state machines that in turn specify the operation of each entity that the autonomic system is governing. The management information that the autonomic system is monitoring signals any context changes, which in turn adjusts the set of policies that are being used to govern the system, which in turn supplies new information to the state machines. The goal of the reconfiguration process is specified by the state machines, hence, new configuration commands are dynamically constructed from these state machines.

2.4.2 The Foundation – Integrating Models and Ontologies

An information model can be thought of as the defining document that is used to model all of the different managed objects in a managed environment. DEN-ng specifies a single information model that is used to derive multiple data models. This is required because of the diversity inherent in management information, which in turn necessitates different types of repositories to facilitate the storage, querying and editing of these data. Deriving different data models from a single common information model provides data coherency and increased manageability [15] [24].

An important point missed by most standards bodies and fora is that information models alone are not enough to capture the semantics and behavior of managed network entities. Information and data models are excellent at representing facts, but do not have any inherent mechanisms to represent semantics required to reason about those facts. Furthermore, most information and data models are designed as current state models using private data models [14]. That is, they provide class and object definitions to model the current state of a managed entity, but do not provide mechanisms to model how that managed entity changes state over its lifecycle. IETF MIBs and ITU-T M, G, and X Recommendations are examples of 'current state' models. Autonomic and cognitive applications require monitoring the lifecycle of a managed entity, not just its current state. UML was designed to do this, but note that a significant amount of management data, such as that produced by the IETF and the ITU-T, is not in UML. The TeleManagement Forum (TMF) and 3GPP/2 are among the few standards bodies and fora that require UML-compliant models to be built. (Note that the Distributed Management Task Force, or DMTF, does not produce UML-compliant models, since it redefines MOF! MOF in the UML sense stands for the Meta Object Facility [10], and is used for modeling and representing metadata, including the UML metamodel; a metamodel is defined as a language for representing a model. In contrast, the DMTF's MOF standards for Managed Object Format, and is a way to describe object-oriented class and instance definitions in textual form. Hence, OMG MOF is *not* the same as DMTF MOF (http://www.dmtf.org/education/mof/), even though the DMTF claims that their models are "MOF-like".). The FOCALE architecture uses UML definitions augmented with ontological information to produce rich data that is then fed into machine-based learning and reasoning algorithms to avoid these problems. Hence, the FOCALE architecture is able to include legacy knowledge while providing means to edit and incorporate new knowledge.

In general, an information or data model may have multiple ontologies associated with it. This is because ontologies are used to represent relationships and semantics that cannot be represented using information languages, such as UML. For example, even the latest version of UML doesn't have the ability to represent the relationship 'is similar to' because it doesn't define logic mechanisms to enable this comparison. (Note that this relationship is critical for heterogeneous end-to-end management, since different devices have different languages, programming models and side effects [20].)

Hence, the approach used in FOCALE relies on the fusion of information models, ontologies and machine learning and reasoning algorithms to enable semantic interoperability between different models. The approach used in FOCALE is to develop a set of ontologies that provide a set of meanings and relationships for facts defined in the information model and data models. This enables the system to *reason* about the facts represented in the information and data models.

The autonomic manager uses the ontologies to analyze sensed data to determine the current state of the managed entities being monitored. Often, this task requires inferring knowledge from incomplete facts. For example, consider the receipt of an SNMP alarm. This is a potentially important fact, especially if the severity of the alarm is assigned as 'major' or 'critical'. However, the alarm in and of itself doesn't provide the *business* information that the system needs. Which customers are affected by the alarm? Which Service Level Agreements (SLAs) of which customers are effected? These and other questions are critical in enabling OSSs and BSSs to decide which problems should be worked on in what order.

Given the above example, FOCALE tries to determine automatically (i.e., without human intervention) which SLAs of which customer are impacted. Once an SLA is identified, it can be linked to business information, which in turn can assign the priority of solving this problem. FOCALE uses a process known as semantic similarity matching [29] to establish additional semantic relationships between sensed data and known facts. This

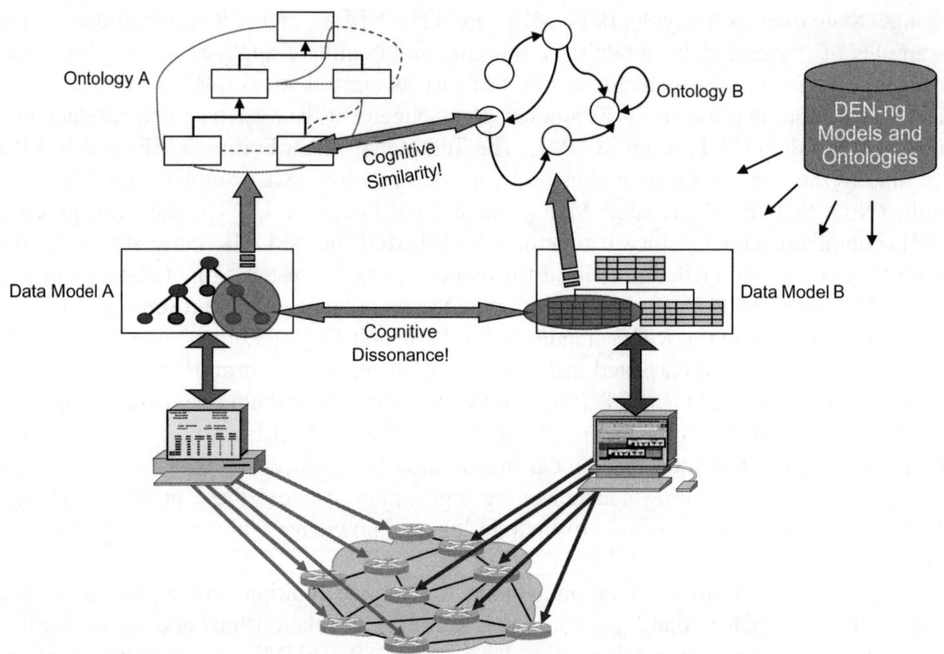

Figure 2.11 Semantic knowledge processing

is required because, in this example, an SLA is *not* directly related in the model to an SNMP alarm. The inference process is thus used to establish semantic relationships between the fact that an SNMP alarm was received and other facts that can be used to determine which SLAs and which customers could be affected by that SNMP alarm.

Figure 2.11 illustrates this process in use. In this figure, two different devices use two different programming languages, represented by the two different data models. They are being used to provide an 'end-to-end' service – hence, two equivalent command sets, one for each device, need to be found. Since the two devices use two different languages, there is a cognitive dissonance present. This is solved by mapping the information in each data model to a set of concepts in each ontology; then, a semantic similarity algorithm is used to define the set of commands in each ontology that best match each other.

2.4.3 Gathering Sensor Data and Sending Reconfiguration Commands

Networks almost always use heterogeneous devices and technologies. Therefore, FOCALE provides an input mediation layer to map different languages and technologies into a single language that can be used by the rest of the autonomic system. For example, it is common to use an SNMP-based system to monitor the network, and a vendor-proprietary CLI, because many devices cannot be configured using SNMP. Hence our dilemma: there is no standard way to know which SNMP commands to issue to see if the CLI command worked as intended or not. The idea of the mediation layer is to enable a legacy managed element with no inherent autonomic capabilities to communicate with

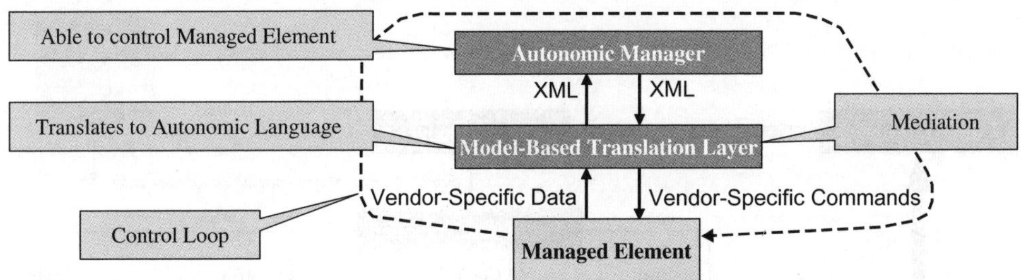

Figure 2.12 Input/output mediation layer of FOCALE

autonomic elements and systems. A model-based translation layer (MBTL), working with an autonomic manager, performs this task.

The MBTL solves this problem by accepting vendor-specific data from the managed element and translating it into a normalized representation of that data, described in XML, for further processing by the autonomic manager. Figure 2.12 shows how the MBTL works. Two assumptions are made: (1) the managed element has no inherent autonomic capabilities; and (2) the managed element provides vendor-specific data and accepts vendor-specific commands. The MBTL enables diverse, heterogeneous managed elements to be controlled by the same autonomic manager by translating their vendor-specific languages and management information into a common form that the autonomic manager uses, and vice versa. Conceptually, the MBTL is a programmable engine, with different blades dedicated to understanding one or more languages.

The MBTL translates vendor-specific data to a common XML-based language that the autonomic manager uses to analyze the data, from which the actual state of the managed element is determined. This is then compared to the desired state of the managed element (as defined by the finite state machine). If the two are different, the autonomic manager directs the issuance of any reconfiguration actions required; these actions are converted to vendor-specific commands by the MBTL to effect the reconfiguration.

2.4.4 The Role of Context and Policy Management

In traditional autonomic systems, the control loop is static. The autonomic manager is limited to performing the state comparison and computation of the reconfiguration action(s), if required, *inside* the control loop. While this meets the needs of static systems, it fails to meet the needs of systems that change dynamically, since the functions inside the control loop cannot change. Thus, FOCALE places the control function (i.e., the autonomic manager) *outside* of the loop. This enables independent control over the constituent elements of the control loop. Figure 2.13 shows an enhanced architecture that uses policy to control the autonomic manager, which then controls each of the architectural components of the control loop. As stated previously, context changes are detected by the system, which in turn change the active policies that are being used at any given time. Note that the use of two different control loops, one for maintenance operations and one for reconfiguration operations, is fundamental to overcoming the limitations of a single static control loop having fixed functionality.

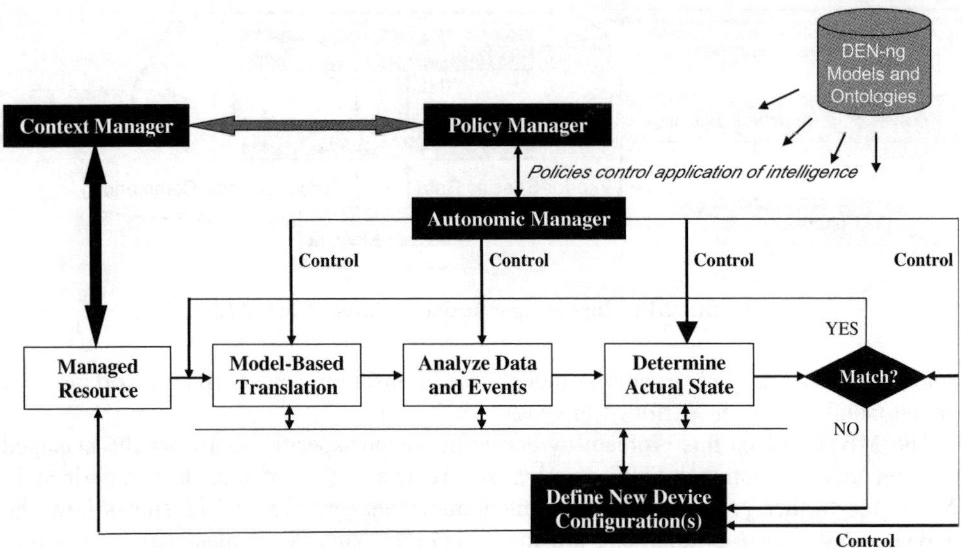

Figure 2.13 Role of context and policy management in FOCALE

The autonomic manager is responsible for controlling and adjusting (as necessary) the functionality of each of the components of the control loop. For example, the autonomic manager may determine that, due to the type of data that needs to be processed, a multilevel threshold match is more appropriate than looking for a particular attribute value.

The policy manager is responsible for translating business requirements written in a restricted natural language (e.g., English) into a form that can be used to configure network resources [18]. For example, the business rule 'John gets Gold Service' will be translated into a set of configuration commands that will be applied to all affected devices. This translation is done via the policy continuum ([14] [23]), which forms a continuum of policies in which each policy captures the requirements of a particular constituency. For example, business people can define an SLA in business terms; this is then transformed into architectural requirements and finally network configuration commands. Note that each policy of the SLA do not 'disappear' in this approach – rather, they function as an integrated whole (or continuum) to enable the translation of business requirements to network configuration commands. Its use is shown in Figure 2.14.

Moving the autonomic manager from inside the loop to outside the loop (so that it can control all loop components) is a significant improvement over the state of the art. For example, assume that computational resources are being expended on monitoring an interface, which suddenly goes down. Once the interface is down, it will stay down until it is fixed. FOCALE recognizes this, stops monitoring this interface, and instead starts monitoring other entities that can help establish the root cause of why the interface went down. This re-purposing of computational resources is directed by the autonomic manager, which controls each architectural component.

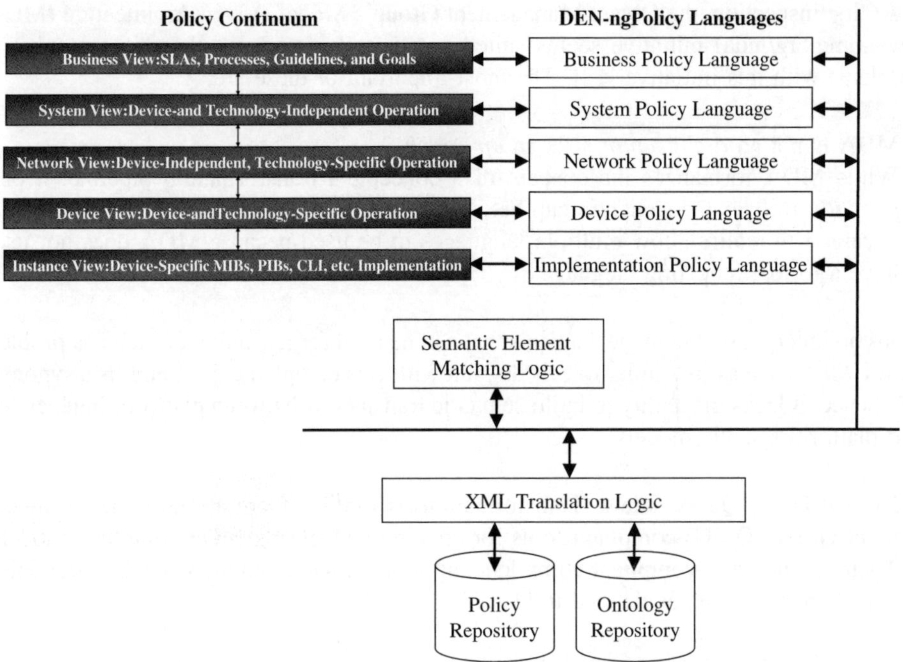

Figure 2.14 Use of the policy continuum in FOCALE

2.4.5 Orchestrating Behavior Using Semantic-Based Reasoning

The mission of any autonomic system is to adapt to change, according to its underlying business rules. Adaptation requires cognition. Once the autonomic manager was moved outside the loop, it made it easier to add cognitive functionality that could help the autonomic manager in its tasks. It also enabled cognition to be used by the autonomic manager to help recognize state and context changes, among other things.

One of the most important novelties of FOCALE is its ability to dynamically adjust policies based on context. Current systems that use policy as part of their autonomic system use it in a static way. Specifically, they assume that the policies that are applied to a system can handle all interactions that the system will have. Current systems are thus examples of 'self-managed' systems. In contrast, FOCALE is a 'self-governed' system. That is, FOCALE uses autonomic mechanisms to modify the policies (or even change them completely) if it determines that those policies are no longer the best policies to use (e.g., because context has changed) to govern the managed element.

Autonomic systems imply a need to generate code, since if the functionality of the system needs to change, then commands have to be issued to realize those changes. Current state-of-the-art systems can generate code statically (in order to reconfigure system components), since its policies and components are defined statically. In contrast, FOCALE must generate code dynamically, according to the current state of the system, to reconfigure one or more managed entities.

At first inspection, the Object Management Group's Model Driven Architecture (MDA; www.omg.org/mda) initiative seems suited to this task. However, there is a number of problems with this initiative [12]. The most important of these are:

- MDA is not an *architecture*, it is an *approach*.
- While MDA formalizes the concept of a conceptual transformation pipeline, it only *partially* bridges the semantic gap between different languages.
- It cannot in reality allow multiple languages to be used because MDA does not itself have a formal modeling language.

It has no inherent semantic mechanisms that can be used for reasoning (which is a problem with UML), such as 'is similar to', 'is disjoint with', 'is complement of' and 'is a synonym of'; hence, it lacks the ability to build semantic translations between platform-independent and platform-specific models.

- Even if QVT (Query–View–Transformation) is ratified, there are no Object Management Group (OMG)-compliant tools for creating and storing MOF-compliant models.
- There are no OCL-compliant compilers and transformation engines (OCL is the object constraint language, as defined in [10]).

Remember that device languages are *programming languages*, and thus are not full-fledged conversational languages. Device vendors will never agree on a common programming language, so FOCALE provides a device 'Esperanto' using the MBTL. This provides a *language translation* between the management system and each device language being used, as well as between different device languages. Ontologies help solve this problem from a linguistics point of view by defining a *shared* vocabulary, along with rules for defining the structure and grammar of the common language. This structure focuses on the structure and formation of words in the vocabulary (morphology), sentence structure (syntax), and meaning and meaning change (semantics).

Figure 2.15 shows an example illustrating this problem. We have a set of M total devices, each having a different number of commands (i commands for Device A, and j commands for Device M). So, our first objective is to build relationships between one command in one device to a set of commands in another device. In this example, command 1 of Device A (denoted as Command 1) is semantically related to the set of commands 2 and 3 of Device B (denoted as Command 2 and Command 3) with 85% equivalence (denoted as the '0.85' annotation under the arrow).

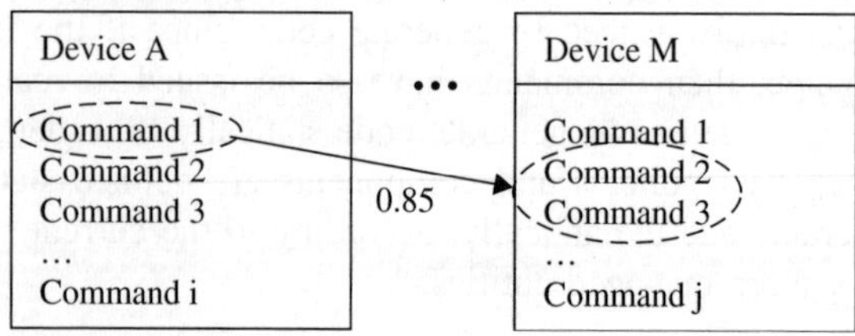

Figure 2.15 Mapping a command from one device to a set of commands in another device

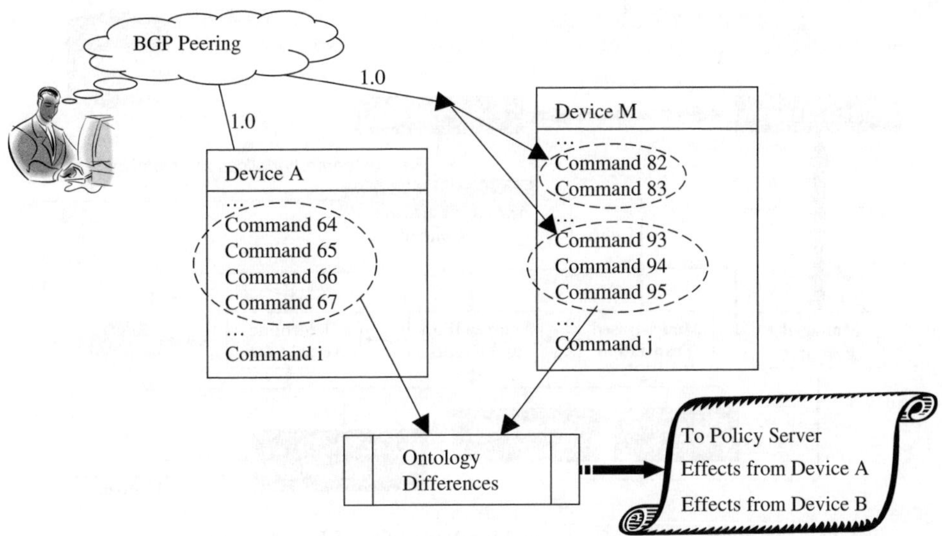

Figure 2.16 Mapping a high-level function to different command sets in different devices

Note that this is a *unidirectional* mapping. Once a set of these relationships is defined, we can build another mapping that abstracts this to a set of common functions. For example, enabling BGP peering, as shown in Figure 2.5, proceeds as first identifying the high-level function (BGP) to be performed, and then mapping this function to the appropriate set of commands in each device. This is shown in Figure 2.16.

In Figure 2.16, we see that the same high-level function (BGP peering) is mapped to two different sets of commands. This reflects the fundamental fact that different devices have different programming structures, and perform the same high-level task using different commands. Significantly, this means that sometimes the semantics of performing these tasks are different, and must be taken into account. Hence, we need an ontology merging tool (or set of tools) to find these semantic differences.

2.4.6 The Resulting Architecture – FOCALE

FOCALE stands for Foundation, Observation, Comparison, Action and Learning Environment. The complete high-level FOCALE architecture is shown in Figure 2.17. This approach assumes that any managed element (which can be as simple as a device interface, or as complex as an entire system or network) can be turned into an autonomic computing element (ACE) by connecting the managed element to the same autonomic manager using the MBTL engine and appropriate blades. By embedding the same autonomic manager and the same MBTL engine in each ACE, uniform management functionality is provided throughout the autonomic system, which simplifies the distribution of its management functionality.

The two new components in this diagram are the 'ontological comparison' and the 'reasoning and learning' functions. The former implements semantic equivalence matching, as shown in Figure 2.15 and Figure 2.16, and is used by other components to find

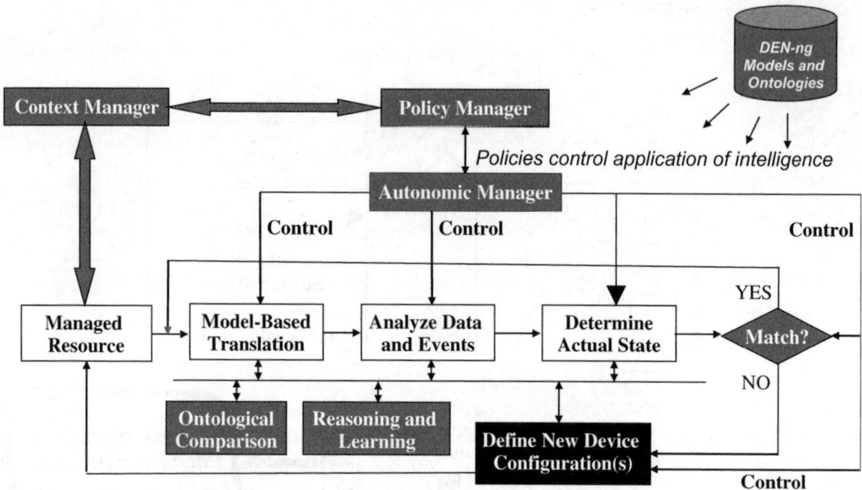

Figure 2.17 Complete high-level FOCALE architecture

equivalent semantic terms in the analysis, learning and reasoning functions; the latter implements reasoning and learning algorithms.

The bus shown in Figure 2.17 enables the reasoning and learning components to 'watch' the current operation being performed, and be used in addition to or instead of that operation. For example, machine learning can examine the operations being performed, and note the effectiveness of the actions taken given the context, policies and input data used. This can be used to build a knowledge base that can help guide future decisions. Similarly, an abductive reasoning algorithm can be used to generate hypotheses as to the root cause of problems sensed, which the autonomic manager then tries to verify by using the models and ontologies to query the system and the environment to gather more data to support each hypothesis. The combination of these learning and reasoning algorithms enable the autonomic system to *adapt* to changing business goals, user needs and environmental conditions.

2.5 Application to Wired and Wireless Cognitive Networks

This section will show how the FOCALE architecture can be used in cognitive networks by looking at Motorola's Seamless Mobility initiative (http://www.motorola.com/content.jsp?globalObjectId=6611-9309). For other connotations of seamless mobility, please see http://w3.antd.nist.gov/seamlessandsecure.shtml. This initiative aims to deliver a continuity of experiences across different domains, networks and devices. FOCALE is used to distribute intelligence by augmenting sensor data with important semantics. These semantics include learning and reasoning mechanisms to enable the system to hypothesize and infer what is happening and what should be done to protect business goals and objectives. The FOCALE architecture is easily scalable by using a common building block (the ACE) as a means to provide an autonomic interface for managed elements that have no inherent autonomic capabilities. Since each ACE has the same management interface, governance actions can be easily distributed.

2.5.1 Seamless Mobility

Seamless mobility is motivated by several forces. First, networks and technologies are converging, and will end up providing a single, inclusive seamless experience. This has given rise to more powerful devices that can themselves create a wide variety of different networks – from personal area networks centered in a car to mobile ad hoc networks that are formed and dissolved when needed. Second, everything is being digitized, which enables – at least in theory! – easier access to a wide variety of data. Third, and most important, the boundary between work, home, entertainment and other periods of life is dissolving, becoming a set of roles that blend as opposed to different physical and logical places and times that have hard demarcations.

The above three reasons drive what Motorola calls an 'experiential architecture', which is the foundation of Motorola's seamless mobility experience.

Motorola defines seamless mobility as 'a set of solutions that will provide easy, uninterrupted access to information, entertainment, communication, monitoring and control – when, where and how we want, regardless of the device, service, network or location'. While handover between different networks and technologies (e.g., cellular to VoIP, or WiMAX to WLAN) is important, the essence of seamless mobility is to *preserve the user's context and session*. This does not imply the use of an omnipotent device or a converged mega-network. Rather, it envisions interoperable devices and networks, with intelligence to manage the boundaries or seams and give the user a consistent experience. This intelligence needs to be coordinated and distributed. For example, a user can have multiple devices that are online but distributed – how are they coordinated, how do they learn user preferences, and so forth.

Motorola's vision of seamless mobility is based on an experiential architecture that is built from standards-based, reusable software elements within a distributed framework. The architecture incorporates a continuum of networks, ranging from wide area (e.g., CDMA, GSM, cable and other current offerings to emerging 4G networks) to shorter range networks (e.g., 802.11, ZigBee and Bluetooth) that may be deployed in businesses, homes, vehicles, public places and even as wearable devices. All are connected to a common IP core network through appropriate gateways. Each network has its own mobility manager that assists with seamless transitions between devices and technologies within its network and to other networks.

The seamless mobility functional architecture is shown in Figure 2.18. It is built to provide a compelling end-user experience by integrating network and application management. The purpose of the connectivity architecture is to provide seamless handover between sessions while providing heterogeneous network access. The purpose of the experience architecture is to guide the provisioning and activation of functionality for end users. Finally, security and manageability are addressed at all levels of the architecture.

Seamless mobility devices have the intelligence and flexibility to respond to changing needs and environmental conditions. Note that this does not mean that the device has to be an autonomic computing element; rather, as long as the device can communicate with the autonomic system, it can be integrated into the autonomic system. This enables the user's goal, personal context and learned personal preferences to determine the type and extent of interaction. This is where the FOCALE architecture is used.

In particular, the MBTL module of FOCALE gains added importance. Sensor information, especially for determining context, is diverse in semantics, data structures, protocols

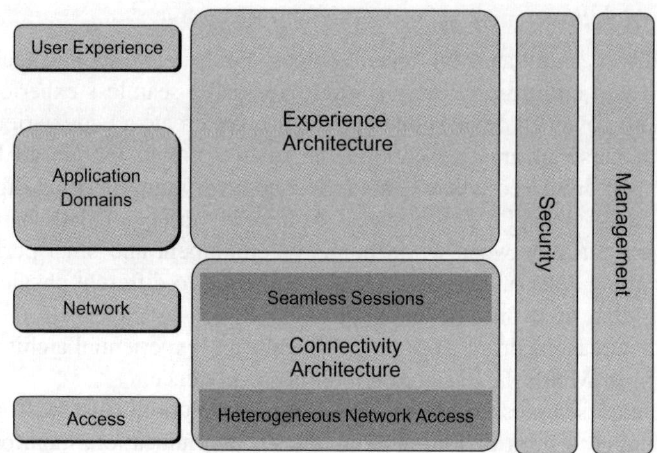

Figure 2.18 Motorola's seamless mobility functional architecture

and data. Most sensor information cannot be naturally combined. Hence, the MBTL serves the important function of accepting the appropriate sensor data and translating it to a form that the rest of the autonomic system can understand.

2.5.2 Application to Wired Networking

FOCALE was originally tested against wired networking scenarios. This is because many aspects of wired networking are in fact more straightforward to compute than similar aspects of wireless networking, due to the simple fact that analysis is often simpler (e.g., wireless signals are affected by other wireless signals, such as those of phones) and quality of service (QoS) is markedly easier to deploy. DEN-ng also had its start in IP networking, and hence has more developed models to define facts for the learning and reasoning processes.

2.5.3 Cognitive Networking Functions

Motorola has defined the following categories that enable intelligent interaction in a cognitive environment:

- **Goal-oriented:** users must be able to focus on their tasks without having to spend effort in telling devices what they want and reconfiguring devices. In other words, users want management to be as invisible as possible. This means that management functionality must be more intuitive and flexible, enabling dynamic tasks that adapt to a user's needs to replace the static, predefined procedures that cannot take a user's varying needs into account.
- **Context-aware:** devices should be able to differentiate between various locations, users and goals, and operate appropriately (e.g., by vibrating instead of ringing in a theater).
- **Adaptive to user needs:** devices should be able to deduce information based on user patterns and responses as well as environmental conditions in order to optimize the behavior of the device at any given time.

- **Interactive:** devices must be able to deal with user input ambiguity and prompt the user for appropriate clarification when required.

An interesting dichotomy exists: while users want intelligent devices, studies have shown that they are unwilling to pay for them. Furthermore, network equipment vendors tend to allocate footprint size to features that are consumer driven. For example, space and power are extremely scarce resources for mobile phones. User-centric features, such as embedded cameras with higher resolution, or more codecs, are much easier features to sell than the nebulous promise of 'more intelligent devices'! Our FOCALE architecture deals with this problem by not mandating devices to be rebuilt. Rather, the use of intelligent agents that implement the ACE façade is used.

By design, the FOCALE architecture makes technological complexity invisible, while enabling complete freedom of movement between devices and environments.

2.5.4 Cognitive Wireless Networking Requirements

The following technologies are some of the means for providing the flexibility and functionality required by cognitive wireless capabilities:

- **Reconfigurable wireless access**, provided by the ability to reconfigure radio access technologies and spectra used.
- **Reconfigurable wireless transmission**, provided by the ability to reconfigure parameters including coding, error control, transmission power and modulation.
- **Optional wireless connectivity to the backbone network**, which is required if mesh or similar network technologies are used.

2.5.5 Cognitive Wireless Networking Extensions to FOCALE

Figure 2.19 shows the modifications made to FOCALE to support wireless cognitive networks. Work on this is currently being done by Motorola Labs. The 'managed resource' in Figure 2.17 is replaced by the combination of a reconfigurable access point (the actual

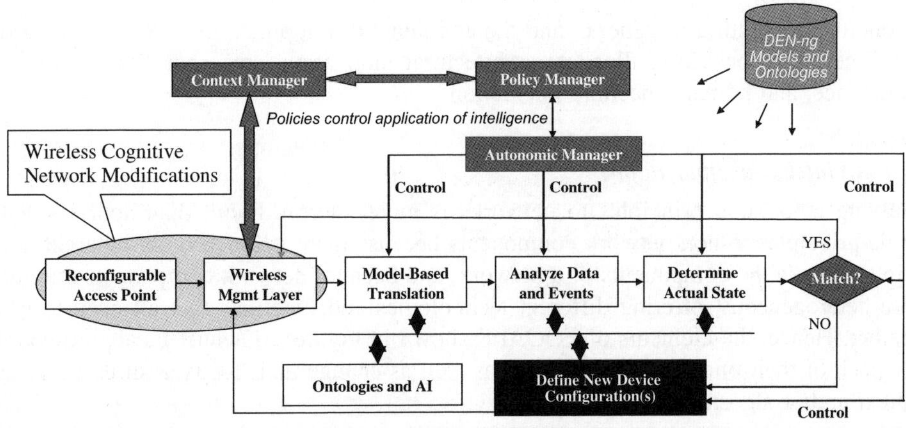

Figure 2.19 FOCALE extended to support wireless cognitive networks

'managed resource') and a wireless management layer (which serves as an application programming interface (API) to FOCALE). The purpose of the wireless management layer is to enable wireless cognitive networking functionality to be integrated with wired networking functionality. The wireless management layer is similar in design to existing cognitive networking projects, and the API enables it to be easily interfaced to FOCALE by building a new MBTL blade that communicates with the API.

The reconfigurable wireless access point uses a completely different protocol stack, and has very different functionality, than the wired networks that FOCALE was designed for. Hence, by using a wireless management layer, this functionality can be translated to a form that is equivalent to functions found in wired networks. This then enables wired and/or wireless networks to be controlled by FOCALE.

As a simple example, consider the adjustment of the radio access technology. In a seamless mobility environment, it is likely that multiple devices used by multiple end users will use different applications, having different service characteristics (e.g., QoS and security). Multimode and reconfigurable devices can connect simultaneously to different wireless networks, and can be reconfigured to change the networks that they connect to. Hence, in a cognitive network, the specific network resources that a device connects to can change as a function of context and the overall business policies of the network.

The wireless management layer presents an API that enables the autonomic system to select the type of radio access technology to use for a given task, and then reconfigure the protocols of the device to enable that technology. There are many different radio access technology selection schemes available: the point of FOCALE is to view these as *resources* that can then be managed using *policies* as a function of the current *context* and *business goals* that are active. In order to do this, FOCALE makes use of its novel semantic analysis using ontologies, models, and learning and reasoning algorithms. One promising implementation is based on game-theoretical techniques, wherein the best strategy is found for each user when their needs conflict with each other (e.g., all users want the 'best' service, but are using different applications having different network resource requirements) [11].

2.6 Challenges and Future Developments

Autonomic computing in general, and the continued development of FOCALE in particular, face three specific challenges: architectural innovations, incorporation of artificial intelligence, and human–machine interaction.

2.6.1 Architectural Innovations

Applying autonomic principles to networks is much more difficult than applying autonomic principles to non-network components because there is much more inherent interaction between the components of a network, and because network components are much more heterogeneous, offering different technologies and functions that all need to work together. Hence, the elements of FOCALE shown in Figure 2.17 must be able to coordinate each of their one or more activities as well as change their focus as directed by any context and/or algorithmic changes.

This will result in more flexible, service-oriented patterns of interaction, as opposed to traditional top-down, hierarchical systems management. Abstraction is especially important in these types of patterns, as gathering information that is seemingly not directly related to each other requires the ability to represent and reason about needs, capabilities, dependencies and constraints of system components, and how they affect protecting business goals.

The FOCALE team is thus interested in defining a set of architectural components and patterns that can support self-knowledge, self-learning and self-reasoning.

2.6.2 Incorporation of Artificial Intelligence

Several different artificial intelligence mechanisms, such as those that support data mining [7] [8], have previously been successfully used to help systems cope with information overload, and more particularly, to determine if information is relevant or not. Machine-learning techniques [9] are used to enable a computer program to learn from experience. More formally, [9] defines machine learning as: 'A computer program is said to learn from experience E with respect to some class of tasks T and performance measure P, if its performance at tasks in T, as measured by P, improves with experience E.' Machine learning is important because the system needs to learn changes to itself as well as to the environment quickly in order to support business goals and objectives. This is complicated by the fact that the environment is very noisy, consisting of a lot of information having different syntax and semantics. Hence, exploration of the environment can be very costly. In addition, there may be hundreds of tunable parameters and adjustments that could each offer help, but choosing the correct ones is difficult to determine due to their interaction and/or attendant cost (e.g., requiring a reboot).

The FOCALE team is thus interested in defining how emergent behavior can be better modeled and understood, so that reasoning and learning may work together to imbue intelligence into the system.

2.6.3 Human–Machine Interaction

The aim of our research in FOCALE is to develop 'invisible management'. Users should not be bound by the user interface of a device, but rather by their needs. Along these lines, we are researching new languages, metaphors and interaction paradigms that enable humans to leverage the capabilities of autonomic networking. For example, if languages can be defined that enable the human to properly specify business goals and objectives to autonomic networks and *visualize* their potential effect, the human can more easily choose among competing alternatives.

2.6.4 Current Progress

FOCALE is currently being prototyped in Motorola Labs in Schaumburg, Illinois. Current research in wired networking has focused on protecting QoS. For example, policies can be defined that instruct the autonomic system what to do when QoS levels are threatened. Detailed models have been built that represent not just the function of different vendor-specific commands (via information and data models), but more importantly, the *effect* of the commands (via ontologies and reasoning algorithms).

In contrast, current research in wireless networking has focused on building reasoning and learning algorithms that are focused on determining how to switch radio access technologies and frequencies as the context changes.

2.7 Conclusions

This chapter has provided a brief introduction to, and motivation for, cognitive wireless networking from a network management viewpoint. Autonomic computing principles were described by first examining IBM's popular autonomic computing architecture. This led to a more in-depth understanding of why existing autonomic approaches were not completely suitable for network management. We then focused on enhancements to the IBM architecture. This built up into an examination of a novel autonomic networking architecture called FOCALE. This architecture was designed specifically to support network management. Specific attention was given to its set of novel features, including the use of multiple control loops, information and data models, ontologies, context awareness and policy management.

FOCALE is a closed loop system, in which the current state of the managed element is calculated, compared to the desired state, and then action taken if the two states aren't equal. Hence, FOCALE uses separate control loops – one for when the state of the managed element is equal to (or converging towards) the desired state, and one for when it isn't. This protects the business goals and objectives of the users as well as the network operators.

FOCALE uses knowledge engineering techniques to manage legacy as well as new (as in inherently autonomic) components. FOCALE orchestrates the behavior of systems and components by using predefined finite state machines. The finite state machines are built from the information and data models, and then augmented with semantics from ontologies. Ontologies are used as a shared vocabulary to enable heterogeneous devices and components to communicate with each other.

After explaining the FOCALE architecture in more detail, the chapter concluded by applying FOCALE to wireless cognitive network management. Motorola's seamless mobility was used as a use case. A very simple modification to FOCALE was made to enable it to manage both wired and wireless networks.

Glossary

The following are definitions of key terms used in this chapter:

- **Autonomic** – Autonomic has many definitions. For the purposes of this chapter, autonomic is defined as self-governing. This is *not* the same as self-managing. Imagine a black box. A self-managing black box is one in which policies that are applied to the black box are able to handle user needs and environmental conditions. Practice has shown that predefined policies are not sufficiently flexible to accomplish this. Therefore, a self-governing system is one in which the system uses self-knowledge and adjusts its behavior to take into account changing user needs, business goals and environmental conditions.
- **Autonomic network** – An autonomic network is one which supports the dynamic delivery of services and resources, as directed by business goals, wherein the nature

of said services and resources *adapt* to the changing environment. Business goals are enforced by using policies to control the *structure* and *operation* of one or more closed control loops, which control the reconfiguration functions.

- **Autonomic system –** An autonomic system is a self-governing system, in which the governance model is expressed using policies. Governance takes the form of a closed control loop, in which the autonomic system senses changes in itself and its environment, analyzes those changes to ensure that business goals and objectives are still being met, plans changes to be made if business goals and objectives are threatened, executes those changes, and observes the result.
- **Cognitive network –** Cognitive networks are networks that can dynamically alter their functionality and/or topology in accordance with the changing needs of its users, taking into account current environmental conditions. This dynamic modification is done in accordance with applicable business rules and regulatory policies.
- **Data model –** The implementation of the characteristics and behavior of a component or system using a particular vendor, platform, language and repository.
- **DEN-ng –** A UML-compliant object-oriented information model that describes managed entities and their relationships. It has three salient characteristics: (1) it uses finite state machines to orchestrate behavior; (2) it is built using patterns and roles (making it inherently extensible); and (3) it consists of static as well as dynamic models.
- **Information model –** A representation of the characteristics and behavior of a component or system independent of vendor, platform, language and repository.
- **Policy –** A set of rules that is used to manage and control the changing and/or maintaining of the state of one or more managed objects.
- **Self-governance –** Self-governance reflects the ability of a system to conduct its own affairs. Self-governance is accomplished through the use of self-knowledge to model system capabilities and environmental constraints as a function of context to: (1) define the desired state of the managed element and (2) determine the actual state of the managed element by comparing data received by the autonomic system to the model of the system.
- **Self-knowledge –** The ability of a system to model the characteristics and behavior of, and the interaction between, the users of the system, the environment in which the system operates, and the components making up the system, as well as the system itself.

References

[1] Agoulmine, N. and Strassner, J. (2004) Gestion de réseaux: de SNMP aux annuaires de politiques. Informatique Professionnelle, 222, 4–10.

[2] Clark, D.D., Partridge, C., Ramming, J. and Wroclawski, J. (2003), *A Knowledge Plane for the Internet.* Sigcomm, pp. 25–9, August.

[3] IBM (2001) *Autonomic manifesto* (2001) www.research.ibm.com/autonomic/manifesto.

[4] IBM (2003) An architectural blueprint for autonomic computing, April.

[5] Jamalipour, A., Wada, T. and Yamazato, T. (2005) A tutorial on multiple access technologies for beyond 3G mobile networks. *IEEE Communications Magazine*, **43**(2), 110–17.

[6] Kephart, J.O. and Chess, D.M. (2003), *The vision of autonomic computing.* In www.research.ibm.com/autonomic/research/papers/AC_Vision_Computer_Jan_2003.pdf.

[7] Larose, D. (2004) *Discovering Knowledge in Data: An Introduction to Data Mining*, John Wiley & Sons, Ltd, Chichester, UK.

[8] Larose, D. (2006) *Data Mining Methods and Models*, John Wiley & Sons, Ltd, Chichester, UK.

[9] Mitchell, T. (1997) *Machine Learning*, McGraw-Hill, New York.
[10] OMG (2003), *OMG Unified Modeling Language Specification*, version 1.5, March.
[11] Osborne, M. (2003), *An Introduction to Game Theory*, Oxford University Press, New York.
[12] Raymer, D., Strassner, J. and Lehtihet, E. (2007) *In search of a model driven reality*. To be submitted to ICSE.
[13] Strassner, J. (2002) *A new paradigm for network management – business driven device management*. In SSGRRs 2002 summer session.
[14] Strassner, J. (2003). *Policy-Based Network Management*. Morgan Kaufman.
[15] Strassner, J. (2004a) *A model-driven architecture for telecommunications systems using DEN-ng*. ICETE 2004 conference.
[16] Strassner, J. (2004b) Building better telecom models through the use of models, contracts, and life cycles. ECUMN, October.
[17] Strassner, J. (2005) *Knowledge management issues for autonomic systems*. In TAKMA 2005 conference.
[18] Strassner, J. (2006) *Seamless mobility – a compelling blend of ubiquitous computing and autonomic computing*. In Dagstuhl Workshop on Autonomic Networking, January.
[19] Strassner, J. (2007) Knowledge engineering using ontologies. *Elsevier Handbook on Network and System Administration*, in press.
[20] Strassner, J. and Kephart, J. (2006) *Autonomic systems and networks: theory and practice*. NOMS 2006 Tutorial.
[21] Strassner, J. and Menich, B. (2005) *Philosophy and methodology for knowledge discovery in autonomic computing systems*. PMKD 2005 conference.
[22] Strassner, J. and Raymer, D. (2006) *Implementing next generation services using policy-based management and autonomic computing principles*. NOMS 2006.
[23] Strassner, J., Raymer, D., Lehtihet, E. and Van der Meer, S. (2006) *End-to-end model-driven policy based network management*. In Policy 2006 Conference.
[24] Strassner, J. and Reilly, J. (2006) *Introduction to the SID*. In TMW University Tutorial, May.
[25] Strassner, J. and Twardus, K. (2004) *Making NGOSS autonomic – resolving complexity*. In Session TECH2, TMW, Nice, May.
[26] TMF (2004a) *The NGOSS lifecycle methodology*. GB927, v1.3, November.
[27] TMF (2004b) The NGOSS technology neutral architecture. TMF053 (and Addenda), June.
[28] TMF (2004c) *Shared Information and Data (SID) model*. GB922 (and Addenda), July.
[29] Wong, A., Ray, P., Parameswaran, N. and Strassner, J. (2005) Ontology mapping for the interoperability problem in network management. *Journal on Selected Areas in Communications*, **23**(10), 2058–68.

3

Adaptive Networks

Jun Lu, Yi Pan, Ryota Egashira, Keita Fujii, Ariffin Yahaya and Tatsuya Suda

Bren School of Information and Computer Sciences, University of California, Irvine

3.1 Introduction

As illustrated in Figure 3.1, modern networks are complex systems composed of many heterogeneous nodes, links and users. Modern networks are often operating in environments where network resources (e.g., node energy and link quality), application data (e.g., the location of data) and user behaviors (e.g., user mobility and user request pattern) experience changes over time. Those changes can degrade network performance and cause service interruption. For example, when the quality of a link being used for data transmission degrades, data being transmitted on the link may be lost. Also, when a user moves to a different location, data that needs to be delivered to the user may be mistakenly sent to the previous location, causing service interruption for the user. In order to maintain performance and service continuity under dynamic environments, networks must provide mechanisms to adapt to changes. Such adaptive mechanisms are implemented within four basic network functions: constructing hop-by-hop connectivity; routing data; scheduling data transmission; and controlling transmission rate. A network adapts to changes by adjusting the behaviors of one or more of those network functions.

This chapter closely examines networks that adapt. It first examines three dynamic factors that trigger network adaptation, namely the dynamic factors of network resources, application data and user behaviors. This chapter then introduces four basic network functions and illustrates typical adaptive techniques for each of the four functions. Typical adaptive techniques are illustrated by describing representative network schemes that adapt to the dynamic factors. For each scheme, this chapter describes the algorithms used to adapt, and their subsequent benefit. This chapter ends with a comprehensive analysis of existing adaptive techniques.

Cognitive Networks: Towards Self-Aware Networks Edited by Qusay H. Mahmoud

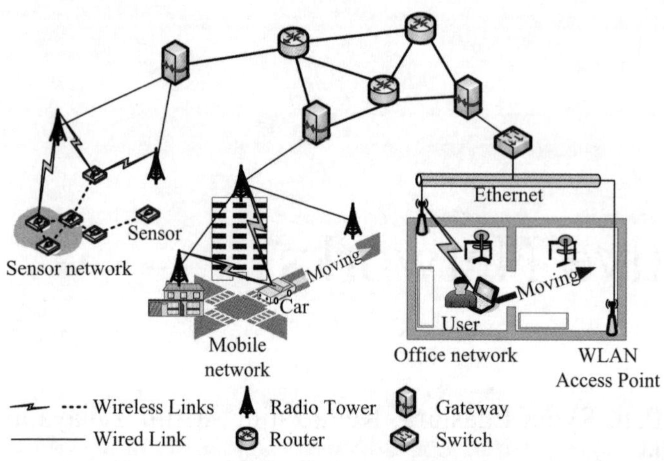

Figure 3.1 Modern network scenarios

The rest of the chapter is organized as follows. Section 3.2 discusses the factors that change and trigger a network to adapt. Section 3.3 explains the four network functions that adapt their behaviors to changes. Section 3.4 presents existing work on networks that adapt by giving a detailed description of representative schemes for each of the four network functions. Section 3.5 provides a review on typical techniques adopted by networks that adapt and Section 3.6 concludes our discussion.

3.2 Dynamic Factors

A network adapts to mitigate the effect of changes in three dynamic factors, namely network resources, application data and user behaviors. This section describes these dynamic factors in detail.

The dynamic factor of network resources refers to both node and link resources. Node resources such as power, processing capability, storage capacity and buffer size may change in their availability and amount. Similarly, link resources such as bandwidth may change in their availability, amount and quality. For example, the availability of node and link resources may change when a node or link fails. The amount of node resources may change when the residual power on a battery-operated node decreases during the operation of the network. The amount of link resources may also change as the available bandwidth of a link changes depending on its traffic load. The quality of link resources may change due to abrupt radio frequency (RF) noise or interference within a wireless network environment.

The dynamic factor of application data refers to data generated by the application in a network such as files in a peer-to-peer network and sampling data collected in a sensor network. Application data may change in their availability and amount. For example, the availability of application data may change when data are newly created or deleted in the network.

The dynamic factor of user behaviors refers to the activities of network users. User behaviors may change in their locations, number, requirements and request patterns. For example, users may move to new locations while sending or receiving data through the network. The number of users may change when new users join or existing users leave the network. User requirements may vary because some users may require data being transmitted with low latency while others may require data being transmitted with low communication overhead. User request patterns may change when users change their preferences on what data to search for and retrieve from the network.

Changes in any of the above dynamic factors may affect the performance and service continuity of the network. A network needs to adapt to mitigate the effects of those changes.

3.3 Network Functions

Given a network and the need to transmit data from a node(s) to another node(s) within the network, the network typically implements the following functions: (1) constructing hop-by-hop connectivity to establish connections between nodes within the network; (2) routing data from a source node to a destination node; (3) scheduling data transmission to avoid collisions or scheduling node duty cycles to conserve energy; and (4) controlling data transmission rate to avoid congestion or buffer overflow. Changes in the dynamic factors described in Section 3.2 may affect the performance or interrupt the service continuity of those network functions. A network needs to adapt its network functions to mitigate the effects of those changes. This section describes those network functions in details.

3.3.1 Constructing Hop-by-Hop Connectivity

Hop-by-hop connectivity refers to either a physical link or a logical connection between two nodes used to exchange data. A network constructs and maintains hop-by-hop connectivity to establish a connected topology that allows data transmission between any pair of nodes in the network. Figure 3.2(a) shows the initial network with the dotted lines representing potential connections between nodes. Figure 3.2(b) illustrates the same network after it has constructed hop-by-hop connectivity to form a connected topology. Figure 3.2(b) also shows that only a subset of potential links is used to construct the connected topology. A network constructs hop-by-hop connectivity only in the data link and the application layers. While constructing hop-by-hop connectivity in the data link layer deals with physically linked neighboring nodes, constructing hop-by-hop connectivity in the application layer deals with logically connected nodes that may be several physical hops away.

3.3.1.1 Data Link Layer

For the data link layer, constructing hop-by-hop connectivity is a key issue mostly in multihop wireless networks such as sensor networks and ad hoc networks. Although there exists a similar function to establish topology in wired networks such as the spanning tree protocol, this chapter focuses on hop-by-hop connectivity construction functions that are performed at the data link layer to construct a physical topology for multihop wireless networks.

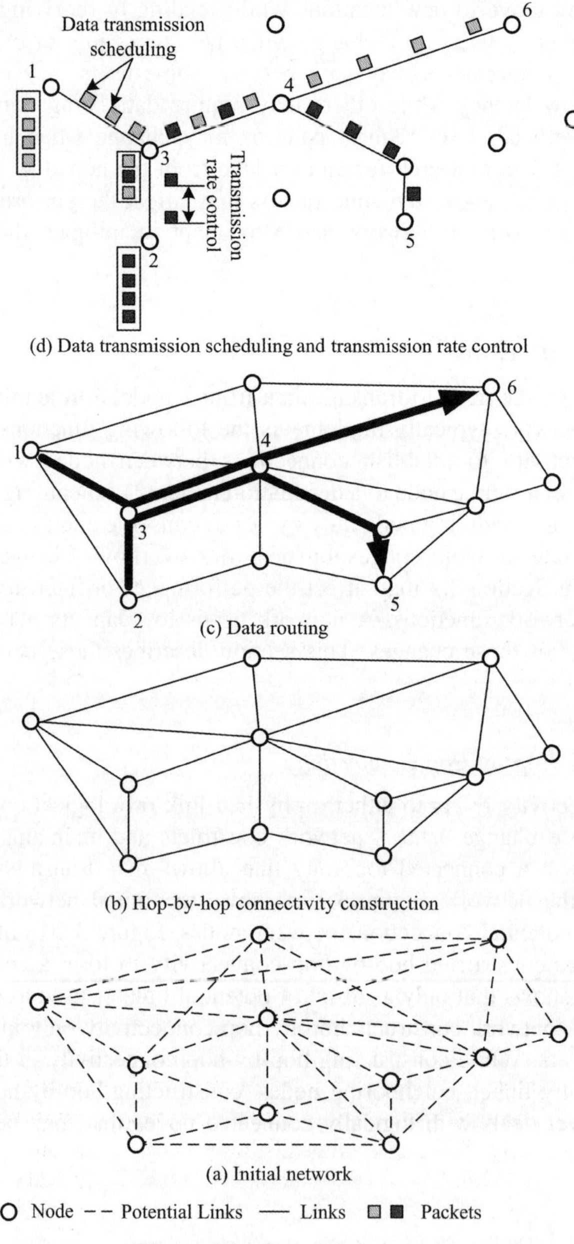

Figure 3.2 Network functions

In multihop wireless networks, it is important for the data link layer to construct hop-by-hop connectivity not only to maintain a connected topology but also to reduce energy consumption and to improve network capacity. The connected topology is maintained collaboratively either by controlling the transmission power of nodes or by controlling the active or inactive states of available links.

3.3.1.2 Application Layer

For the application layer, constructing hop-by-hop connectivity is a key issue mostly in peer-to-peer (P2P) networks. Other types of networks either establish topology through the data link layer or are outside the scope of this chapter.

In P2P networks, it is important for the application layer to maintain a connected topology to support P2P network operations, namely, data publishing and data discovery/retrieval. Data publishing refers to the process where nodes in a P2P network store data published by a source node. Data discovery and retrieval refers to the process where a node searches for and retrieves the stored data. In contrast to the data link layer, a link within a P2P network refers to a logical connection that could exist between nodes that may be several physical hops away. The connected topology is maintained by identifying the logical links that are most beneficial to the application. For example, logical links may be established among the nodes that store more data than others in order to ensure successful data discovery and retrieval.

3.3.2 Routing Data

To route data within a connected topology, a network first finds a path from a source to a destination, and then starts transmitting the data through the path. Figure 3.2(c) illustrates that a path is found (shown as arrows) between two pairs of nodes where nodes 1 and 2 are the sources and nodes 5 and 6 are the destinations, respectively. A network routes data only in the network layer and the application layer.

3.3.2.1 Network Layer

Given a physical topology constructed in the data link layer, the network layer is responsible for routing data from a source node to a destination node. Routing data from a source node to a destination node typically requires finding a path that starts from the source node, ends at the destination node, and satisfies a certain requirement such as the shortest path or the path with enough bandwidth.

3.3.2.2 Application Layer

Similar to routing data in the network layer, the goal of routing data in the application layer is to find a suitable path between two communicating nodes for a given logical topology. Routing data in the application layer is required to support data publishing and data discovery/retrieval in P2P applications. To publish data, a source node determines a path to the data storage locations and uses the path to transmit the data. To discover data, a node needs to find a path to send a query to the location where the corresponding data are stored. To retrieve the discovered data, the nodes storing the data need to find a path to transmit the data back to the query originator.

3.3.3 Scheduling Data Transmission

After finding a path to route data from a source node to a destination node, the source node (and/or intermediate nodes) needs to schedule when to transmit data onto the path. Scheduling data transmission is important because a path may not always be available.

For example, some nodes on the path may be temporarily off to save their energy, or cannot forward data because their communication channel is used by other nodes. To guarantee that data are transmitted only at the appropriate time, the source node (and/or intermediate nodes) needs to adapt its data transmission schedule. For example, assume the scenario where nodes 1, 2 and 3 are in one broadcasting domain (e.g., nodes 1, 2 and 3 are connected through wireless LAN or Ethernet), as illustrated in Figure 3.2(d). When both nodes 1 and 2 transmit data to node 5, they must schedule their data transmission to avoid collision at node 3. Scheduling data transmission is implemented either as node duty cycle management or as medium access control in the data link layer.

3.3.3.1 Node Duty Cycle Management

For power-constrained nodes such as nodes powered by batteries in multihop wireless networks, it is critical to reduce node activity to conserve energy. Node duty cycle management controls when to send/receive data and when to activate/deactivate a network interface for each node in order to save energy while guaranteeing satisfactory performance in throughput, data delivery latency and energy efficiency. Node duty cycle management is not required for nodes with constant power supply (e.g., nodes in a wired network).

3.3.3.2 Medium Access Control

In networks utilizing a broadcast channel, nodes need to schedule data transmission to resolve contention and avoid collision. Medium access control decides which node is allowed to use the channel at a certain time.

3.3.4 Controlling Transmission Rate

When a network transmits data from a source node to a destination node, it is important for the source node to control the transmission rate so that the traffic load does not overwhelm the intermediate links or the destination node. Transmitting too much data may cause congestion at an intermediate link or buffer overflow at the destination node, which may lead to wasted bandwidth, high latency of data transmission or failure of data delivery. Figure 3.2(d) illustrates a scenario where nodes 1 and 2 may drain the buffer at node 3 and deplete the bandwidth of the link between nodes 3 and 4, if they do not control their transmission rate. The transmission rate is controlled by the flow control and the congestion control in different layers.

3.3.4.1 Flow Control

Flow control adjusts the transmission rate according to the destination node's capabilities. Flow control can be performed at the data link layer between two physically connected neighboring hops or at the transport layer between two end nodes that are multiple hops away.

3.3.4.2 Congestion Control

Congestion control adjusts the transmission rate according to the network's current capacity. Congestion control can be performed at the network layer or the transport layer.

3.4 Representative Adaptation Techniques

Changes in dynamic factors can adversely affect the four common network functions described in Section 3.3. To mitigate the impact of those changes, researchers have proposed and developed various mechanisms addressing changes in different dynamic factors. This section introduces typical adaptive techniques by presenting and discussing representative schemes that adapt to various dynamic factors for each of the four common network functions.

3.4.1 Constructing Hop-by-Hop Connectivity

Constructing hop-by-hop connectivity establishes a connected network topology, which allows data to be routed between any pairs of nodes in the network. A connected topology should enable data routing that satisfies a set of user requirements, such as transmission latency, load balance and energy efficiency. However, changes in a network environment may cause connectivity between nodes to degrade or fail. Therefore, a network needs to adapt to changes in dynamic factors while constructing hop-by-hop connectivity.

Nodes adapt to changes in dynamic factors that affect topology connectivity in the following manner. First, nodes obtain information regarding their neighboring nodes. For example, nodes may acquire residual energy of a neighboring node, capacity of a link to a neighboring node, and/or distance to a neighboring node. Second, based on the obtained information, nodes evaluate their neighboring nodes according to certain criteria. For example, nodes may evaluate their neighboring nodes based on their energy consumption and the quality of the connectivity between them. Third, based on the evaluation, nodes establish links to new neighboring nodes, maintain the current connectivity to neighboring nodes, or disconnect from neighboring nodes. Constructing hop-by-hop connectivity adapts to changes in three dynamic factors: network resources; application data; and user behaviors. The following introduces representative schemes that adapt to those changes.

3.4.1.1 Adaptation to Changes in Network Resources

The first example of how a network adapts in constructing hop-by-hop connectivity is a scheme that establishes a connected topology for a P2P network. Gianduia (Gia) [6] constructs hop-by-hop connectivity for P2P file sharing by adapting to dynamic node resources such as computation power, bandwidth and storage capacity. First, each node obtains information from neighboring nodes regarding their resources. The node then evaluates neighboring nodes using the obtained information. Based on its preference (e.g., a node may prefer neighboring nodes with more resources), a node selects neighboring node(s) and establishes link(s) to them. Each node continues this process of obtaining information, evaluating neighboring nodes, and establishing link(s) to them until the number of links at the node reaches an upper bound. This upper bound of the number of links that a node can maintain is determined by the resources of the node. With this scheme, nodes with more (less) resources can have more (less) links to neighboring nodes. A node regularly monitors the resources of each neighboring node and decides whether to keep a link to that node accordingly. Because more links implies more traffic and higher workload, Gia improves the efficiency and scalability of a P2P network by allowing nodes with more resources to establish and maintain more links.

The next example of how a network adapts in constructing hop-by-hop connectivity is hybrid energy-efficient distributed (HEED) clustering [32]. In this scheme, hop-by-hop connectivity is constructed for a sensor network by adapting to changes in the amount of node energy. HEED organizes nodes (sensors) into clusters led by head nodes, as shown in Figure 3.3. In HEED, a node decides to become a head node based on its own residual energy. Therefore, nodes are not required to communicate with neighboring nodes to obtain information. Each node computes the probability to become the head node according to its residual energy such that nodes with more residual energy have higher probability to become head nodes. Based on this probability, each node may elect itself to be a head node and inform its decision to its neighboring nodes. A node that has determined to stay as a non-cluster head selects a head node in its neighborhood (if one exists) with the preference for low intra-cluster communication cost (e.g., the transmission power of each cluster member to reach the head node). The process continues until every node determines either to become a head node or to participate in a cluster. The whole clustering process is executed periodically to adapt the changes in node residual energy due to power consumption and the changes in intra-cluster communication cost caused by node movements. By constructing hop-by-hop connectivity (i.e., organizing nodes into clusters) adaptively to changes in node residual energy and intra-cluster communication cost, HEED decreases and balances energy consumption over different nodes, resulting in longer lifetime for a sensor network.

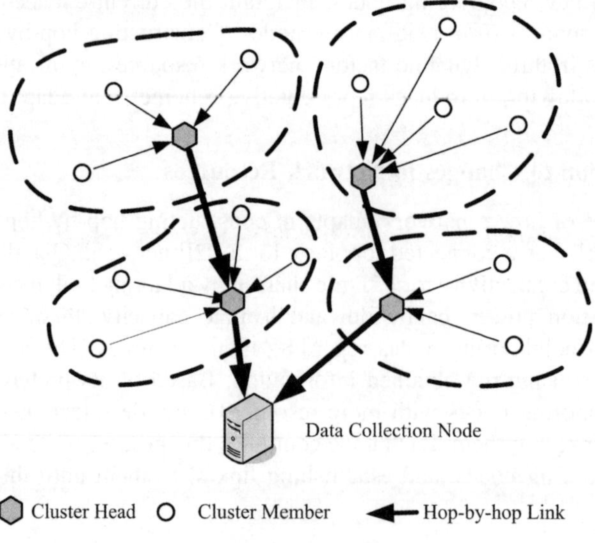

Figure 3.3 Node clustering

Another example is configurable topology control (CTC) proposed in [29], which constructs the hop-by-hop connectivity of a multihop wireless network by adapting to changes in link quality. CTC is executed in two phases: first, nodes exchange beacon messages with their two-hop neighbors to collect link quality and link power consumption information. Then, based on the collected information, each node computes its transmission power. Specifically, each node identifies the links with maximum transmission power and replaces them with alternative paths composed of a sequence of low-power links while guaranteeing link quality. By adapting hop-by-hop connectivity construction to changes in link resource, CTC conserves node energy while providing satisfactory path quality and communication performance.

3.4.1.2 Adaptation to Changes in Application Data

Application data in the network change in terms of their location distribution. One such example is a sensor network, where data are generated by sensors upon detecting phenomena. Roaming hub-based architecture (RoamHBA) [10] provides a solution for tracking multiple targets moving in a wireless sensor network. In RoamHBA, multiple monitoring processes (called agents) migrate from one node to another following the estimated locations of the targets being monitored. To facilitate agent collaboration for in-network processing, those agents need to construct hop-by-hop connectivity to form a connected topology through which they can communicate to each other. Since both agents and targets are moving, agents construct hop-by-hop connectivity while adapting to changes in the locations of agents, which reflect target movements. To construct a connected topology, agents first obtain information regarding the locations of other agents. Then, a roaming hub composed of backbone nodes is formed by considering the locations of all agents to enable efficient communication among them. Once a roam hub is determined, agents establish a path to the roaming hub and send data through it. Agents continue to monitor the locations of other agents, and adapt the roaming hub according to the changing agent locations. By adaptively constructing hop-by-hop connectivity among agents according to changes in their locations, RoamHBA provides an efficient way to support group communication among agents for multiple target tracking. There also exists a scheme in mobile sensor networks [31] that adaptively constructs a topology to monitor a moving target in the network. This scheme adapts to changes in the location of the target and the number of available mobile sensors close to the target.

3.4.1.3 Adaptation to Changes in User Behaviors

Adapting connectivity according to changes in user behaviors such as user distribution, user requirements and user request pattern improves network performance because data communication usually ends at users who actually consume the data. For example, the interest-based locality scheme [25] constructs hop-by-hop connectivity in a P2P network by adapting to changes in user behaviors. In this scheme, nodes obtain information regarding the files that other nodes store by actively searching and passively observing the passing traffic. Nodes manage two types of links to neighboring nodes: base links and extended links (interest-based shortcuts). A node forms base links to randomly selected nodes to construct the base topology. On top of the base topology, a node forms interest-based shortcuts to other nodes storing the same files. The formed shortcuts are used to forward user requests for a file. It has been observed that a node having a file a user is interested in is also likely to have other files of interest to the same user; using this premise, the aforementioned shortcuts improve search efficiency in P2P file-sharing systems. Nodes regularly establish and disconnect shortcuts according to dynamically changing files that other nodes store. Nodes also monitor the searching performance through the formed shortcuts in order to discard less effective shortcuts. When users change their interests, the interest-based shortcuts among nodes also change accordingly. In this manner, the interest-based locality scheme improves search efficiency by adapting to dynamically changing user interests.

3.4.1.4 Analysis

A network adapts to changes in dynamic factors when constructing hop-by-hop connectivity in order to improve the quality of the formed network topology so as to better support data transmission upon it.

Adapting to changes in network resources to construct hop-by-hop connectivity helps to achieve load balancing over different nodes. For example, to mitigate the impact of the changes in network resources on maintaining a connected topology, HEED adapts by allowing nodes with more residual energy to become head nodes with a higher probability, thus balancing the work load over different nodes and consequently prolonging the lifetime of the network. Gia adapts by adjusting the number of links to/from other nodes according to changes in node capabilities, and as the result, nodes with higher capabilities bear more workload, which improves the throughput (i.e., how many search queries can be processed in a unit time) of the whole P2P network. Adapting to changes in network resources to maintain a connected topology also achieves better link quality among nodes. For example, CTC maintains a connected topology while providing guaranteed communication performance by adapting to changes in link quality.

Adapting to changes in application data when constructing hop-by-hop connectivity improves the efficiency of data transmission. For example, to mitigate the impact of the changes in application data while maintaining a connected topology, RoamHBA adapts connectivity among multiple mobile agents according to their changing locations in order to support efficient group communication among mobile agents that track multiple moving targets.

Adapting to changes in user behaviors when constructing hop-by-hop connectivity results in better user connectivity and thus improves the efficiency of data transmission to the users. For example, the interest-based locality scheme adaptively establishes shortcuts between nodes to adapt to changes in user interests. The shortcuts guide user requests to nodes storing the file being searched for, thus reducing the communication overhead of file search.

3.4.2 Routing Data

Routing data from a source to a destination is an essential function of a network and is directly related to the quality of data delivery. Data should be routed through the high-quality fulfilling certain performance criteria such as energy consumption, load balancing and throughput. However, a path may degrade in performance or become unusable due to changes in dynamic factors. Therefore, adaptation of data routing to changes in dynamic factors is critical for network performance.

The quality of a path refers to different meanings in different networks and applications. For example, in a network supporting real-time applications, latency of a path will be the major measurement of path quality; while in a wireless network, the energy of wireless nodes and the bandwidth of wireless links on a path may be the main concern. This implies that an all-in-one solution for adapting data routing to changes is not feasible. Thus, in order to gain an understanding of the current approaches for adaptive data routing, this section presents representative schemes that adapt data routing to changes in various dynamic factors.

Adaptive data routing usually involves three phases: obtaining information about the available resources at the nodes and links on available paths; evaluating the quality of the available paths based on the obtained information; and choosing the optimal path to route data. Routing data usually adapts to changes in network resources, application data and user behaviors. The following introduces the representative schemes that adapt to those changes.

3.4.2.1 Adaptation to Changes in Network Resources

AntNet [5] is a data routing scheme that adapts data routing to the changes in the availability and quality of network resources. In AntNet, each node keeps a routing table storing the probabilities of reaching known destinations through each outgoing link. Each node periodically generates an agent and sends it to a randomly selected destination. While traversing the network to the destination, the agent keeps track of the time it takes to get to the destination. After reaching the destination, the agent returns to the original node following the reverse path and updates the routing table at each node on the path based on the measured time to reach the destination – the path with lower latency is marked with higher probability. Upon receiving a packet addressed to a specific destination, a node probabilistically routes it onto an outgoing link in accordance to the routing table. In this way, data are routed along high-quality paths of low latency with high probability.

Resilient overlay network (RON) [2] constructs an overlay network topology over the current Internet and enables adaptive data routing between domains on the constructed topology. In RON, data routing adapts to changes in the availability, amount and quality of network resources by varying data routing paths. Each router collects information about the quality of overlay links (measured by latency and throughput) by sending probing packets. When routing data, nodes prefer a path that consists of high-quality links. By adapting data routing to changes in the quality of a path, RON routes data adaptively through a path with low latency and high throughput.

Inter-domain resource exchange architecture (iREX) [30] is an adaptive inter-domain routing scheme for data needing special quality of service (QoS) on the Internet. iREX adapts to changes in the availability and quality of network resources by applying economics to create a market for inter-domain network resources. iREX provides automation to a previously manual methodology for QoS policy deployment by empowering each domain to independently advertise and acquire network resources in deploying an inter-domain path. In iREX, each domain decides on a price for the links that it owns and advertises known prices to its neighboring domains. Upon receiving the advertisements from other domains, a domain filters them to keep only the cheapest high-quality resources. In deploying a path to route data, a source domain chooses a set of resources that make up the cheapest path and reserves the resources directly with the resource owners. Data are sent using source routing along the deployed path. iREX also has a reputation system that adapts to quality by excluding low-quality resource from being used, and a fault-tolerance system that adapts to availability. By adapting inter-domain routing to changes in resource prices that directly translates to the availability and quality of network resources, a domain can select the cheapest path with low price and quality guarantee.

Greedy perimeter stateless routing (GPSR) [15] is a geographical routing scheme that routes data to a specific geographical location instead of a destination ID such as an IP address. GPSR adapts to changes in the availability of network resources in terms of failure/recovery of a link and a node. In GPSR, each node collects and stores information of neighboring nodes such as their IDs and locations by listening to beacons periodically broadcasted by neighboring nodes. A node routes data to a geographical location by forwarding the data to a neighboring node with the shortest geographical distance to the destination location. If no neighbor is closer to the destination location than the current node, the node routes the data to a neighboring node identified by the right-hand rule, where each incoming packet is routed to the neighboring node that is encountered first by a counterclockwise scan starting from the direction from which a packet arrived. Packet forwarding stops when the data packet reaches a node located at the destination location. Since routing decisions are merely based on locations of neighbors, which can be obtained locally, nodes can change their routing decisions immediately when the locations of their neighbors change. Therefore, GPSR accommodates changes in the availability of nodes and links in a scalable manner.

Geographical and energy aware routing (GEAR) [33] is another geographical routing scheme. Unlike GPSR, GEAR adapts to changes in the amount of network resources (i.e., the amount of energy consumed at a node). Node energy is regarded as a scarce resource in sensor networks, and GEAR considers the amount of consumed energy at a node in selecting data routing paths. In GEAR, nodes obtain the locations and the energy consumption of neighboring nodes, and each node routes data by forwarding them to a neighbor with a shorter distance to the destination location and with less energy consumption. By considering node energy consumption in routing, GEAR distributes energy consumption of data transmission over different sensors and thus prolongs the lifetime of a sensor network.

3.4.2.2 Adaptation to Changes in Application Data

Information-directed routing [17] is a data-routing scheme for wireless sensor networks that adapts to changes in the availability and amount of application data. Two different routing algorithms are proposed for two specific tasks: routing a query to where data being searched for are and routing a query to a node that collects the data acquired by the query. The first algorithm searches for a path with maximum amount of data, without specifying the destination. Nodes are aware of the amount of data contributed by a neighboring node (referred as the information value) and the topology information within multiple hops. To forward a query, a node selects the node with the highest information value within multiple hops as the destination and chooses the path with minimum hops and maximal amount of data to reach the selected destination. The second algorithm searches for a path with maximum amount of data and limited total communication cost (i.e., energy consumption) to a given destination. Each node estimates the amount of data on each path that is within the communication cost limit. The path with the lowest communication cost and the most data is chosen to route the query. Both information-directed routing algorithms maximize the amount of data retrieved by adapting routing to the availability and the amount of application data.

Intelligent search [14] adapts data query routing to the availability of application data (i.e., data distribution). In intelligent search, each node obtains a profile of a neighboring node by recording query hits received from the neighboring node and messages broadcasted by that neighboring node when replying a query. Upon receiving a new query, a node refers to the profiles of its neighbors and forwards the incoming query to the neighbor with the highest chance of returning a query hit for the incoming query. In order to determine how likely a neighboring node returns a hit for the query, a node evaluates similarity between an incoming query and a past query by comparing keywords contained in both queries; the more keywords they share, the more similar they are. A node then identifies a neighboring node that has answered in the past a query that is the most similar to the incoming one. When the availability of application data changes, the profile information about the data stored at neighboring nodes also changes accordingly, and the change in profile information further impacts query routing. A node learns of the change through the success/failure of recent queries and updates its profiles. In this manner, intelligent search reduces the number of hops to discover data of interest and improves the success rate in data discovery.

In addition to intelligent search, there exists other research [9] [18] addressing the adaptability to application data in P2P resource discovery. They mainly differ on what information is collected for a profile for a neighboring node, and how to choose a neighboring node as a next hop based on the profile. For example, one scheme [9] allows users to return their preferences for received search results, and stores information about user preference into a profile for a neighboring node. The profile is used to forward subsequent queries to nodes that are most likely to return preferred search results.

3.4.2.3 Adaptation to Changes in User Behaviors

Reverse path multicast [8] is a multicast routing scheme that adapts to the changes in user behaviors, such as user movements and users joining and leaving a multicast group. To join a multicast group, a user sends a 'join' message towards the source. Upon receiving a join message, an intermediate node records the link from which the join message arrived. The paths between users and the source node form a multicast tree rooted at the source node and ended at the users. Data from the source node are routed along the multicast tree from the source node to users. Reverse path multicast is also able to detect and truncate the branches that do not include any user of multicast. Reverse path multicast adaptively maintains a multicast tree, and multicasting data on top of such a multicast tree results in less bandwidth occupation.

Mobile IP [20] is an IP routing scheme that supports routing data to mobile users roaming over multiple networks. In mobile IP, a mobile user is assigned a permanent home IP address. As a mobile user moves to a new network, the user obtains a new local IP address in the new network. Both the permanent home IP address and the new local IP address are stored at a centralized entity called the home agent. The home agent receives the data addressed to the mobile user and redirects them to the new local IP address of the mobile user. In this manner, mobile IP allows data to reach a mobile user regardless of which network the mobile user is currently in.

3.4.2.4 Analysis

Adapting to changes in network resources when routing data results in routing paths with high quality. For example, AntNet adapts to changes in path latency by using agents to obtain the latency of random paths to a destination and assigning lower latency paths with higher probabilities for routing data. To adapt to changes in link quality, RON nodes first collect information about the latency and throughput of links on a path using probing packets. Data are then routed through a path with low latency and/or high throughput. iREX chooses a path for inter-domain data routing by adapting to changes in available bandwidth and quality of inter-domain links – the inter-domain path with more bandwidth and better quality is favored. Adapting to changes in network resources when routing data also achieves better load balancing among nodes and links. For example, GEAR distributes traffic load over different nodes by adapting to the amount of energy consumed at a node, while iREX adaptively distributes traffic load over different inter-domain paths according to the available bandwidth of inter-domain links. Routing schemes such as GPSR use only information about one-hop neighborhoods, and thus realize scalable routing of data while adapting to changes in link and node availability caused by node movements, nodes leaving and joining the network.

Adapting to changes in application data when routing data improves data transmission performance by choosing routing paths according to the distribution of data within a network. For example, information-directed routing adapts the routing path of a query to maximize the amount of information that can be obtained on the path when routing a query. In intelligent search, queries are routed adaptively based on data availability at different nodes. To improve the success rate of data discovery, queries are forwarded toward the node that most likely has the data of interest.

Adapting to changes in user behaviors when routing data allows the network to support applications where the distribution and availability of users fluctuate, caused by user movements, users leaving and joining the network. For example, reverse path multicast adaptively forms a multicast group by accommodating users joining and leaving the multicast group, thus reduces the bandwidth occupied by multicasting. Mobile IP adapts to user mobility in routing data to ensure delivery to a user even when the user moves to a new network.

3.4.3 Scheduling Data Transmission

After identifying the receiver node as the next hop to forward data, a source node needs to decide the most appropriate time to actually transmit the data onto the link to the receiver node – referred to as data transmission scheduling. Because multiple nodes may share the same link, a source node may only transmit data when the link is available. A source node also needs to determine whether the receiver node is ready to receive the data. Changes in network resources, application data and user behaviors greatly impact the performance of data transmission scheduling. For example, as a wireless link becomes unreliable, continuing data transmission through the link simply wastes energy. To mitigate the impact of changes in these dynamic factors, a node must adapt data transmission scheduling accordingly in order to decide the most appropriate time to transmit data.

Adaptive data transmission scheduling usually involves two phases: obtaining information about network resources and traffic in a network, and deciding when to send data through the link based on the obtained information. Data transmission scheduling usually adapts to changes in network resources and application data. The following introduces the representative schemes that adapt to those changes.

3.4.3.1 Adaptation to Changes in Network Resources

Channel state dependent packet scheduling (CSDPS) [4] is a data transmission scheduling scheme for wireless LAN where a base station communicates to multiple mobile nodes. CSDPS adapts to changes in network resources in terms of the quality of links. In CSDPS, the quality of a wireless link is measured by bit error rate and packet loss rate. A base station detects a burst link error when it does not receive an ACK after multiple retransmissions of a packet. When a base station detects a burst error in the communication to a mobile station, it marks the queue of that mobile node (a base station has a separate queue for each mobile node). Packets in the marked queue are assigned lower priority than those in an unmarked queue, and packets are transmitted according to their priority levels. A marked queue will be unmarked after a certain period. In this manner, CSDPS adaptively schedules packets to be transmitted to different mobile nodes according to their priority levels (which reflect the quality of a link that a packet is transmitted onto) and thus avoids a packet destined to a mobile node with a low-quality channel to block other packets bound for a mobile node with a high-quality channel (which is called the head-of-line (HOL) blocking problem). Consequently, CSDPS improves the network throughput.

Opportunistic packet scheduling and media access control, or OSMA [28], adapts data transmission scheduling to the changes in network resources (i.e., the quality of a link) in order to solve the HOL blocking problem. Unlike CSDPS, OSMA uses multicast control messages and priority-based election to schedule data transmission. In OSMA, when a source node has data for different receivers, it transmits a multicast control message containing the priorities of those receivers. Upon receiving a multicast control message, each candidate receiver evaluates the instantaneous link quality through physical layer analysis of the received control message. The candidate receiver having good link quality and the highest priority replies with a message to indicate the intention of data transmission.

3.4.3.2 Adaptation to Changes in Application Data

Timeout-MAC (T-MAC) [7] is a MAC layer scheme for wireless sensor networks, which adapts data transmission scheduling to changes in the amount and availability of application data (traffic load). In T-MAC, nodes switch between active and sleep states for the purpose of energy conservation. Neighboring nodes exchange their schedule information and synchronize the start of the active state so that they can communicate to each other during the active period. Nodes transmit all data packets in bursts and sleep

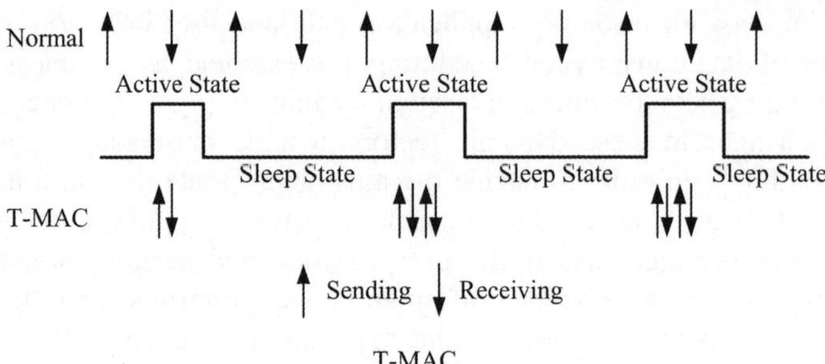

Figure 3.4 Timeout-MAC

between transmission bursts by adaptively ending the active period when no activation event has occurred for a certain period of time. As illustrated in Figure 3.4, T-MAC has a variable active period and adjusts the data transmission time by sending data packets in bursts. Adapting node transmission scheduling to traffic load reduces the amount of energy wasted on idle listening while still maintaining a reasonable throughput.

Traffic-adaptive medium access (TRAMA) [22] also adapts data transmission scheduling to changes in the amount and the availability of application data (traffic load). In TRAMA, time is divided into periods, and each period is composed of time slots. Each node learns about its two-hop neighboring nodes by propagating one-hop neighbor information among neighboring nodes. Nodes also obtain information regarding the transmission schedules of one-hop neighboring nodes. With the information about two-hop neighboring nodes and the transmission schedule of one-hop neighboring nodes, each node determines whether it has the highest priority in a time slot among two-hop neighboring nodes using a hash function with its node ID and the current time slot ID as the seeds. If it has data to send, the node with the highest priority among two-hop neighboring nodes will transmit data in the time slot. If the node with the highest priority has no data to transmit, another node may be elected to transmit in the time slot. A node switches to the receive state if it is the intended receiver of the node that will transmit data in the current time slot. Otherwise, the node switches to the sleep state for energy conservation. In TRAMA, a node adjusts its data transmission scheduling according to the application data to be transmitted in its one-hop neighborhood.

Cross-layer scheduling [24] is another data transmission scheduling scheme that adapts to changes in the application data (traffic load). In cross-layer scheduling, each node schedules its data transmission based on the current traffic load on the path to the destination. When a node has data to send to the destination, it sets up a transmission schedule along the path towards the destination. During the path set up, intermediate nodes on the path agree on their working schedules such that data are forwarded continuously toward the destination in order to minimize data delivery latency. When there are multiple data paths passing a node, the final working schedule of the node is the union of the schedules for all the paths. In the case of a schedule conflict between two paths, an error message is returned to the source node so that the source node sets up a new path. Nodes stay active according to their working schedules for data transmission and sleep at other times to conserve energy. By adapting data transmission scheduling to traffic load, the cross-layer scheduling scheme conserves energy while reducing packet delivery latency.

3.4.3.3 Analysis

Adapting to changes in network resources when scheduling data transmission reduces the latency of data transmission. For example, CSDPS schedules data transmission by keeping separate queues for different mobile nodes and giving priority to mobile nodes with high-quality channels. Thus, CSDPS avoids traffic being blocked by packets bound for a mobile node with a low-quality link. As the result, data transmission latency is reduced. OSMA approaches the same problem using multicast control messages and priority-based election.

Adapting to changes in application data when scheduling data transmission reduces energy consumption at a node by scheduling nodes to sleep when there is no need to transmit and receive application data. For example, T-MAC reduces energy consumption at each node by having nodes transmit messages in bursts and sleep between transmission bursts. In TRAMA, each node schedules data transmission based on the information about the application data to be transmitted within its neighborhood and sleeps when it is not scheduled to transmit or receive. In the cross-layer scheduling scheme, a node establishes its schedule for each path passing the node in order to conserve energy and minimize data delivery latency.

3.4.4 Controlling Transmission Rate

After data transmission begins, a source node needs to decide how much data to send per unit time – referred to as controlling transmission rate. Transmission rate control is important because links and nodes on the path and/or the receiver have a capacity limit, and sending data that exceeds the limit causes transmission failure. Controlling transmission rate is performed at the MAC, network, transport and application layers to limit the transmission rate to the link capacity and the receiver capabilities.

Adaptive transmission rate control usually involves two phases: obtaining information about changes in the dynamic factors that affect network resources and application data, and adapting the transmission rate to those changes. Transmission rate control adapts to changes in network resources and application data. The following introduces the representative schemes that adapt to those changes.

3.4.4.1 Adaptation to Changes in Network Resources

In wireless networks, transmission rate control needs to adapt to changes in the dynamic factors that are specific for wireless networks, such as dynamic link capacity and quality.

Collision-aware rate adaptation (CARA) [16] is a MAC layer transmission rate control scheme designed for wireless LANs. In CARA, each node adapts the transmission rate to changes in the amount and quality of network resources in terms of available link capacity and link quality. To detect the changes in the available capacity and the quality of a wireless link, a node monitors the transmission of control packets (RTS/CTS frames) and data packets through RTS probing and CCA (clear channel assessment), respectively. The node identifies the link to be 'collision' if the transmission of a control packet fails, or 'quality degradation' if the transmission of the control packet succeeded but the transmission of the data packet following the control packet fails. The node then decides the transmission rate based on the current state of the link – reducing the transmission rate when the link is in the collision state and keeping the current transmission rate when the link is in the quality degradation state. Since CARA avoids unnecessary data transmission throttling in response to random errors on a wireless link, it improves link utilization for wireless networks.

TCP Westwood [12] is another example where a node adapts its transmission rate to changes in the amount and quality of network resources in terms of available link capacity and link quality. Unlike CARA which is implemented at the MAC layer, TCP Westwood is implemented at the transport layer – it obtains the information about data transmission on an end-to-end path and adapts transmission rate according to the obtained information. TCP Westwood performs adaptive rate estimation based on the arrival rate of ACK packets. Two rate estimation algorithms, the bandwidth estimation (BE) algorithm and the rate estimation (RE) algorithm, are employed to measure the appropriate transmission rate of an end-to-end path. The BE algorithm gives better estimation on the transmission rate when the dominant cause of packet loss is random link error, while the RE algorithm yields a better estimation of the transmission rate when the dominant cause of packet loss is persistent congestion. A node identifies the network to be congested when the BE rate is much higher than the RE rate, and adjusts its transmission rate to the RE rate. If a node identifies that the network is not congested, it adjusts transmission rate to the BE rate. In this manner, TCP Westwood limits transmission rate only when the network is congested and thus improves the utilization of network capacity when packet loss is due to random link error.

Reception control protocol (RCP) [13] is another transport layer scheme that adapts to changes in the capacity and quality of a link by regulating transmission rate. Given the observation that a wireless network is often used to connect a user terminal to the Internet, and a user terminal often acts as a receiver which constantly receives data from the Internet but sends few data, RCP assumes that receivers have direct access to wireless links. Thus, in RCP, it is the receiver node, not the sender node, that acquires the wireless link state and performs transmission rate control. Upon detecting packet loss, a receiver node examines whether the packet loss is due to congestion or wireless link error through physical layer assessment. If the packet loss is due to congestion, the receiver adjusts (lowers) the transmission rate and informs the rate to the sender. In this manner, a node only adjusts its transmission rate when the network is congested. Therefore, RCP improves the utilization of network capacity.

Source-adaptive multilayered multicast (SAMM) [26] is an adaptive transmission rate control scheme for video multicast applications. A sender of multicast video in SAMM employs a layered coding technique in encoding video streams, and it adjusts the transmission rates in multiple video layers to adapt to changes in available bandwidth at different receivers. SAMM employs two different mechanisms to acquire the information regarding the amount of bandwidth available at multicast receivers: a network-aided explicit bandwidth assessment mechanism and a receiver-estimated available bandwidth mechanism. In the network-aided explicit bandwidth assessment mechanism, each of the intermediate nodes on a multicast path monitors how much bandwidth is available for the multicast transmission, and inserts the information into a control packet periodically sent by the sender node to the receiver nodes. Upon receiving a control packet, receiver nodes become aware of the amount of bandwidth available on the path from the sender node to the receiver nodes. Each receiver node then reports the amount of available bandwidth to the sender node. In the receiver-estimated available bandwidth mechanism, each receiver node monitors the incoming traffic and estimates the amount of available bandwidth through observing the amount of incoming traffic. Each receiver node then reports the amount of available bandwidth to the sender node. With either mechanism, the sender adjusts the transmission rates on multiple video layers based on the amount of available bandwidth at multicast receiver nodes, allowing different receivers to receive different layers of video in order to maximize video throughput.

There are some schemes that adapt to changes in the available resources of wireless networks where users frequently move. For example, a sender node in the adaptive scheme [19] for a video streaming application adapts to changes in the amount of available bandwidth when a mobile receiver node moves into a new network. The sender node employs a layered video coding technique, and adjusts transmission rates of multiple video layers during the handoff according to the amount of available bandwidth in different wireless networks (Figure 3.5). In order to adjust the video transmission rate, a sender node uses a rate estimation algorithm (e.g., TFRC [11]) to estimate the available bandwidth on the new path to the mobile receiver node during the handoff. A sender node then adjusts the transmission rates of different video layers based on the estimated available bandwidth so that the important base layer video stream is transmitted without loss when a mobile receiver node moves into a wireless network with less amount of bandwidth. Through adaptive transmission rate control, the scheme delivers smooth video to mobile nodes.

There also exist adaptive transcoding schemes (e.g., the scheme proposed in [3]) that change the data transmission rate for different receiver nodes to adapt to changes in the amount of available bandwidth on the path to the receiver nodes. In these schemes, an intermediate node on a multicast path transcodes a video stream if the transmission rate of the incoming stream surpasses the transmission rate of the outgoing stream. In this manner, adaptive transcoding schemes customize video quality for different receiver nodes, and thus improve the utilization of link bandwidth.

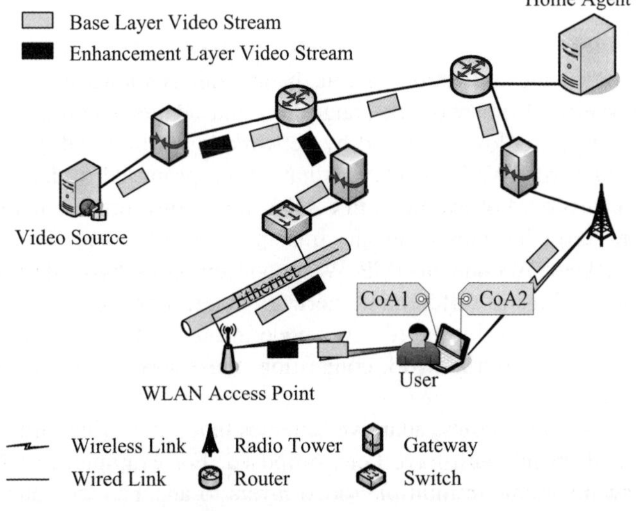

Figure 3.5 Multipath handoff

3.4.4.2 Adaptation to Changes in Application Data

Congestion detection and avoidance (CODA) [27] adapts to changes in the amount and availability of application data by controlling the rate of data transmission in wireless sensor networks. In wireless sensor networks, data are usually transmitted from multiple sender nodes (sensors) that detected the same event to a receiver node (a sink). In CODA, each node monitors its message queue length, which reflects how much application data it needs to send, and senses the channel to obtain the information about how busy the surrounding channel is. If there are messages in the queue and the channel is busy, the node regards the network as congested and reduces transmission rate using two different algorithms, the open-loop algorithm and the closed-loop algorithm. The open-loop algorithm employs backpressure messages propagated from a receiver node that detected congestion towards the sender nodes to allow the intermediate nodes to throttle their sending rates or drop packets. In the closed-loop algorithm, if a sender node detects congestion, it triggers the receiver node to regulate transmission rates of other sender nodes. In CODA, sender nodes adjust transmission rate by adapting to changes in the amount and availability of application data surrounding the receiver node and/or sender nodes. Another work [23] shares similar adaptive features with CODA.

The TCP congestion control algorithm [1] [21] adapts to changes in the amount of application data by controlling data transmission rate. A TCP source node decides its transmission rate in response to changing traffic loads on the end-to-end path. In TCP, a source node monitors the round trip time and the packet loss through acknowledgement from the receiver. The TCP source node adjusts transmission rate by changing the sending window size in each round trip time using the following algorithm: when no packet loss is detected, the TCP source node increases the congestion window size by one transmission unit size for each round trip time; if packet loss is detected, the TCP source identifies the network to be overloaded and the congestion window size is reduced by half. By applying this rate adaptation algorithm, TCP regulates its transmission rate when the network is overloaded with application data.

3.4.4.3 Analysis

In wireless networks, the utilization of link bandwidth is improved if the sender node differentiates link error from network congestion and adapts accordingly (i.e., reducing transmission rate if packet loss is caused by network congestion and ignoring packet loss due to random link error). Various schemes have been proposed to distinguish link error from network congestion. For example, in CARA, nodes discriminate network congestion from random link error by monitoring the transmission failures of short control packets and long data packets. A node in TCP Westwood employs two different transmission rate estimation algorithms to differentiate network congestion from random link error. In RCP, congestion is detected by the receiver node through physical layer assessment. By distinguishing link error from network congestion, those schemes improve the utilization of link bandwidth.

In video multicast applications, adaptive schemes that adjust video transmission rate to changes in available bandwidth have been proposed. For example, in SAMM, a sender node encodes video streams in multiple video layers to adapt to the changes in available bandwidth at different receivers. The adaptive scheme in [19] adjusts the transmission rates of video layers by adapting to the change in the amount of available bandwidth when a mobile receiver node moves into a new wireless network. Adaptive transcoding schemes customize the data transmission rates for different receiver nodes according to changes in the amount of available bandwidth on the path to those receiver nodes. By adapting the transmission rate to changes in available bandwidth, those schemes improve the throughput of video multicasting.

Adaptive schemes are widely used for congestion control. Adapting to changes in application data allows nodes to avoid network congestion by reducing transmission rate. For example, in CODA, sender nodes detect network congestion by monitoring their message queues and sensing the load of channel. Upon detecting network congestion, the sender nodes reduce the transmission rate to avoid network congestion. In TCP, a sender node detects changes in application data (i.e., network congestion) through packet loss. When a packet is detected to be lost, the sender node limits its data transmission rate to prevent network congestion. Through congestion control, data from multiple sender nodes are transmitted through a network without overloading the network capacity.

3.5 Discussion

The process of adapting to changes in dynamic factors that affect networks involves first detecting and then reacting to the changes. Detecting a change refers to collecting information about a dynamic factor and reacting to a change refers to the action taken by a network in order to improve network performance or maintain service continuity after detecting the change.

Detecting changes is critical to the effectiveness of a network's adaptation scheme because a network's ability to adapt greatly depends on the information collected while detecting a change. The accuracy of the collected information determines the correctness of the network's reaction, and the latency and cost of the information collection impacts various characteristics such as the convergence time and overhead of the whole adaptive scheme. Generally, a node may collect the information either actively or passively. A node may actively collect information about dynamic factors by explicitly sending messages to other nodes. For example, a node in AntNet uses agents to probe the quality of a specific path and updates the probability of forwarding along that path based on its quality. A node may passively collect information about dynamic factors by listening to the network to infer or estimate network state without explicitly requesting such information. For example, intelligent search learns about the availability of application data (i.e., data distribution) and the user behaviors (i.e., query pattern) only through routing history such as query hits received from a neighboring node. Another example of passively collecting information is the TCP congestion control algorithm, in which a node infers information on network congestion from packet loss. Compared with the active collecting of information, passive collecting of information does not incur communication overhead when collecting information; however, passive collecting may result in a slower learning process. The scope of information collecting also impacts various characteristics of the adaptive scheme such as the time to converge, and the performance and overhead of the scheme. Generally, a node may either use information locally available without communicating with any other nodes, or collect information from its neighbors. Both active and passive collecting information can collect either local or multihop information. AntNet, RON and TCP congestion control algorithm are examples of multihop information collection while HEED only requires local information. Collecting information by communicating with neighbors incurs more communication overhead and higher latency than simply using information locally available at a node. However, an adaptive scheme may need to react to changes multiple hops away, making multihop information collection inevitable. For example, as a transport layer protocol, TCP needs to adjust transmission rate based on network congestion detected from packet loss, which is usually several hops away from the sender.

Reacting to changes is also critical to the effectiveness of a network's adaptation scheme because it reflects how well a network is able to adjust its behavior and this directly impacts network performance. Once a node detects a change in a network, it either adjusts its own behavior or tries to influence the behaviors of other nodes based on the detected changes. An example of changing its own behavior is found in CARA. In CARA, a sender monitors the network to differentiate congestion from link degradation and adjust its own behavior by changing the transmission rate accordingly. RCP is an example of influencing the behavior of other nodes. In RCP, a receiver node monitors packet loss to determine the proper transmission rate for rate control at the receiver side, and then informs the sender about the transmission rate to use. Usually, causing behavior adjustments in other nodes incurs more overhead and takes longer to take effect when compared to a node adjusting its own behavior. However, based on the system design, causing behavior adjustments in others allows the decoupling of the detection of a change from the reaction to the change and increases flexibility and efficiency of information collection. In adjusting behaviors, a node may adjust behaviors in a probabilistic or deterministic manner. While most of the schemes are deterministic, there are some schemes that adjust behaviors probabilistically. For instance, AntNet relies on probabilistically choosing a path to forward data. In HEED, cluster head nodes are elected probabilistically. With probabilistic behavior adjustment, there is a chance that the adjustment may not be optimal, and thus, probabilistic behavior adjustment may not provide the most efficient results. However, probabilistic behavior adjustment makes a network more robust against inaccurate information on and rapid changes in network state.

3.6 Conclusion

This chapter includes a comprehensive discussion on current network schemes that adapt to changes in various dynamic factors. Dynamic factors that affect a network are resources, application data and user behaviors. Adaptive schemes employed in four basic network functions are presented and analyzed in detail.

In conclusion, networks that employ adaptive schemes are usually more efficient and can provide better service continuity under dynamic environments. Due to the foreseeable wide deployment of wireless networks and new types of services, networks in the future will exhibit even higher degrees of dynamics and adaptive schemes in networks will become even more important.

References

[1] Allman, M., Paxson, V. and Stevens, W. (1999) *TCP congestion control*. RFC 2581, Internet Engineering Task Force (IETF), April.

[2] Andersen, D.G., Balakrishnan, H., Kaashoek, M.F. *et al.* (2001) *Resilient overlay networks.* Proceedings of 18th ACM Symposium on Operating Systems and Principles (SOSP), October 21–4, Banff, Canada.

[3] Assuncao, P.A.A. and Ghanbari, M. (1997) *Optimal transcoding of compressed video.* Proceedings of International Conference on Image Processing (ICIP), October 26–29, Santa Barbara, CA, USA.

[4] Bhagwat, P., Bhattacharya, P., Krishna, A. *et al.* (1996) *Enhancing throughput over wireless LANs using channel state dependent packet scheduling.* Proceedings of INFOCOM, March 24–28, San Francisco, USA.

[5] Caro, G.D. and Dorigo, M. (1998) AntNet: distributed stigmergetic control for communications networks. *Journal of Artificial Intelligence Research*, **9**, 317–65.

[6] Chawathe, Y., Ratnasamy, S., Breslau, L. *et al.* (2003) *Making Gnutella-like P2P systems scalable.* Proceedings of ACM SIGCOMM, 25–29 August, Karlsruhe, Germany.

[7] Dam, T.V. and Langendoen, K. (2003) *An adaptive energy-efficient MAC protocol for wireless sensor networks.* Proceedings of the First International Conference on Embedded Networked Sensor Systems (SenSys), November 5–7, Los Angeles, USA.

[8] Deering, S.D. and Cheriton, D.R. (1990) Multicast routing in datagram internetworks and extended LANs. *ACM Transactions on Computer Systems*, **8**(2), 85–111.

[9] Egashira. R., Enomoto, A. and Suda, T. (2005) *Distributed service discovery using preference.* Proceedings of the IEEE/ICST CollaborateCom.

[10] Fang, Q., Li, L., Guiba, L. *et al.* (2004) *RoamHBA: maintaining group connectivity in sensor networks.* Proceedings of the Third International Symposium on Information Processing in Sensor Networks (IPSN), April 26–27, Berkeley, CA, USA.

[11] Floyd, S., Handley, M., Padhye, J. *et al.* (2000) *Equation-based congestion control for unicast applications.* Proceedings of Sigcomm, August 28–September 1, Stockholm, Sweden.

[12] Gerla, M., Ng, B.K.F., Sanadidi, M.Y. *et al.* (2004) TCP Westwood with adaptive bandwidth estimation to improve efficiency/friendliness tradeoffs. *Computer Communications*, **27**(1), 41–58.

[13] Hsieh, H.Y. and Sivakumar, R. (2003) *A receiver-centric transport protocol for mobile hosts with heterogeneous wireless interfaces.* Proceedings of ACM MOBICOM, September 14–19, San Diego, USA.

[14] Kalogeraki, V., Gunopulos, D. and Zeinalipour-Yazti, D. (2002) *A local search mechanism for peer-to-peer networks.* Proceedings of the 11th International Conference on Information and Knowledge Management (CIKM), November 4–9, McLean, VA, USA.

[15] Karp, B. and Kung, H.T. (2000) *Greedy perimeter stateless routing for wireless networks.* Proceedings of the Sixth Annual ACM/IEEE International Conference on Mobile Computing and Networking (MobiCom), August 6–11, Boston, MA, USA.

[16] Kim, J., Kim, S., Choi, S. *et al.* (2006) *Collision-aware rate adaptation for IEEE 802.11 WLANs.* Proceedings of IEEE INFOCOM, April 23–29, Barcelona, Spain.

[17] Liu, J., Zhao, F. and Petrovic, D. (2005) Information-directed routing in ad hoc sensor networks. *IEEE Journal on Selected Areas in Communications*, **23**(4), 851–61.

[18] Moore, M. and Suda, T. (2002) *Distributed discovery in peer-to-peer networks.* Proceedings of the First Annual Symposium on Autonomous Intelligent Networks and Systems.

[19] Pan, Y., Lee, M., Kim, J.B., *et al.* (2004) An end-to-end multi-path smooth handoff scheme for stream media. *IEEE Journal of Selected Areas of Communications, Special-Issue on All-IP Wireless Networks*, **22**(4), 653–63.

[20] Perkins, C. (2002) *IP mobility support for IPv4.* RFC 3344, Internet Engineering Task Force (IETF), August.

[21] Postel, J. (1981) *Transmission control protocol.* RFC 793, Internet Engineering Task Force (IETF), September.

[22] Rajendran, V., Obraczka, K. and Gracia-Luna-Aceves, J.J. (2003) *Energy-efficient, collision-free medium access control for wireless sensor networks.* Proceedings of the First International Conference on Embedded Networked Sensor Systems (SenSys), November 5–7, Los Angeles, USA.

[23] Sankarasubramaniam, Y., Akan, Ö.B. and Akyildiz, I.F. (2003) *ESRT: event-to-sink reliable transport in wireless sensor networks.* Proceedings of the 4th ACM International Symposium on Mobile Ad Hoc Networking and Computing (MobiHoc), June 1–3, Annapolis, Maryland, USA.

[24] Sichitiu, M.L. (2004) *Cross-layer scheduling for power efficiency in wireless sensor networks.* Proceedings of IEEE INFOCOM, March 7–11, Hong Kong, China.

[25] Sripanidkulchai, K., Maggs, B. and Zhang, H. (2003) *Efficient content location using interest-based locality in peer-to-peer systems.* Proceedings of IEEE INFOCOM, March 30–April 3, San Francisco, USA.

[26] Vickers, B.J., Albuquerque, C. and Suda, T. (2000) Source-adaptive multilayered multicast algorithms for real-time video distribution. *IEEE/ACM Transactions on Networking (TON)*, **8**(6), 720–33.

[27] Wan, C.Y., Eisenman, S.B. and Campbell, A.T. (2003) *CODA: congestion detection and avoidance in sensor networks*. Proceedings of ACM SENSYS, November 5–7, Los Angeles, USA.

[28] Wang, J., Zhai, H. and Fang, Y. (2004) *Opportunistic packet scheduling and media access control for wireless LANs and multi-hop ad hoc networks*. Proceedings of IEEE Wireless Communications and Networking Conference, March 21–25, Atlanta, GA, USA.

[29] Xing, G., Lu, C. and Pless, R. (2005) Configurable topology control in lossy multi-hop wireless networks. Technical Report, Washington University at St. Louis, CSE-34, July 27.

[30] Yahaya, A. and Suda, T. (2006) *iREX: Inter-domain resource exchange architecture*. Proceedings of IEEE INFOCOM, April 23–29, Barcelona, Spain.

[31] Yang, M. and Suda, T. (2007) *Barrier coverage using mobile sensors*. Technical Report, UCI.

[32] Younis, O. and Fahmy, S. (2004) *Distributed clustering in ad-hoc sensor networks: a hybrid, energy-efficient approach*. Proceedings of IEEE INFOCOM, March 7–11, Hong Kong, China.

[33] Yu, Y., Govindan, R., Estrin, D. (2001) *Geographical and energy aware routing: a recursive data dissemination protocol for wireless sensor networks*. Technical Report UCLA/CSD-TR-01-0023, May.

4

Self-Managing Networks

Raouf Boutaba and Jin Xiao
School of Computer Science, University of Waterloo

4.1 Introduction: Concepts and Challenges

Human reliance on advanced digital communications and distributed applications has reached a phenomenal stage, where everyday life and business processes hinge on the proper and efficient operation of computer networks and applications. It is apparent that networks and software systems are becoming larger in scales, more complex in design, more pervasive in distribution, and the runtime environment more dynamic – they have grown beyond the limit of manual administration. Yet at the same time, there is rising pressure in making networks and systems more manageable, their operations more efficient, and their deployment and maintenance more cost effective. In 2002, IBM stated that the IT industry is facing a crisis and network and system management is becoming the grand research challenge [30]. The immensity and immediacy of this problem is perhaps best illustrated with today's stark IT business reality [52]:

- 33% of the total cost of ownership is spent on network and software recovery and 50% on failure protection.
- 40% service outages are caused by human error.
- 80% IT expenditure is on operations and maintenance.

As a result, researchers have been actively seeking solutions to this management crisis. Although advancements in hardware and software technologies (e.g., decreases in computing and network capacity cost, emergence of distributed software, web service technology) will ameliorate the problem to some extent, the ultimate solution lies in a fundamental shift in the design philosophy of network and system management. No longer can the networks remain primitive and the management systems omnipotent and all encompassing – a large degree of self-awareness and self-governess must be realized

Cognitive Networks: Towards Self-Aware Networks Edited by Qusay H. Mahmoud

in the networks and systems. In response, there has been a major push for self-managing networks and systems in the last five years. Although management automaton has been the ultimate goal of network and systems researchers for decades, never before has there been such a strong surge of concerted efforts from both academia and industry, and the need for effective solutions so immediate.

In addressing the topic of self-managing networks, readers should be conscious of the fact that self-management is really in essence another aspect of the cognitive network concept. It envisions a system that can understand and analyze its own environment, and can plan and execute appropriate actions with little or no human inputs. And it considers the problem from a functional perspective, focusing on how self-management concepts and designs could be used to solve various management problems in fault, security, performance, configuration, and so on. We will continue our discussion of this topic in Section 4.8. While putting together the materials for writing this chapter, we realized that as there are many different research topics and theories in self-management that are ongoing today, it is quite difficult to present them to the readers in a comprehensive and coherent manner, yet still cover enough depth. In the end, we adopted a tutorial-like organization where we first introduce the concepts, frameworks and general theories of self-management and then present topics in specific problem domains. Thus the remainder of this chapter is organized as follows: we first give a short introduction to the visions and challenges of self-management (Section 4.2), then we explore the theories and concepts that support autonomic architecture and self-management design, especially for large-scale networks (Section 4.3); of strong relation to cognitive networks, we examine how knowledge could be represented and acquired by autonomic components, and the theories and foundations for automated reasoning and analysis (Section 4.4); this concludes our examination of the general self-management scope. We then present research advances and promising explorations in self-management with respect to specific problem domains (Section 4.5); followed by two key topics in self-management that have not been a focus of traditional management research: system validation and benchmarking (Section 4.6) and self-stabilization (Section 4.7). Finally, we conclude the chapter with summary and discussions (Section 4.8). A fairly extensive reference of literature works has been included to provide further coverage of the concepts or problems discussed in this chapter.

4.2 The Vision and Challenges of Self-Management

4.2.1 The Self-Management Vision

IBM research [34] has used the term ‘autonomic computing’ as a distributed system where a set of software/network components can regulate and manage themselves in areas of configuration, fault, performance and security to achieve some common user-defined objectives. The word ‘autonomic’ originates from the autonomic nervous system that acts as the primary conduit of self-regulation and control in human bodies. Four ‘self’ properties of autonomic computing are defined:

- Self-configuration: the entities can automate system configuration following high-level specifications, and can self-organize into desirable structures and/or patterns.
- Self-optimization: the entities constantly seek improvement to their performance and efficiency, and are able to adapt to changing environment without direct human input.

- Self-healing: the entities can automatically detect, diagnose and recover from faults as the result of internal errors or external inconsistencies.
- Self-protection: the entities can automatically defend against malicious attacks or isolate the attacks to prevent system-wide failures.

Although the autonomic computing concept was first proposed for distributed systems and software components, it is equally relevant to self-managing networks, especially with the increasing associations between networks and distributed applications and the extensive use of network management applications. In general, we consider a self-managing network as a particular type of autonomic system. Hence, the terms 'autonomic systems' and 'self-managing networks' are used interchangeably in our discussion. The left-hand side of Figure 4.1 shows the proposed anatomy of an autonomic component whereby the autonomic manager interacts with the managed elements and its surroundings by taking inputs from the external environment, applying analysis and reasoning logic, generating corresponding actions, and executing these actions as output. This workflow fits well with the classic monitor and control loop of network management (right-hand side of Figure 4.1), where monitored data from the networks are used to determine the appropriate management decisions and then translated into corresponding control actions. Today's network management applications require extensive human involvement in management decision making.

One important question to ask is whether self-managing networks are realizable. In fact, various degrees of automation are prevalent in today's networks and systems, such as the automatic CPU scheduling and memory allocation in operating systems, the TCP congestion control, optical network fault protection and automatic recovery. The operations of today's networks and systems are highly dependent on these automated processes. Their success story gives hope that self-managing networks are indeed possible. Nevertheless, automation of networks and systems thus far has been applied to specific problem domains and to limited degree. To enable full automation of the networks and systems, as we envision the future to be, there are many unresolved problems and issues.

4.2.2 Challenges in Self-Managing Networks

In this section, we briefly outline some of the challenges in self-managing networks. Further readings on the challenges of developing autonomic systems can be found in [29], [33], [34], [43], [47], [50], including some research direction.

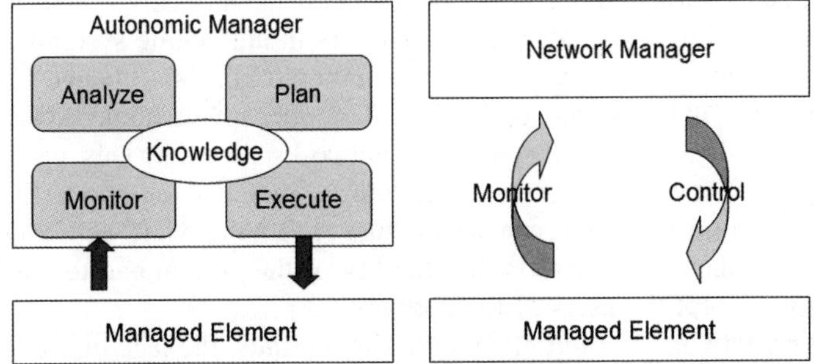

Figure 4.1 The anatomy of autonomic computing and the monitor and control loop of network management

- The lifecycle of a self-managing network must follow a rigorous engineering approach to ensure the validity, consistency and correctness of the system from its development, to operation and removal. This process must be firmly established in order to ensure the resulting system is reliable and robust. In addition, there must be independent validation and benchmarking methodologies and mechanisms in place that not only can evaluate the performance of autonomic systems in general, but also can provide safeguard against unforeseeable errors.
- Monitoring is of paramount importance to self-managing networks. Because of the closed-loop interaction pattern an autonomic component exhibits, the component must be able to determine what, when and where to monitor. The monitoring data may be diverse in content and context, and the monitoring parameters are likely to be subject to frequent changes. It is not clear how such a process should be structured in a distributed and efficient manner. What overhead the monitoring activities should place on the networks. And to what granularity, in time and in volume, should the monitoring activities be bounded by.
- Knowledge is critical to autonomic reasoning, analysis and decision making. The knowledge an autonomic system is required to gather and formulate far exceeds the level of data representation and interpretation capability of today's network management applications and distributed systems. There is a need for knowledge representation of the environment, various processes, services and objects, the autonomic component, the states the system could take on, the interactions a component could have, the properties the system and components are associated with, and the features/services the system or component provides. There is also a need for representation of high-level objectives, system behaviors and rules. The problem is further complicated by the wide range of context such knowledge could be taken from, the need for common information representation, and the possibility of varied semantics and interpretations.
- Interactions among autonomic components or even autonomic systems pose another challenge. How do they coordinate to achieve system-wide objectives? How do they negotiate services and requirements? How can the overall system behavior be determined from the interactions of individual components? These are some of the important problems that require well-formulated solutions.
- System stability, robustness, resilience and correctness are some of the key properties of self-managing networks. Not all of them are well understood or easily analyzable, especially in the self-management context. For example, how does localized reasoning based on limited information lead to stable global behavior? How does one avoid global inconsistencies and undesirable system behaviors in autonomic systems?
- In a dynamic environment, processes must undergo constant change to adapt to the new environment and maintain optimality. The concept of homeostasis [29], central to biology, is prevalent to self-managing networks as well. This need for constant adaptation poses a new challenge: as the autonomic system evolves overtime in response to changed environments, how can we be sure it remains valid and consistent? What effect does this adaptation have on the stability of the system and to what extent does the adaptation disrupt the users of the network?
- Self-managing networks exhibit many undesirable traits: the lack of conformity among autonomic components, the complexity of the autonomic systems, the constant evolution

of the network environment, and the implicit designs and architectural assumptions the designers are prone to make. All of these factors make interoperability among components and systems difficult.

Furthermore, the lessons we have learned from past network design and automation [43] suggest that successful network design should:

- make assumptions explicit
- be able to handle common failures
- understand the interactions between different control loops
- make goals explicit
- make systems more user friendly and understandable
- have built-in validation

These are additional requirements that we must consider in designing self-managing networks.

4.3 Theories for Designing Self-Managing Networks

To date, many autonomic computing and/or networking architectures and frameworks have been proposed in the literature. Rather than present their architectural details, we focus on the promising theories that aid in the design of self-managing networks.

4.3.1 Design Patterns

Design patterns offer an effective method of capturing human knowledge in how to cope with specific problems from an architectural standpoint. Applying appropriate design patterns in self-managing networks not only results in intelligent organization of autonomic components at runtime, but also provides a concrete guideline to component interactions. More importantly, autonomic systems developed using design patterns are guaranteed to yield desirable system output and correct system states. In a broad sense, design patterns could be general design practices, such as the patterns proposed in [62]:

- Goal-driven self-assembly: this pattern is useful for self-configuration. System configuration decisions are made *a priori* and are assigned to the system components as goals. When the component joins an autonomic system, it knows how to contact a service registry to obtain resources and services based on its goal description.
- Self-regenerating cluster: two or more instances of a particular type of autonomic component are tied together in a cluster. They share the same input element and process external requests assigned by some scheduling algorithm. Instances in a cluster monitor others to ensure their proper operations.
- Market-control model: autonomic components compute the utility of candidate services or resources and make local purchase decisions for resource acquisition. A resource-arbiter element could be added to compute system-wide optimal allocation of resources among autonomic components.

Design patterns could also be tailored to solve specific problems and could be combined together to form larger patterns. For example, Wile [63] outlines some problem-specific design patterns:

- Resource reallocation pattern: probes are associated with resource consumers to monitor resource usage and gauges are used to compute average or maximum resource usage, any violations of threshold are reported through an alarm.
- Model comparator pattern: two identical copies of an autonomic entity coexist – an actual copy and a simulated copy. Environmental events that a component responds to are fed to the simulator. A comparator gauge compares the output of the actual copy to the output of the simulated copy to determine any inconsistencies.

Similar to design patterns, architectural descriptions and models could be used to facilitate runtime binding of autonomic components and resources. Work on architectural prescriptions [28] captures the functionalities, constraints, resource requirements and operational states of an autonomic component as activities, roles and intents using architectural description language and state change models. This allows the resulting system to find suitable components at runtime and be able to reason about a component's ability to fulfill a particular task.

The primary problem with design patterns is their rigidity. Although design patterns are highly effective in addressing specific problems, they are only effective under the architect's envisioned context. When the environmental conditions change that render the old design obsolete, autonomic components that follow design patterns cannot evolve to cope with the new environment and may even produce erroneous executions.

4.3.2 Multi-Agent Systems

As the anatomy of an autonomic component corresponds closely to the structure of an agent in multi-agent systems, research in multi-agent systems could find application in autonomic systems, particularly with respect to agent interactions and collaborations. The COUGAAR [27] agent system is one such example. COUGAAR agents are arranged into collaborative societies with common and specific goals. Communication between agents is unique in that agents do not directly communicate with each other. Rather a task is formed and agents capable of performing the task are associated with the task through an information channel. Thus, agents interact with each other for the purpose of achieving a specific task and complex composition of agents is possible depending on the complexity of the task. Specific work on multi-agent systems for automated network management [36] adapts management domain specific models to traditional agency concept and describes management specific agent interactions. In this work, not only are agents assigned roles (e.g. managed element, manager) but also all agent-to-agent interactions are typed according to their roles and task dependencies.

Multi-agent systems offer a structured way of defining agent interactions and coordinating agent activities, but they do not guarantee system-wide correctness or consistency as in the case of design patterns.

4.3.3 Theories in Closed-Loop Control

Control theory was first established as a general reference architecture for adjusting system behaviors to achieve desired objectives. Because adjustments are made incrementally

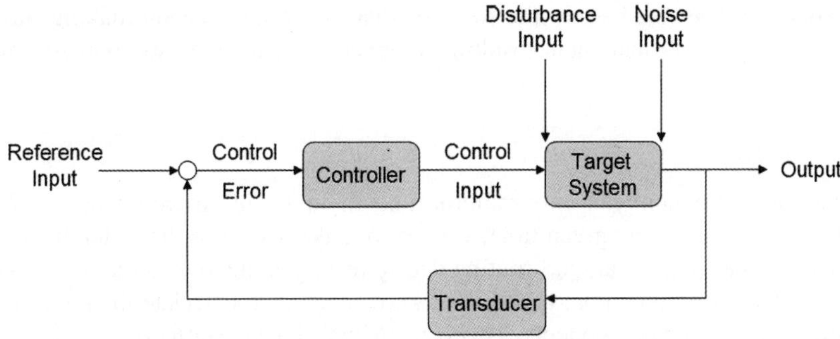

Figure 4.2 Block diagram of the control system

based on environmental feedbacks, this process fits well with the execution flow of an autonomic component, and has thus been proposed as a methodology for developing autonomic components [16]. Figure 4.2 depicts the basic elements of a control system. The controller is the autonomic component and the target system is the environment. The controller affects the target system by feeding it control input. The output from the target system is then measured. Uncertainties in the environment and the monitoring devices are represented by disturbance and noise. The transducer smoothes the measured output to a form comparable to the desired input. The difference between the measurement and the objective produces a control error that serves as feedback to the controller.

Stochastic adaptive control is a self-optimization theory from real-time control systems. Its adaptation to network self-optimization [56] also follows a similar execution flow of monitor, plan and execute. Measurements from the environment are used to build an optimization model. The model produces a series of optimization steps and only the immediate step is carried out, thus providing some degree of look ahead. The resulting changes in the environment are further monitored to refine the model. Due to the locality of the information on the environment and the possibility of multiple variables required for monitoring, the optimization model does not always yield global optimal solutions.

Although closed-loop controls are very well-formed self-optimization mechanisms, they require clear and simple environment models (e.g. identifying the control input, understanding the dependencies among control variables) to be most effective.

4.3.4 Other Theories for Designing Self-Managing Networks

Design patterns from biology [4] such as diffusion, replication and stigmergy offer additional guidance to autonomic component design. Operational knowledge could be used to record component functions and dependencies to increase their usability in autonomic systems [67]. Situation calculus and the belief–desires–intention framework [45] can provide autonomic components with additional adaptability to new environments and help them reason about the degree of collaboration they require to achieve their goals.

4.4 Self-Management Intelligence: To Know and to Act

In this section we examine how knowledge can be acquired and represented in self-managing networks to facilitate the understanding of the environment and the autonomic components. Then we will study the theories that drive the decision-making process of autonomic components and their resulting interactions with the environment and each other.

4.4.1 Knowledge

An autonomic component can only monitor a small subset of environmental variables in its local environment at any given time, yet has to make decisions that often have system-wide consequences (e.g. local QoS routing decisions impact the overall end-to-end QoS of user traffic). One way to combat this lack of knowledge is to correlate the monitored data among other autonomic components. This is nontrivial since the component must be aware of what information the others have, how the information is represented and whether it can be accessed in a timely fashion. Intercomponent negotiations and interactions must then take place to solicit the required data. The amount of network and system resource consumed to acquire such data remotely may also be a problem. QMON [41] attempts to address this issue by offering differentiated monitoring classes, hence prioritized monitor traffic based on the QoS class of the service an autonomic entity is supporting. Alternatively, some decision-making algorithms are designed to deal with imperfect information or localized data, such as the use of learning algorithms and market-based approaches. In some cases, it is more important for an autonomic system to understand the dependencies among the monitored data. Research in data mining and data reduction may be highly relevant here. Concepts such as rough set theory and principal component analysis could offer effective means of deriving such dependencies based on limited data input [66]. While attribute reduction can yield insights into the preciseness of the monitored variable and the significance of a variable by using a dicernibility matrix, principal component analysis can efficiently reduce the dimension of the variable space by retaining only high variance variables.

It is also important to have knowledge representation of the autonomic components themselves. The definition and specification of self is fundamental to automatic configuration and runtime reconfiguration. The work of [7] focuses on externalizing an autonomic component based on its defined services and SLAs. This is somewhat analogous to how in object-oriented programming the interfaces of an object could be exposed without exposing its implementation details. The work of Supadulchai and Aagesen [53] defines an autonomic component's capacity as the resource, functions or data it represents, and studies how such capacity could be dynamically generated for a composite consisting of multiple autonomic components. To facilitate common understanding among autonomic components that may have been developed for different purposes and in different problem domains, work on semantic representation of service (e.g., [20]) provides an effective mechanism of bridging the difference in representation and vocabulary. The particular work of Fujii and Suda [20] assumes an autonomic component to have a set of operations and focuses on modeling the semantics of each operation's input/output and the associated operational properties.

4.4.2 Decision Making in Autonomic Components

Policy-based control is an effective method of defining the behaviors of an autonomic component and can be used to drive the condition–response interactions between an autonomic component and its environment. The work of [8] proposes such a policy-based self-management system that leverages the traditional policy-based management architecture. An extensive database of polices and their associated action sets are stored in a knowledge base. An event handler interprets monitored environmental variables to determine whether a particular enforced policy condition is met. The decision of applying specific polices is performed at the policy decision point, translated into executions at the policy enforcement point and carried out by the effectors. An event analyzer incorporating information processing and pattern recognition is suggested to help the policy decision point in deriving long-term trends in observed environment variables. Although elaborate, such policy architecture may not be able to cope with environmental conditions that are not prescribed in its knowledge base. Other work ([3], [8], [40], [48]) explores various ways of dynamic policy generation. The Accord programming system [40] is designed to provide rule-based descriptions of a component's runtime behavior and the language also supports rule combinations for component composition. The system is further augmented with some look-ahead control [6] providing predictions on the operations of a component in the current environment and deducing any necessary adaptation the component should undertake. Samman and Karmouch [48] examine how high-level user preferences and business goals could be translated into appropriate network-level objectives. The work recognizes the disparity in QoS representation across different layers (e.g. user, application, system, network) and seeks automated mapping of QoS attributes. The result of this mapping forms the basis for their assembly of policies and actions. A reassessment mechanism is also proposed to monitor the actual performance of the established policies through environmental feedbacks. Work on self-optimization using policies [3] similarly uses business-level policies to control autonomic system behavior. However, a different approach is taken in that prescribed models are used to assist in the translation of policies to system behaviors. The work also suggests that reinforcement learning and statistical techniques could be useful in providing specialized models such as user behavior prediction and user attribute modeling.

Emergence properties inherent to biological insect colony may also be of interest to self-managing networks, as not only do they exhibit the desired 'self' properties, but also each component is simplistic in design and yet together they perform complex global tasks. For example, work on bio-networking architecture [54] provides a middleware model of a biological insect colony. Each autonomic component in BioNet, called a cyber-entity (CE), is a simple component with diverse roles. Each CE has a set of defined behaviors: communication, energy exchange and storage, lifecycle regulation, relationship management, discovery, pheromone emission and environmental sensing. Two key biological concepts that are used to facilitate CE interactions are energy exchange and pheromone emission.

- Energy exchange: each CE has an energy value to spend on acquiring services and resources. Energy is earned by providing service to others.
- Pheromone emission: a CE will periodically emit pheromones whose strength is determined by the CE's configuration or energy level.

With such simple constructions, self-regulated behaviors such as reproduction/death, migration and collaboration could be realized. For example, when the energy level is high in the environment, CEs are more attracted to the site due to the high demand for services, and CEs may even copy themselves. When the energy level in the environment is low, CEs may terminate themselves due to lack of energy. In general, pheromones provide an indication of a CE's presence and the population density. Pheromones may also be used for service discovery. Preliminary experimentation indicates that the simple construction of CEs does not add significant performance overhead to the networks.

Market-based approaches to autonomic decision making and component interactions constitute another promising direction of research. Such a minimalist approach allows the components to freely make their own self-maximizing decisions regardless of the overall system welfare and results in significant design and control simplification. Two groups exist in the market: buyers and sellers. The sellers offer services at particular prices, which may vary depending on the availability of the resources and the demand for them. The buyers purchase the services based on their specific needs at purchase time. The decision process itself is driven by the evaluation of utility functions. Price may be an input to the function or as a comparator to the utility. In [60], such a market approach is applied to media streaming in overlay networks. The buyer maintains a list of high utility sellers of a desired service while the seller determines the amount of throughput to allocate to each buyer based on utility computation. The service prices are updated periodically based on utility maximization for the future. It is provable that game theoretical equilibrium of such system could be reached when seller/buyer has perfect information about the market and there exists some stable equilibrium points. In reality, a component may not have perfect information about the system and environment. The concept of Catallaxy economy offers an alternative solution. The Catallaxy economy model is similar to the aforementioned free-market model. It assumes that agents operate under self-interest to gain utility. However, the Catallaxy model acknowledges that decision making in markets is subject to 'constitutional ignorance': it is impossible to know or obtain data on each and every environmental variable that is relevant to utility maximization. Some work, such as [19] explores this branch of market science in the context of self-managing networks.

Reinforcement learning is effective in dealing with dynamic environment and uncertainties. The techniques do not require exact modeling of the system and they assume the environment is constantly changing and hence requires constant adaptation. These two factors make this technique quite valuable to autonomic computing and networking. However, the main drawbacks of reinforcement learning are: the learning process may be slow; a large volume of consistent training data is required; and it may not be able to capture complex multi-variable dependencies in the environment. Tesauro *et al.* [55] remedy some of these drawbacks by seeking a hybrid approach where initial models are used to guide the learning process and in turn the models are refined over time using reinforcement learning. Some experimentation with the hybrid approach suggests its effectiveness over pure reinforcement learning techniques.

4.5 Self-Management Advances in Specific Problem Domains

4.5.1 Self-Optimization

In [59] utility functions are used for the self-optimization of data center servers. The purpose of the utility function is to provide to data service requests the best resource allocation among a group of data servers. The utility computation itself is performed in two levels. At the local level, the autonomic application manager determines the utility of servicing a specific selection of requests based on its fixed resources and demand requirements. At the global level, allocation of resources to application managers is determined by a global resource arbiter.

Collaborative reinforcement learning is used in a k-component model to generate self-adaptive distributed systems [18]. The k-component is a modeling framework for creating distributed components and specifying their interfaces. Collaborative reinforcement learning is a modification of the reinforcement learning approach, which allows the components to not only learn from environmental feedbacks but also from the experiences of neighboring components. Global optimization problems are thus tackled in a distributed manner by first having the individual component perform its local reasoning using reinforcement learning and then advertise the results to the neighbors. Based on the results, the neighbors may initiate another round of distributed optimization process. The dynamic control of behavior based on learning (DCBL) middleware [58] uses reinforcement learning for QoS management. The middleware implements a wide range of reinforcement learning algorithms.

Threshold-based performance management [9] is used to ensure autonomic services can uphold their promised service-level objectives. The authors observe that traditional threshold setting often produces ineffective solutions because the exact threshold level at the component level cannot be predefined at the design time and is rather sensitive to the changing environment conditions and the runtime configuration of the component. The work proposes to conduct online adjustments of component level thresholds based on high-level service level objectives using generic algorithms and logistic regression.

Other self-optimization related research include QoS performance management using online control and predictive filtering [1]; combining queuing modeling, long-term predictive provisioning and short-term reactive provisioning for the management of multi-tier network applications [57]; and resource scheduling with automated demand and QoS requirement estimation and allocation in shared resource pools [46].

4.5.2 Self-Healing and Self-Protection

Efficient self-healing and protection mechanisms require the autonomic system to be predictive about the status of its environment and itself, and be able to reason about the changing environment for the purpose of fault localization and recovery. Much of the knowledge and foresight have to come from the runtime observation of the environment and system operations, with very little predefined designs or patterns to follow. Thus far, machine-learning techniques have found much applicability in this problem domain. For example, the decision tree learning approach is used in [12] to find faulty components. A decision tree is trained to classify the operation requests during a faulty period and the results are used to extrapolate and predict fault locations based on the paths of success and failed requests. Reinforcement learning is applied to automatic network repair in [39] to study its feasibility and efficiency. Steinder and Sethi [51] use belief networks to provide fault localization for communication networks. Two Bayesian inference algorithms, belief-updating and most-probable-explanation, are used to compute queries to the belief network. The resulting network is able to tolerate uncertainties in system observations, diagnose and isolate multiple simultaneous faults.

Abnormal trace detection can be used to identify fault or attacks in networks. However, the technique is difficult to apply in practice due to the variance in normal user behavior. Jiang *et al.* [32] constructed n-gram automata for abnormal trace detection leveraging the fact that n-gram automata's reachability structure is not changed by users' dynamic access behaviors. Compared with traditional probabilistic modeling, which is prone to sudden changes in user behavior (e.g. flash crowd effect), the n-gram automata are able to detect abnormal network behaviors in some cases. The extraction of n-grams from abnormal traces uses a similar technique to sequential pattern mining.

Improving system robustness through self-induced disruptions is a radical concept explored in [5]. The work suggests that controlled randomization is an effective mean of improving system performance and tolerance to faults. The technique represents the network and system dependencies among components as a connected graph and then randomly rewires existing edges or inserts new edges at random. Their findings show that this random perturbation not only increases the system performance when used in conjunction with self-optimization techniques but also reduces the likelihood of system-wide failures (e.g. due to system organization moving towards an unbalanced structure through gradual adaptations).

4.5.3 Self-Organization

Design patterns for autonomic system are used in the Unity architecture [13]. The goal-driven self-assembly pattern is implemented in Unity to realize self-organization. On start up, each autonomic component only knows about its high-level description of objectives and is given a point of contact. The required services and resources are automatically discovered at runtime through service discovery mechanisms (e.g. at the service registry). A policy registry is also provided as a common service from where various configurations, polices and rules could be retrieved by the autonomic components and then applied. The goal-driven self-assembly pattern in Unity, as described, is at a preliminary stage. The designers are still required to specify the essentials of the system's functions and relationships.

In order to accomplish a common task or for the sake of improving operational efficiency, self-structuring among the autonomic components may be necessary. Some research ([44], [31]) deals with the dynamic clustering of autonomic components. In a cluster, a 'cluster head' is often elected as a special-purpose node, performing coordination functions such as inter-cluster coordination, intra-cluster administration, etc. Ragusa *et al.* [44] propose self-managed clustering through the use of mobile agents. The cluster is formed via a clone-and-migrate process of the mobile agent. The agent examines the local routing table to determine the relative distance to each node in the cluster. If the cluster size exceeds threshold, the agent clones itself and partitions the cluster in two. If the cluster size is reasonable, the agent migrates among the cluster nodes until the central point in the cluster is found. Aggregation is also possible when neighboring clusters are both small in size. After the formation of a cluster, the mobile agent periodically rechecks the network distance to cluster nodes to determine whether re-election of a cluster head is necessary. Jesi *et al.* [31] propose an SG-2 protocol for cluster-head election. The protocol does not require distance measurement between cluster nodes as in traditional approaches. Rather, a virtual coordinate is selected on an n-dimensional space by a virtual coordinate service. The Euclidean distance between the coordinates of two nodes is used to predict the actual network roundtrip time between the two cluster nodes. Although this work is conducted in the scope of P2P networks, the technique is generally applicable to the dynamic clustering of autonomic components.

Emergence from biology describes how simple local behaviors by entities without global knowledge result in global behavior. In [2] the emergence concept is used to design an 'emergent' election algorithm. A key feature of emergence is that interactions are generally between the components and the environment (e.g. emitting pheromone trails), and the communications are one way and asynchronous. Individual messages in emergent systems have low values on their own and the system behavior is nondeterministic. These features of emergence make their designs extremely simple on the component level and render component validation tractable. Thus far, it is uncertain whether emergent designs could lead to complex and yet stable system behaviors in self-managing networks. Another related concept is stigmergy: insects coordinate their behaviors by using environmental modifications as cue. Work in collaborative construction [61] explores this promising direction. Swarms of robots are able to construct complex building structures from blocks by following simple rules and observing local environmental stimuli. The similarities between the collaborative construction and the self-organizing properties of autonomic systems suggest that indeed it is possible to achieve similar constructs, such as clustering and election, in networks using stigmergy concepts.

4.5.4 Other Problem Domains in Networks

How to best route traffic in networks based on user's QoS requirements is an important area of research in networking. The fluctuations in network performance and the large number of source–destination pairs in the network make the problem quite challenging and difficult to tackle. Research on cognitive packet network (CPN) ([22], [24]) seeks to provide adaptive distributed routing in the network using neural network and reinforcement learning techniques. Packets are classified into three types in CPN: smart packet, dumb packet and acknowledgement packet. Each router runs an independent random neural network to determine the best path from current node to destination based on learned path information, and the packets are routed using the output of the neural network. A smart packet remembers the list of routers it visits on its way to the destination and when it reaches the destination, an acknowledgement packet is sent back in reverse path order after removing cyclic subpaths. QoS measurements are also included in the ACK packet. Thus the system is able to route user traffic in a distributed way while seeking the best QoS path for the user. Gelenbe *et al.* [22] also propose an alternative source-based routing technique using a genetic algorithm. Since CPN is able to trace the routes user traffic takes and conduct reverse path routing of ACK packets, it is also able to fight against DDoS attacks [23]. Leibnitz *et al.* [37] propose accommodating user traffic using distributed multipath routing where activity functions from biological attractor selection are used to find multiple paths to carry the user traffic. The algorithm is globally stable and can withstand sporadic disruptions in the network.

With the networks pushing toward user and service centric paradigms, the management of users and their services is becoming an added concern. Xiao and Boutaba [65] use QoS-aware routing algorithms to facilitate the automatic selection of service providers and their associated service classes that jointly form an end-to-end service composition for the user traffic. Service providers and their service classes are abstracted into a weighed service domain graph upon which QoS path selection heuristics are applied to find the lowest cost path that satisfies the user's QoS requirement. The same technique is used for the automatic adaptation of the service composition in the presence of faults and performance degradations. Swarm intelligence is applied in SwarmingNet [14] to facilitate automatic service configuration. Using the multimedia messaging service as an example, the work shows that swarm intelligence could in some cases outperform traditional agent-based service configuration.

4.6 Benchmarking and Validation

4.6.1 Benchmarking

Because of the reliance we must place on automation and the fact that the complexity of the autonomic systems and the network environment will be far beyond human tractability, every piece of autonomic component in the network must be validated and benchmarked. Benchmarking is an age-old science in engineering and software productions. Thus far, traditional benchmarking involves subjecting the object under test to a tightly controlled benchmarking environment and injecting representative workloads. Performance data is then gathered for comparison. However, a number of problems arise when applying this approach to benchmarking autonomic systems. What constitutes a representative workload? First, the majority of autonomic systems are strongly influenced by their environment and they are as interaction heavy as a pure input–output system. Secondly, given the diversity of the functions and self-properties of autonomic systems, it becomes quite difficult to select a general set of representative workloads for comparison. What constitutes a good benchmarking environment? This challenge is similar to the one faced by the networking community at large. It is difficult to create a satisfying and sizable simulated environment under which the system benchmarking could be carried out with conviction. There are simply too many variables to control and their dependencies too complex to comprehend. What are good autonomic performance metrics? There simply isn't sufficient groundwork laid in this area to even formulate a reasonable set of performance metrics. Some preliminary attempts can be found in [42], where general categorization of benchmarking metrics for autonomic systems are proposed: QoS, cost, granularity/flexibility, failure avoidance/robustness, degree of autonomy, adaptivity, time to react and adapt, stability and sensitivity. However, these metrics are far from being in concise measurable forms. Additional discussions on benchmarking autonomic systems can be found in [10] and [38].

4.6.2 Validation

Validation of autonomic systems can be carried out at both design time and operation time. At design time, depending on the particular theory or architectural model used, validity and correctness constraints may be incorporated to ensure the resulting system is consistent and valid. Mechanisms that facilitate runtime validity checking could also be incorporated into the system design. For example, Desmet *et al.* [15] show the importance of maintaining consistency and correctness in component composition and adaptation. They propose employing architectural descriptions and implementation dependency specifications to ensure composition validity. Soft state fail-safe models are used for component adaptation. Gjorven *et al.* [26] propose the use of mirror-based reflections to encapsulate information about the implementation of a service and to include developer's knowledge about the service specifics.

One approach to runtime validation uses model-based inference. Essentially, a known 'good configuration' model of the system is compared with its runtime counterpart to determine the validity of the running system. In [21], architectural models of the running system are maintained as the basis for comparison against the running system. The Glean system [35] does this by first formulating the 'good configuration' classes of a system and then extracts correctness constraints from these classes. The inferred correctness constraints can then be used to determine the correctness of the current system configuration.

Wolf *et al.* [64] analyze system-wide behaviors in steady scenarios and conduct macroscopic numerical analysis to determine the system-wide behaviors of distributed and emergent systems. Chaparadza *et al.* [11] discusses the challenges and approaches to autonomic self-monitoring as a tool for runtime self-validation.

4.7 Self-Stabilization

4.7.1 Concept of Self-Stabilization

A primary characteristic of any distributed system is that the global system behavior is a manifestation of individual component behaviors and their effects on the environment. As expected, in large scale such complex interactions are difficult to abstract and control, and could indeed result in undesirable global system states. When considering a distributed system as a set of states (current configuration and system properties) and the transitions among states (system behaviors), then the property of stabilization categorizes the set of states as legitimate or illegitimate states. A legitimate state is one where the configuration of the system is consistent and its functions desirable.

The concept of self-stabilization was first introduced in 1973 by Dijkstra [17]. He describes a self-stabilizing system as a distributed system where individual component behaviors are determined by a subset of global system state that is known to the component and the global system must exhibit the property of 'regardless of its initial state, it is guaranteed to arrive at a legitimate state in a finite number of steps'. Two key properties are: (1) the system could initiate in any state and (2) the system can always recover from transient faults. When discussing self-stabilization, the concept of closure and convergence [49] are often mentioned. The closure property states that if the system is stable, after a number of executions it will remain stable unless perturbed by external force. The convergence property states that if the system is unstable, it has a tendency to move towards a stable state over time. The length of time that must elapse for this process to terminate is sometimes referred to as the convergence span. It is apparent that self-stabilization exhibits many of the desired self-properties of self-managing networks and hence should be an important area of investigation in autonomic research. Although there are some existing self-stabilizing algorithms proposed for specific network problems (e.g. the maximum flow problem [25]), thus far little research has been done in studying the feasibility of applying self-stabilization to self-managing networks or analyzing the self-stabilization properties of existing self-management theories. However, we believe this is a promising area of research. For example, many of the proposed theories for the design of self-managing networks exhibit certain degrees of self-stabilization, such as the dynamic equilibrium concept from game theory, and the emergence concept in biological modeling. In Schneider (1993), a number of theoretical factors that are detrimental to self-stabilization, are suggested: symmetry, termination, isolation and look-alike configurations.

4.7.2 Stability of a Network Path

As the primary function of a network is to transport traffic between source and destination nodes with efficiency, it is important to examine the concept of stability in the context of network paths. Consider the following scenario: assuming a self-stabilizing network with two legitimate states, each of which has a completely different way of routing traffic through the network. The convergence property states that if the system is in an illegitimate state, it will eventually return to one of the two legitimate states. The closure property assures us that regardless of the time elapsed; the system will stay in legitimate states, unless disrupted by external force. However, these two properties alone are not strong enough to guarantee the stability of a path, since constant oscillation between the two legitimate states is then possible, which results in undesirable route flapping. Hence a much stronger sense of self-stabilization is needed in self-managing networks, especially with respect to the time dimension.

4.8 Conclusion

In this chapter, we have presented an overview of the concepts, theories and advances in self-managing networks, and discussed specific self-management problems. There is a strong tie between self-awareness and self-management, where self-awareness must be present to enable self-management and self-management in turn makes the system or network more self-aware. In fact, a large body of research in self-management deals with the problems of knowledge acquisition, representation, dissemination and analysis (Section 4.4). In the end, the road to self-management is full of difficult challenges both from general system designs and from the specific network management domains. In particular, we have brought forth two particular issues in self-management: benchmarking and validation; and the concept of self-stabilization. They will become major research focus in the coming years, because in contrast to the traditionally human-intensive management solutions, a much stronger emphasis must now be placed on system reliance and trustworthiness if the vision of self-management is to become a reality. Also because of this degree of reliance and trust we must place on self-managed systems, their design has to follow a rigorous and formal approach that emphasizes verifiable theories and formal design methodologies. The largely distributed nature of self-managing systems offers a rich ground for the exploration of novel theories that favor simple local interactions leading to complex global behaviors.

To date, the movement towards self-management is still largely in the research exploration stage. Some industry bodies, chiefly IBM, have advocated a gradual phasing of self-management properties into their products, as suggested in the IBM autonomic evolution roadmap [30]. We are likely to see initial experimentation results of self-management in the next few years and the actual application in the industry in much longer term.

References

[1] Abdelwahed, S., Kandasamy, N. and Neema, S., (2004) *A control-based framework for self-managing distributed computing systems*. ACM Workshop on Self-Healing Systems, October.

[2] Anthony, R. (2004) *Emergence: a paradigm for robust and scalable distributed applications*. IEEE 1st International Conference on Autonomic Computing, May.

[3] Aiber, S., Gilat, D., Landau, A. *et al.* (2004) *Autonomic self-optimization according to business objectives*. IEEE 1st International Conference on Autonomic Computing, May.

[4] Babaoglu, O., Canright, G., Deutsch, A. *et al.* (2006) Design patterns from biology for distributed computing. *ACM Transactions on Autonomous and Adaptive Systems*, **1**(1), 26–66.

[5] Beygelzimer, A., Grinstein, G., Linsker, R. *et al.* (2004) *Improving network robustness*. IEEE 1st International Conference on Autonomic Computing, May.

[6] Bhat, V., Parashar, M., Liu, H. *et al.* (2006) *Enabling self-managing applications using model-based online control strategies*. IEEE International Conference on Autonomic and Autonomous Systems, July.

[7] Bahati, R., Bauer, M., Vieira, E. *et al.* (2006a) *Mapping policies into autonomic management actions*. IEEE International Conference on Autonomic and Autonomous Systems, July.

[8] Bahati, R., Bauer, M., Vieira, E. *et al.* (2006b) *Using policies to drive autonomic management*. IEEE 7th International Symposium on a World of Wireless Mobile and Multimedia Networks, June.

[9] Breitgand, D., Henis, E. and Shehory, O. (2005) *Automated and adaptive threshold setting: enabling technology for autonomy and self-management*. IEEE 2nd International Conference on Autonomic Computing, June.

[10] Brown, A., Hellerstein, J., Hogstrom, M. *et al.* (2004) *Benchmarking autonomic capabilities: promises and pitfalls*. IEEE 1st International Conference on Autonomic Computing, May.

[11] Chaparadza, R., Coskun, H. and Schieferdecker, I. (2005) *Addressing some challenges in autonomic monitoring in self-managing networks*. IEEE 13th International Conference on Networks, November.

[12] Chen, M., Zheng, A., Lloyd, J. *et al.* (2004) *Failure diagnosis using decision trees*. IEEE 1st International Conference on Autonomic Computing, May.

[13] Chess, D., Segal, A., Whalley, I. *et al.* (2004) *Unity: experience with a prototype autonomic computing system*. IEEE 1st International Conference on Autonomic Computing, May.

[14] Chiang, F. and Braun, R. (2006) *A nature inspired multi-agent framework for autonomic service management in ubiquitous computing environments*. IEEE/IFIP Network Operations and Management Symposium, April.

[15] Desmet, L., Janssens, N., Michiels, S. *et al.* (2004) *Towards preserving correctness in self-managed software systems*. ACM Workshop on Self-Healing Systems, October.

[16] Diao, Y., Hellerstein, J., Parekh, S. *et al.* (2005) A control theory foundation for self-managing computing systems. *IEEE Journal on Selected Areas in Communications*, 23(12), 2213–22.

[17] Dijkstra, E.W. (1974) Self-stabilization in spite of distributed control. *Communications of the ACM*, **17**(11), 643–4.

[18] Dowling, J. and Cahill, V. (2004) *Self-managed decentralised systems using k-components and collaborative reinforcement learning*. ACM Workshop on Self-Healing Systems, October.

[19] Eymann, T., Reinicke, M., Ardaiz, O. *et al.* (2003) *Self-organizing resource allocation for autonomic networks*. IEEE 14th International Workshop on Database and Expert Systems Applications, September.

[20] Fujii, K. and Suda, T. (2005) Semantics-based dynamic service composition. *IEEE Journal on Selected Areas in Communications*, **23**(12), 2361–72.

[21] Garlan, D. and Schmerl, B. (2002) *Model-based adaptation for self-healing systems*. ACM Workshop on Self-Healing Systems, November.

[22] Gelenbe, E., Gellman, M., Lent, R. *et al.* (2004) *Autonomous smart routing for network QoS*. IEEE 1st International Conference on Autonomic Computing, May.

[23] Gelenbe, E., Gellman, M. and Loukas, G. (2005) *An autonomic approach to denial of service defense*. IEEE 6th International Symposium on a World of Wireless Mobile and Multimedia Networks, June.

[24] Gelenbe, E. and Liu, P. (2005) *QoS and routing in the cognitive packet network*. IEEE 6th International Symposium on a World of Wireless Mobile and Multimedia Networks, June.

[25] Ghosh, S., Gupta, A. and Pemmaraju, S. (1995) *A self-stabilizing algorithm for the maximum flow problem*. IEEE 14th Annual International Phoenix Conference on Computers and Communication.

[26] Gjorven, E., Eliassen F., Lund, K. *et al.* (2006) *Self-adaptive systems: a middleware managed approach.* IFIP/IEEE International Workshop on Self-Managed Systems & Services, June.

[27] Gracanin, D., Bohner, S. and Hinchey, M. (2004) *Towards a model-driven architecture for autonomic systems.* IEEE 11th International Conference and Workshops on the Engineering of Computer-Based Systems, May.

[28] Hawthorne, M. and Perry, D. (2004) *Exploiting architectural prescriptions for self-managing, self-adaptive systems: a position paper.* ACM Workshop on Self-Healing Systems, October.

[29] Herrmann, K., Muhl, G. and Geihs, K. (2005) Self-management: the solution to complexity or just another problem? *IEEE Distributed Systems Online*, **6**(1).

[30] IBM (2002) *Autonomic computing architecture: a blueprint for managing complex computing environments.* IBM and Autonomic Computing, October.

[31] Jesi, G., Montresor, A. and Babaoglu, O. (2006) *Proximity-aware superpeer overlay topologies.* IFIP/IEEE International Workshop on Self-Managed Systems & Services, June.

[32] Jiang, G., Chen, H., Ungureanu, C. *et al.* (2005) *Multi-resolution abnormal trace detection using varied-length n-grams and automata.* IEEE 2nd International Conference on Autonomic Computing, June.

[33] Kephart, J. (2005) *Research challenges of autonomic computing.* ACM 27th International Conference on Software Engineering, May.

[34] Kephart, J. and Chess, D. (2003) The vision of autonomic computing. *IEEE Computer*, January.

[35] Kiciman, E. and Wang, Y. (2004) *Discovering correctness constraints for self-management of system configuration.* IEEE 1st International Conference on Autonomic Computing, May.

[36] Lavinal, E., Desprats, T. and Raynaud, Y. (2006) *A generic multi-agent conceptual framework towards self-management.* IEEE/IFIP Network Operations and Management Symposium, April.

[37] Leibnitz, K., Wakamiya, N. and Murata, M. (2005) *Biologically inspired adaptive multi-path routing in overlay networks.* IFIP/IEEE International Workshop on Self-Managed Systems & Services, May.

[38] Lewis, D., O'Sullivan, D. and Keeney, J. (2006) *Towards the knowledge-driven benchmarking of autonomic communications.* IEEE 7th International Symposium on a World of Wireless Mobile and Multimedia Networks, June.

[39] Littman, M., Ravi, N., Fenson, E. *et al.* (2004) *Reinforcement learning for autonomic network repair.* IEEE 1st International Conference on Autonomic Computing, May.

[40] Liu, H. and Parashar, M. (2006) Accord: a programming framework for autonomic applications. *IEEE Transactions on Systems, Man and Cybernetics*, **36**(3), 341–52.

[41] Mbarek, N. and Krief, F. (2006) *Service level negotiation in autonomous systems management.* IEEE 3rd International Conference on Autonomic Computing, June.

[42] McCann, J.A. and Huebscher, M.C. (2004) *Evaluation issues in autonomic computing.* Grid and Cooperative Computing Workshops, October.

[43] Mortier, R. and Kiciman, E. (2006) *Autonomic network management: some pragmatic considerations.* ACM SIGCOMM Workshop, September.

[44] Ragusa, C., Liotta, A. and Pavlou, G. (2005) An adaptive clustering approach for the management of dynamic systems. *IEEE Journal on Selected Areas in Communications*, **23**(12), 2223–35.

[45] Randles, M., Taleb-Bendiab, A., Miseldine, P. *et al.* (2005) *Adjustable deliberation of self-managing systems.* IEEE 12th International Conference and Workshops on the Engineering of Computer-Based Systems, April.

[46] Rolia, J., Cherkasova, L., Arlitt, M. *et al.* (2005) *An automated approach for supporting application QoS in shared resource pools.* IFIP/IEEE International Workshop on Self-Managed Systems & Services, May.

[47] Salehie, M. and Tahvildari, L. (2005) *Autonomic computing: emerging trends and open problems.* ACM Design and Evolution of Autonomic Application Software, May.

[48] Samman, N. and Karmouch, A. (2005) An automated policy-based management framework for differentiated communication systems. *IEEE Journal on Selected Areas in Communications*, **23**(12), 2236–47.

[49] Schneider, M. (1993) Self-stabilization. *ACM Computing Surveys*, **25**(1), 45–67.

[50] Smith, D., Morris, E. and Carney, D. (2005) *Interoperability issues affecting autonomic computing.* ACM Design and Evolution of Autonomic Application Software, May.

[51] Steinder, M. and Sethi, A. (2004) Probabilistic fault localization in communication systems using belief networks. *IEEE/ACM Transactions on Networking*, **12**(5), 809–822.

[52] Sterritt, R. and Hinchey, M., (2005) *Why computer-based systems should be autonomic.* IEEE 12th International Conference and Workshops on the Engineering of Computer-Based Systems, April.

[53] Supadulchai P. and Aagesen F. (2005) *A framework for dynamic service composition*. IEEE 6th International Symposium on a World of Wireless Mobile and Multimedia Networks, June.

[54] Suzuki, J. and Suda, T. (2005) A middleware platform for a biologically inspired network architecture supporting autonomous and adaptive applications. *IEEE Journal on Selected Areas in Communications*, **23**(12), 249–60.

[55] Tesauro, G., Jong, N., Das, R. *et al.* (2006) *A hybrid reinforcement learning approach to autonomic resource allocation*. IEEE 3rd International Conference on Autonomic Computing, June.

[56] Tianfield, H. (2003) *Multi-agent based autonomic architecture for network management*. IEEE International Conference on Industrial Informatics, August.

[57] Urgaonkar, B., Shenoy, P., Chandra, A. *et al.* (2005) *Dynamic provisioning of multi-tier Internet applications*. IEEE 2nd International Conference on Autonomic Computing, June.

[58] Vienne, P. and Sourrouille, J. (2005) *A middleware for autonomic QoS management based on learning*. ACM Software Engineering and Middleware, September.

[59] Walsh, W., Tesauro, G., Kephart, J. *et al.* (2004) *Utility functions in autonomic systems*. IEEE 1st International Conference on Autonomic Computing, May.

[60] Wang, W. and Li, B. (2005) Market-based self-optimization for autonomic service overlay networks. *IEEE Journal on Selected Areas in Communications*, **23**(12), 2320–32.

[61] Werfel, J. and Nagpal, R. (2006) Extended stigmergy in collective construction. *IEEE Intelligent Systems*, **21**(2), 20–8.

[62] White, S., Hanson, J., Whalley, I. *et al.* (2004) *An architectural approach to autonomic computing*. IEEE 1st International Conference on Autonomic Computing, May.

[63] Wile, D., (2004) *Patterns of self-management*. ACM Workshop on Self-Healing Systems. October.

[64] Wolf, T., Samaey, G., Holvoet, T. *et al.* (2005) *Decentralised autonomic computing: analysing self-organising emergent behaviour using advanced numerical methods*. IEEE 2nd International Conference on Autonomic Computing, June.

[65] Xiao, J. and Boutaba, R. (2005) QoS-aware service composition and adaptation in autonomic communication. *IEEE Journal on Selected Areas in Communications*, **23**(12), 2344–60.

[66] Zeng, A., Pan, D., Zheng, Q. *et al.* (2006) Knowledge acquisition based on rough set theory and principal component analysis. *IEEE Intelligent Systems*, **21**(2), 78–85.

[67] Zenmyo, T., Yoshida, H. and Kimura, T. (2006) *A self-healing technique based on encapsulated operation knowledge*. IEEE 3rd International Conference on Autonomic Computing, June.

5

Machine Learning for Cognitive Networks: Technology Assessment and Research Challenges

Thomas G. Dietterich
School of Electrical Engineering and Computer Science, Oregon State University, Oregon, USA

Pat Langley
Institute for the Study of Learning and Expertise, Palo Alto, California, USA

5.1 Introduction

Clark, Partridge, Ramming, and Wroclawski [9] [10] [36] recently proposed a new vision for computer network management – the Knowledge Plane – that would augment the current paradigm of low-level data collection and decision making with higher level processes. One key idea is that the Knowledge Plane would learn about its own behavior over time, making it better able to analyze problems, tune its operation, and generally increase its reliability and robustness. This suggests the incorporation of concepts and methods from machine learning [23] [28], an established field that is concerned with such issues.

Machine learning aims to understand computational mechanisms by which experience can lead to improved performance. In everyday language, we say that a person has 'learned' something from an experience when he can do something he could not, or could not do as well, before that experience. The field of machine learning attempts

Cognitive Networks: Towards Self-Aware Networks Edited by Qusay H. Mahmoud

to characterize how such changes can occur by designing, implementing, running and analyzing algorithms that can be run on computers. The discipline draws on ideas from many other fields, including statistics, cognitive psychology, information theory, logic, complexity theory and operations research, but always with the goal of understanding the computational character of learning.

There is general agreement that representational issues are central to learning. In fact, the field is often divided into paradigms that are organized around representational formalisms, such as decision trees, logical rules, neural networks, case libraries and probabilistic notations. Early debate revolved around which formalism provided the best support for machine learning, but the advent of experimental comparisons around 1990 showed that, in general, no formalism led to better learning than any other. However, it also revealed that the specific features or representational encodings mattered greatly, and careful feature engineering remains a hallmark of successful applications of machine learning technology [26].

Another common view is that learning always occurs in the context of some *performance* task, and that a learning method should always be coupled with a performance element that uses the knowledge acquired or revised during learning. Figure 5.1 depicts such a combined system, which experiences the environment, uses learning to transform those experiences into knowledge and makes that knowledge available to a performance module that operates in the environment. Performance refers to the behavior of the system when learning is disabled. This may involve a simple activity, such as assigning a label or selecting an action, but it may also involve complex reasoning, planning or interpretation. The general goal of learning is to improve performance on whatever task the combined system is designed to carry out.

We should clarify a few more points about the relations between learning, performance and knowledge. Figure 5.1 suggests that the system operates in a continuing loop, with performance generating experiences that produce learning, which in turn leads to changes in performance, and so on. This paradigm is known as *online* learning, and characterizes some but not all research in the area. A more common approach, known as *offline* learning, instead assumes that the training experiences are all available at the outset, and that learning transforms these into knowledge only once. The figure also includes an optional link that lets the system's current knowledge influence the learning process. This idea is not widely used in current research, but it can assist learning significantly when relevant knowledge is available.

In this chapter, we examine various aspects of machine learning that touch on cognitive approaches to networking. We begin by reviewing the major problem formulations that

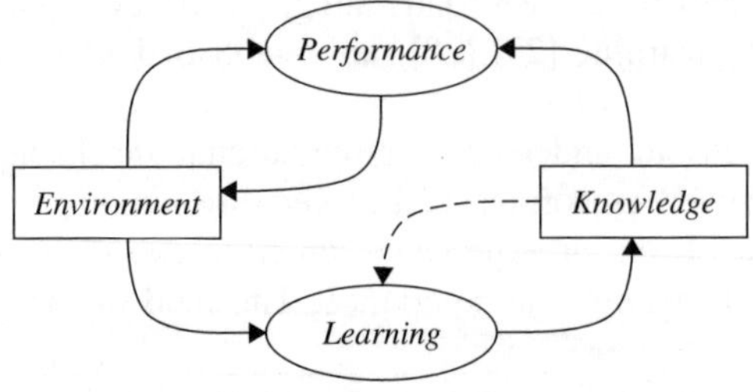

Figure 5.1 Relationship between learning, performance, knowledge and the environment

have been studied in machine learning. Then we consider three tasks that the Knowledge Plane is designed to support and the roles that learning could play in them. Next we discuss some open issues and research challenges that the Knowledge Plane poses for the field of machine learning. Finally, we propose some methods and criteria for evaluating the contribution of machine learning to cognitive networking tasks.

5.2 Problem Formulations in Machine Learning

Treatments of machine learning (e.g., [22] [38]) typically organize the field along representational lines, depending on whether one encodes learned knowledge using decision trees, neural networks, case libraries, probabilistic summaries, or some other notation. However, a more basic issue concerns how one *formulates* the learning task in terms of the inputs that drive learning and the manner in which the learned knowledge is utilized. This section examines three broad formulations of machine learning.

5.2.1 Learning for Classification and Regression

The most common formulation focuses on learning knowledge for the performance task of *classification* or *regression*. Classification involves assigning a test case to one of a finite set of classes, whereas regression predicts the case's value on some continuous variable or attribute. In the context of network diagnosis, one classification problem is deciding whether a connection failure is due to the target site being down, the target site being overloaded, or the ISP service being down. An analogous regression problem might involve predicting the time it will take for the connection to return. Cases are typically described as a set of values for discrete or continuous attributes or variables. For example, a description of the network's state might include attributes for packet loss, transfer time and connectivity. Some work on classification and regression instead operates over relational descriptors. Thus, one might describe a particular situation in terms of node connections and whether numeric attributes at one node (e.g., buffer utilization) are higher than those at an adjacent node.

In some situations, there is no special attribute that one knows at the outset will be predicted from others. Instead, one may need to predict the value of any unobserved attributes in terms of others that have been observed. This performance task, often called *pattern completion* or *flexible prediction*, can be used for symbolic attributes, continuous attributes, or a mixture of them. For example, given information about some network variables that are measured easily and cheaply, one might want to predict the values of other network variables that are more expensive to measure. A related task involves predicting the conditional probabilities that different values will occur for unknown variables given observed values for others. Alternatively, one may want to predict the joint probability distribution over the entire set of variables.

One can formulate a number of distinct learning tasks that produce knowledge for use in classification or regression. The most common, known as *supervised learning*, assumes the learner is given training cases with associated classes or values for the attribute to be predicted. For example, one might provide a supervised learning method with 200 instances of four different types of connection failure, say 50 instances of each class, with each instance described in terms of the attributes to be used later during classification. The analogous version for regression would provide instead the time taken to restore the connection for each training instance.

There exist a variety of well-established paradigms for supervised learning, including decision-tree and rule induction [38] [11], neural network methods [39], support-vector machines [12], nearest-neighbor approaches [1] and probabilistic methods [7]. These frameworks differ in the formalisms they employ for representing learned knowledge, as well as their specific algorithms for using and learning that knowledge. What these methods hold in common is their reliance on a target class or response variable to direct their search through the space of predictive models. They also share a common approach to evaluation, since their goal is to induce predictive models from training cases that have low error on novel test cases.

A second broad class of tasks, *unsupervised learning*, assumes that the learner is given training cases without any associated class information or any specific attribute singled out for prediction. For example, one might provide an unsupervised method with the same 200 instances as before, but not include any information about the type of connection failure or the time taken to restore the connection.

As with supervised learning, there exist many techniques for learning from unsupervised data, but these fall into two broad classes. One approach, known as *clustering* [8] [17], assumes the goal of learning is to assign the training instances to distinct classes of its own invention, which can be used to classify novel instances and make inferences about them, say through pattern completion. For example, a clustering algorithm might group the 200 training instances into a number of classes that represent what it thinks are different types of service interruption. Another approach, known as *density estimation* [37], instead aims to build a model that predicts the probability of occurrence for specific instances. For example, given the same data about service interruptions, such a method would generate a probability density function that covers both the training instances and novel ones.

A third formulation, known as *semi-supervised learning* [4], falls between the two approaches we have already discussed. In this framework, some of the training instances come with associated classes or values for predicted attributes, but others (typically the majority) do not have this information. This approach is common in domains such as text classification, where training cases are plentiful but class labels are costly. The goal is similar to that for supervised learning, that is, to induce a classifier or regressor that makes accurate predictions, but also to utilize the unlabeled instances to improve this behavior. For example, even if only 20 of the 200 training instances on service interruption included class information, one might still use regularities in the remaining instances to induce more accurate classifiers.

Classification and regression are the most basic capabilities for which learning can occur. As a result, the field has developed robust methods for these tasks, and they have been applied widely to develop accurate and useful predictive models from data. Langley and Simon [26] review some early successes of these methods, and they have since formed the backbone for many commercial applications within the data-mining movement. Methods for classification and regression learning can also play a role in more complex tasks, but such tasks also introduce other factors that require additional mechanisms, as discussed in the next section.

5.2.2 Learning for Acting and Planning

A second formulation addresses learning of knowledge for selecting actions or plans for an agent to carry out in the world. In its simplest form, action selection can occur in a purely reactive way, ignoring any information about past actions. This version has a straightforward mapping onto classification, with alternative actions corresponding to distinct classes from which the agent can choose based on descriptions of the world state. One can also map it onto regression, with the agent predicting the overall value or utility of each action in a given world state.

Both approaches can also be utilized for problem solving, planning and scheduling. These involve making cognitive choices about future actions, rather than about immediate actions in the environment. Such activities typically involve search through a space of alternatives, which knowledge can be used to constrain or direct. This knowledge may take the form of classifiers for which action to select or regression functions over actions or states. However, it can also be cast as larger-scale structures called *macro-operators* that specify multiple actions that should be carried out together.

As with classification and regression, one can formulate a number of learning tasks that produce knowledge for action selection and search. The simplest approach, known as a *learning apprentice* [29] or an *adaptive interface* [25], embeds the learner within a larger system that interacts with a human user. This system may accept directions from the user about what choices to make or it may make recommendations to the user, who can then accept them or propose other responses. Thus, the user gives direct feedback to the system about each choice, effectively transforming the problem of learning to select actions into a supervised learning task, which can then be handled using any of the methods discussed earlier. A related paradigm, known as *programming by demonstration* [13] [34], focuses on learning macro-operators for later invocation by the user to let him accomplish things in fewer steps.

For example, one might implement an interactive tool for network configuration that proposes, one step at a time, a few alternative components to incorporate into the configuration or connections to add among existing components. The human user could select from among these recommendations or reject them all and select another option. Each such interaction would generate a training instance for use in learning how to configure a network, which would then be used on future interactions. One can imagine similar adaptive interfaces for network diagnosis and repair.

A closely related formulation of action learning, known as *behavioral cloning* [40], collects traces of a human acting in some domain, but does not offer advice or interact directly. Again, each choice the human makes is transformed into a training case for use by supervised learning. The main difference is that behavioral cloning aims to create autonomous agents for carrying out a sequential decision-making task, whereas learning apprentices and adaptive interfaces aim to produce intelligent assistants. For example, a system could watch a human expert execute a sequence of commands in configuring a computer network, and then transform these into supervised training cases for learning which actions to select or for estimating the value of the available choices. However, one might also attempt to extract, from the same trace, recurring sets of actions for composition into macro-operators that would let one solve the same problem in fewer steps.

A somewhat different formulation involves the notion of learning from delayed reward, more commonly known as *reinforcement learning*. Here the agent typically carries out actions in the environment and receives some reward signal that indicates the desirability of the resulting states. However, because many steps may be necessary before the agent reaches a desirable state (e.g., reestablishing a service connection), the reward can be delayed. Research in the reinforcement learning framework falls into two main paradigms. One represents control policies indirectly in terms of *value functions* that map state descriptions and available actions onto expected values (i.e., expected total future reward) [20] [44]. This approach involves propagating rewards backward over action sequences to assign credit and may invoke a regression method to learn a predictor for expected values. Another paradigm instead encodes control policies more directly as a mapping from state descriptions onto actions, with learning involving search through the space of such policies [31] [46].

One might apply either approach to learning policies for dynamic network routing [6]. The reward signal here might be based on the standard metrics for route performance. The system would try establishing different routes, each of which involves a number of decision-making steps, and learn routing policies based on the observed performance. Over time, the routes selected by the learned policy would change, giving improved behavior for the overall network.

Another formulation is closely related to reinforcement learning, but involves *learning from problem solving* and mental search [42], rather than from actions in the environment. Here the agent has some model of the effects of actions or the resources they require which it can use to carry out mental simulations of action sequences. However, there typically exist many possible sequences, which introduces the problem of search through a problem space. Such search can produce one or more sequences that solve the problem, but it can also generate dead ends, loops and other undesirable outcomes. Both successes and failures provide material for learning, in that they distinguish between desirable and undesirable choices, or at least suggest relative desirability.

Research on learning from problem-solving traces occurs within three broad paradigms. Some work focuses on learning local search-control knowledge for selecting, rejecting or preferring actions or states. This knowledge may be cast as control rules or some related symbolic representation, or it may be stated as a numeric evaluation function. The latter approach is closely related to methods for estimating value functions from delayed reward, which has occasionally been used for tasks such as scheduling [47] and integrated circuit layout [5]. Another paradigm emphasizes the formation from solution paths of macro-operators that take larger steps through the problem space in order to reduce the effective depth of search. A third framework, analogical problem solving, also stores large-scale structures, but utilizes them in a more flexible manner by adapting them to new problems.

For example, one might apply any of these approaches to tasks like network routing and configuration. Such an application would require some model of the effects that individual choices would produce, so that the agent can decide whether a given state is desirable before actually generating it in the world. Thus, the system would start with the ability to generate routes or configurations, but it might do this very inefficiently if the search space is large. After repeated attempts at routing or configuration, it would acquire heuristic knowledge about how to direct its search, letting it produce future solutions much more efficiently without loss in quality.

A final formulation involves the *empirical optimization* of a complex system. Consider the problem of adjusting a chemical plant's parameters to improve its performance (e.g., reduce energy consumption, reduce waste products, increase product quality, increase rate of production, and so forth). If a predictive model of the plant is not available, the only recourse may be to try various settings of the parameters and see how the plant responds.

One example of this idea, *response surface methodology* ([33]) attempts to find the optimal operating point of a system by measuring system behavior at various points. The classic method designs and executes an experiment (e.g., some form of factorial design) about the current operating point and fits the results with a quadratic function to estimate the local shape of the objective function surface. Then it chooses a new operating point at the optimum of that quadratic surface and repeats the process.

Machine learning researchers have studied methods that make weaker assumptions and require fewer training examples. One approach [30] employs regression to analyze the results of previous experiments and determine a region of interest in which the objective function can be approximated well, then chooses a new test point that is distant from other test points while still lying within this region. An alternative approach [2] is more appropriate for searching discrete parameter spaces such as those that arise in network configuration. Given a set of parameter settings (configurations) for which the performance has been measured, one fits a probability distribution to predict where additional 'good' points are located, then samples a new set of configurations according to that distribution, measures their performance, and continues until convergence.

Before closing, it is worth making two other points about learning for action selection and planning. First, in many domains, sensing requires active invocation, so that one can view it as a kind of action. Thus, an agent can learn policies for sensing, say to support efficient network diagnosis, just as it can for effectors, such as closing down a link in response to a suspected attack. Second, some methods for plan learning assume the availability of action models that describe the expected effects when actions are invoked, which leads in turn to the task of learning such action models from observations. This has many similarities to the problem of classification and regression learning, but aims to support higher level learning about policies for acting and planning.

5.2.3 Learning for Interpretation and Understanding

A third formulation focuses on learning knowledge that lets one interpret and understand situations or events. Classification can be seen as a simple example of this idea, since one can 'understand' an instance as being an example of some class. However, more sophisticated approaches attempt to interpret observations in a more constructive manner, by combining a number of separate knowledge elements to explain them. The key difference is that classification and regression are content with models that make accurate predictions, whereas interpretative approaches require models that explain the data in terms of deeper structures. This process of explanation generation is often referred to as *abduction*.

The explanatory or abductive approach is perhaps most easily demonstrated in natural language processing, where a common performance task involves parsing sentences using a context-free grammar or some related formalism. Such a grammar contains rewrite rules that refer to nonterminal symbols for types of phrases and parts of speech, and a parse tree specifies how one can derive or explain a sentence in terms of these rules. One can apply similar ideas to other domains, including the interpretation and diagnosis of network behavior. For example, given anomalous data about the transfer rates between various nodes in a network, one might explain these observations using known processes, such as demand for a new movie that is available at one site and desired by others.

One can state a number of different learning tasks within the explanatory framework. The most tractable problem assumes that each training case comes with an associated explanation cast in terms of domain knowledge. This formulation is used commonly within the natural language community, where the advent of 'tree banks' has made available large corpora of sentences with their associated parse trees. The learning task involves generalizing over the training instances to produce a model that can be used to interpret or explain future test cases. Naturally, this approach places a burden on the developer, since it requires hand construction of explanations for each training case, but it greatly constrains the learning process, as it effectively decomposes the task into a set of separate classification or density estimation tasks, one for each component of the domain knowledge.

A second class of learning task assumes that training instances do not have associated explanations, but provides background knowledge from which the learner can construct them. This problem provides less supervision than the first, since the learner must consider alternative explanations for each training case and decide which ones are appropriate. However, the result is again some model that can be applied to interpret or explain future instances. This formulation is less burdensome on the developer, since he need not provide explanations for each training case, but only a domain theory from which the learner can construct them itself. Flann and Dietterich [18] have referred to this learning task as *induction over explanations*, but it is also closely related to some work on *constructive induction* [16] and *explanation-based generalization* [15].

A final variant on learning for understanding provides training cases with neither explanations nor background knowledge from which to construct them. Rather, the learner must induce its own explanatory structures from regularities in the data, which it can then utilize to interpret and understand new test instances. An example from natural language involves the induction of context-free grammars, including both nonterminal symbols and the rewrite rules in which they occur, from legal training sentences [43]. Clearly, this task requires even less effort on the developer's part, but places a greater challenge on the learning system. This approach has gone by a variety of names in the machine learning literature, including *term generation, representation change* and *constructive induction* (though this phrase has also been used for the second task).

Because learning tasks that produce explanatory models are generally more difficult than those for classification and regression, some researchers have formulated more tractable versions of them. One variant assumes the qualitative structure of the explanatory model is given and that learning involves estimating numeric parameters from the data. Examples of this approach include determining the probabilities in a stochastic context-free grammar, tuning the parameters in sets of differential equations and inferring conditional probabilities in a Bayesian network. Another variation, known as *theory revision*, assumes an initial explanatory model that is approximately correct and utilizes training data to alter its qualitative structure [35]. Examples include revising Horn clause programs from classified training cases, improving sets of equations from quantitative data, and altering grammars in response to training sentences.

5.2.4 Summary of Problem Formulations

In summary, one can formulate machine learning tasks in a variety of ways. These differ in both the manner in which learned knowledge is utilized and, at a finer level, in the

Table 5.1 Summary of machine learning problem formulations

Formulation	Performance task
Classification and regression	Predict y given x
	Predict rest of x given part of x
	Predict $P(x)$ given x
Acting and planning	Iteratively choose action a in state s
	Choose actions $\langle a_1, \ldots, a_n \rangle$ to achieve goal g
	Find setting s to optimize objective $J(s)$
Interpretation and understanding	Parse data stream into tree structure of objects or events

nature of the training data that drives the learning process. Table 5.1 summarizes the main formulations that have been discussed in this section. However, it is important to realize that different paradigms have received different degrees of attention within the machine learning community. Supervised approaches to classification and regression have been the most widely studied by far, with reinforcement learning being the second most common. Yet their popularity in the mainstream community does not imply they are the best ways to approach problems in computer networking, and research on the Knowledge Plane should consider all the available options.

Another important point is that one can often formulate a given real-world problem as a number of quite different learning tasks. For example, one might cast diagnosis of network faults as a classification problem that involves assigning the current network state to either a normal condition or one of a few prespecified faulty conditions. However, one could instead formulate it as a problem of understanding anomalous network behavior, say in terms of unobservable processes that, taken together, can explain recent statistics. Yet another option would be to state diagnosis as a problem of selecting active sensors that narrow down alternatives. Each formulation suggests different approaches to the diagnostic task, to learning knowledge in support of that task, and to criteria for evaluating the success of the learning component.

5.3 Tasks in Cognitive Networking

The vision for the Knowledge Plane [9] [10] [36] describes a number of novel capabilities for computer networks. This section reviews three capabilities that the vision assumes in terms of the cognitive functionalities that are required. These include anomaly detection and fault diagnosis, responding to intruders and worms, and rapid configuration of networks.

5.3.1 Anomaly Detection and Fault Diagnosis

Current computer networks require human managers to oversee their behavior and ensure that they deliver the services desired. To this end, the network managers must detect unusual or undesirable behaviors, isolate their sources, diagnose the fault, and repair the problem. These tasks are made more challenging because large-scale networks are managed in a distributed manner, with individuals having access to information about, and control over, only portions of the system. Nevertheless, it will be useful to examine the activities in which a single network manager engages.

The first activity, *anomaly detection*, involves the realization that something unusual or undesirable is transpiring within the network. One possible approach to this problem, which applies recent advances in Bayesian networks, is to formulate it as a density estimation problem. Individual components, larger regions of the network, or, at some level, the entire Internet could be modeled as the joint probability distribution of various quantities (queue lengths, traffic types, round-trip times, and so on). An anomaly is defined as a low probability state of the network.

Another possible approach is sometimes called one-class learning or learning a characteristic description of a class. A classifier can be learned that attempts to find a compact description that covers a target percentile (e.g., 95%) of the 'normal' traffic. Anything classified as 'negative' by this classifier can then be regarded as an anomaly.

There are several issues that arise in anomaly detection. First, one must choose the level of analysis and the variables to monitor for anomalies. This may involve first applying methods for interpreting and summarizing sensor data. In the Knowledge Plane, one can imagine having whole hierarchies of anomaly detectors looking for changes in the type of network traffic (e.g., by protocol type), in routing, in traffic delays, in packet losses, in transmission errors, and so on. Anomalies may be undetectable at one level of abstraction but easy to detect at a different level. For example, a worm might escape detection at the level of a single host, but be detectable when observations from several hosts are combined.

The second issue is the problem of false alarms and repeated alarms. Certain kinds of anomalies may be unimportant, so network managers need ways of training the system to filter them out. Supervised learning methods could be applied to this problem.

The second activity, *fault isolation*, requires the manager to identify the locus of an anomaly or fault within the network. For example, if a certain route has an especially heavy load, this may be due to changes at a single site along that route rather than to others. Hence, whereas anomaly detection can be performed locally (e.g., at each router), fault isolation requires the more global capabilities of the Knowledge Plane to determine the scope and extent of the anomaly.

The activity of *diagnosis* involves drawing some conclusions about the cause of the anomalous behavior. Typically, this follows fault isolation, although in principle one might infer the presence of a specific problem without knowing its precise location. Diagnosis may involve the recognition of some known problems, say one the network manager has encountered before, or the characterization of a new problem that may involve familiar components.

One can apply supervised learning methods to let a network manager teach the system how to recognize known problems. This could be a prelude to automatically solving them, as discussed below.

Both fault isolation and diagnosis may require active measurements to gather information. For example, an anomaly found at a high level of aggregation would typically require making more detailed observations at finer levels of detail to understand the cause. In the 'Why?' scenario, one can imagine active probes of both the local computer (e.g., its configuration) and the Internet (e.g., 'pings' to see if the destination is reachable and up). Diagnosis usually must balance the cost of gathering information against the potential informativeness of the action. For example, if the ping succeeds, it requires little time, but otherwise it can take much longer to time out. If the goal is to diagnose the problem as quickly as possible, then ping might be a costly action to perform.[1]

Fault isolation and diagnosis also typically require models of the structure of the system under diagnosis. Much recent effort in network research has sought to provide better ways of understanding and visualizing the structure of the Internet. Machine learning for interpretation could be applied to help automate this process. The resulting structural and behavioral models could then be used by model-based reasoning methods to perform fault isolation and diagnosis.

Once a network manager has diagnosed a problem, he is in a position to repair it. However, there may exist different courses of action that would eliminate the problem, which have different costs and benefits. Moreover, when multiple managers are involved in the decision, different criteria may come into play that lead to negotiation. Selecting a repair strategy requires knowledge of available actions, their effects on network behavior, and the trade-offs they involve.

Supervised learning methods could be applied to learn the effects of various repair actions. Methods for learning in planning could be applied to learn repair strategies (or perhaps only to evaluate repair strategies suggested by a human manager). There may be some opportunity here for 'collaborative filtering' methods that would provide an easy way for managers to share repair strategies.

As stated, the 'Why' problem [9] [36] requires diagnosis of an isolated fault, but one can imagine variations that involve answering questions about anomalies, fault locations and actions taken to repair the system. Each of these also assumes some interface that lets the user pose a specific question in natural language or, more likely, in a constrained query language. Defining the space of Why questions the Knowledge Plane should support is an important research task.

5.3.2 Responding to Intruders and Worms

Responding to intruders (human, artificial or their combination) and keeping networks and applications safe encompass a collection of tasks that are best explained depending on the time at which the network manager performs them. We can group them into tasks that occur before, during, or after the occurrence of an intrusion, as the temporal model in Figure 5.2 depicts.

5.3.2.1 Prevention Tasks

Network managers try to minimize the likeliness of future intrusions by constantly auditing the system and eliminating threats beforehand. A network manager proactively performs security audits testing the computer systems for weaknesses – vulnerabilities or exposures. However, scan tools (e.g., Nessus, Satan and Oval) used for penetration or vulnerability testing only recognize a limited number of vulnerabilities given the ever increasing frequency of newly detected possibilities for breaking into a computer system or disturbing its

[1] Recent work in an area known as 'cost-sensitive learning' addresses this trade-off between cost and informativeness.

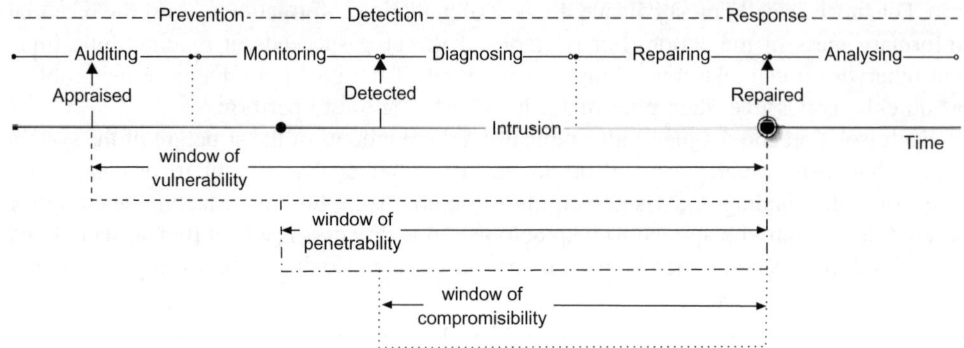

Figure 5.2 Time axis model of incident prevention, detection and response tasks

normal operation. Thus, network managers continually update scan tools with new plug-ins that permit them to measure new vulnerabilities. Once the existence of a vulnerability or exposure is perceived, network managers assess the convenience of discontinuing the service or application affected until the corresponding patch or intrusion detection signature is available. A trade-off between risk level and service level is made in every assessment.

Network managers aim at shrinking the *window of vulnerability*, the time gap between when a new vulnerability or exposure is discovered and a preventative solution (patch, new configuration, etc.) is provided. A basic strategy to accomplish that objective is based on two conservative tasks: first, minimizing the number of exposures (i.e., disable unnecessary or optional services by configuring firewalls to allow only the use of ports that are necessary for the site to function) and, second, increasing awareness of new vulnerabilities and exposures (e.g., the subscription model that Partridge discusses with relation to worms).

Finally, network managers continuously monitor the system so that pre-intrusion behavioral patterns can be understood and used for further reference when an intrusion occurs. *Monitoring* is an ongoing, preventive task.

5.3.2.2 Detection Tasks

The sooner an intrusion is detected, the more chances there are for impeding an unauthorized use or misuse of the computer system. Network managers monitor computer activities at different levels of detail: system call traces, operating system logs, audit trail records, resource usage, network connections, etc. They constantly try to fuse and correlate real-time reports and alerts stemming from different security devices (e.g., firewalls and intrusion detection systems) to stop suspicious activity before it has a negative impact (i.e., degrading or disrupting operations). Different sources of evidence are valuable given the evolving capabilities of intruders to elude security devices. The degree of suspicion and malignancy associated with each report or alert still requires continuous human oversight. Consequently, network managers are continually overwhelmed with a vast amount of log information and bombarded with countless alerts. To deal with this onslaught, network managers often tune security devices to reduce the number of false alerts even though this increases the risk of not detecting real intrusions.

The time at which an intrusion is detected directly affects the level of damage that an intrusion causes. An objective of network managers is to reduce the *window of penetrability*, the time span that starts when a computer system has been broken into and extends until the damage has been completely repaired. The correct diagnosis of an intrusion allows a network manager to initiate the most convenient response. However, a trade-off between quality and rapidness is made in every diagnostic.

5.3.2.3 Response and Recovery Tasks

As soon as a diagnostic on an intrusion is available, network managers initiate a considered response. This response tries to minimize the impact on the operations (e.g., do not close all ports in a firewall if only blocking one IP address is enough). Network managers try to narrow the *window of compromisibility* of each intrusion – the time gap that starts when an intrusion has been detected and ends when the proper response has taken effect – deploying automatic intrusion response systems. Nevertheless, these systems are still at an early stage and even fail at providing assistance in manual responses. Therefore, network managers employ a collection of ad-hoc operating procedures that indicate how to respond and recover from a type of intrusion. The responses to an attack range from terminating a user job or suspending a session to blocking an IP address or disconnecting from the network to disable the compromised service or host. Damage recovery or *repairing* often requires maintaining the level of service while the system is being repaired, which makes this process difficult to automate. Once the system in completely recovered from an intrusion, network managers collect all possible data to thoroughly analyze the intrusion, trace back what happened, and evaluate the damage. Thus, system logs are continuously backed up. The goal of *post-mortem analysis* is twofold. On the one hand, it gathers forensic evidence (contemplating different legal requirements) that will support legal investigations and prosecution and, on the other hand, it compiles experience and provides documentation and procedures that will facilitate the recognition and repelling of similar intrusions in the future.

Ideally, the ultimate goal of a network manager is to make the three windows (vulnerability, penetrability and compromisibility) of each possible intrusion converge into a single point in time. Tasks for responding to intruders (human, artificial or a combination of both) should not differ significantly from those tasks needed to recover from non-malicious errors or failures (Sections 5.3.1 and 5.3.2).

5.3.3 Network Configuration and Optimization

Network configuration and optimization can be viewed as an instance of the general problem of designing and configuring a system. In this section, we review the space of configuration problems and briefly describe the methods that have been developed in artificial intelligence (AI) and machine learning to solve these problems.

5.3.3.1 A Spectrum of Configuration Tasks

The problem of the design and configuration of engineered systems has been studied in artificial intelligence since the earliest days [45]. Configuration is generally defined as a form of routine design from a given set of components or types of components (i.e.,

Table 5.2 Configuration tasks in increasing order of complexity

Problem	Global parameters	Local parameters	Topology	Components
Global parameter configuration	XX			
Compatible parameter configuration	XX	XX		
Topological configuration	XX	XX	XX	
Component selection and configuration	XX	XX	XX	XX

as opposed to designing the components themselves). As such, there is a spectrum of configuration problems of increasing difficulty, as shown in Table 5.2.

The simplest task is *parameter selection*, where values are chosen for a set of global parameters in order to optimize some global objective function. Two classic examples are the task of setting the temperature, cycle time, pressure and input/output flows of a chemical reactor and the task of controlling the rate of cars entering a freeway and the direction of flow of the express lanes. If a model of the system is known, this becomes purely an optimization problem, and many algorithms have been developed in operations research, numerical analysis and computer science to solve such problems.

The second task is *compatible parameter selection*. Here, the system consists of a set of components that interact with one another to achieve overall system function according to a fixed topology of connections. The effectiveness of the interactions is influenced by parameter settings which must be compatible in order for sets of components to interact. For example, a set of hosts on a subnet must agree on the network addresses and subnet mask in order to communicate using IP. Global system performance can depend in complex ways on local configuration parameters. Of course, there may also be global parameters to select as well, such as the protocol family to use.

The third task is *topological configuration*. Here, the system consists of a set of components, but the topology must be determined. For example, given a set of hosts, gateways, file servers, printers and backup devices, how should the network be configured to optimize overall performance? Of course, each proposed topology must be optimized through compatible parameter selection.

Finally, the most general task is *component selection and configuration*. Initially, the configuration engine is given a catalog of available types of components (typically along with prices), and it must choose the types and quantities of components to create the network (and then, of course, solve the topological configuration problem of arranging these components).

5.3.3.2 The Reconfiguration Process

The discussion thus far has dealt only with the problem of choosing a configuration. However, a second aspect of configuration is determining how to implement the configuration efficiently. When a new computer network is being installed (e.g., at a trade show), the usual approach is to install the gateways and routers; then the file and print servers; and finally individual hosts, network access points, and the like. The reason for this is that this order makes it easy to test and configure each component and it minimizes the amount of rework. Automatic configuration tools (e.g., DHCP) can configure the individual hosts if the servers are in place first.

A different challenge arises when attempting to change the configuration of an existing network, especially if the goal is to move to the new configuration without significant service interruptions. Most configuration steps require first determining the current network configuration, and then planning a sequence of reconfiguration actions and tests to move the system to its new configuration. Some steps may cause network partitions that prevent further (remote) configuration. Some steps must be performed without knowing the current configuration (e.g., because there is already a network partition, congestion problem or attack).

We now review some of the existing work on configuration within the artificial intelligence and machine learning communities.

5.3.3.3 Parameter Selection

As we discussed above, parameter selection becomes optimization (possibly difficult, nonlinear optimization) if the model of the system is known. Statisticians have studied the problem of empirical optimization in which no system model is available.

5.3.3.4 Compatible Parameter Configuration

The standard AI model of compatible parameter configuration is known as the *constraint satisfaction problem* (CSP). This consists of a graph where each vertex is a variable that can take values from a set of possible values and each edge encodes a pairwise constraint between the values of the variables that it joins. A large family of algorithms have been developed for finding solutions to CSPs efficiently [14] [22]. In addition, it is possible to convert CSPs into Boolean satisfiability problems, and very successful randomized search algorithms, such as WalkSAT [41], have been developed to solve these problems.

The standard CSP has a fixed graph structure, but this can be extended to include a space of possible graphs and to permit continuous (e.g., linear algebraic) constraints. The field of *constraint logic programming* (CLP) [19] has developed programming languages based on ideas from logic programming that have a constraint solver integrated as part of the run-time system. The logic program execution can be viewed as conditionally expanding the constraint graph, which is then solved by the constraint system. Constraint logic programming systems have been used to specify and solve many kinds of configuration problems.

To our knowledge, there has been no work on applying machine learning to help solve compatible parameter configuration problems. There is a simple form of learning that has been applied to CSPs called 'nogood learning', but it is just a form of caching to avoid wasting effort during CSP search. There are many potential learning problems, including learning about the constraints relating pairs of variables and learning how to generalize CSP solutions across similar problems.

5.3.3.5 Topological Configuration

Two principal approaches have been pursued for topological configuration problems: refinement and repair. Refinement methods start with a single 'box' that represents the entire system to be configured. The box has an attached formal specification of its desired behavior. Refinement rules analyze the formal specification and replace the single box with two or more new boxes with specified connections. For example, a small office network might initially be specified as a box that connects a set of workstations, a file server, and two printers to a DSL line. A refinement rule might replace this box with a local network (represented as a single box connected to the various workstations and servers) and a router/NAT box. A second refinement rule might then refine the network into a wireless access point and a set of wireless cards (or alternatively, into an ethernet switch and a set of ethernet cards and cables). There has been some work on applying machine learning to learn refinement rules in the domain of VLSI design [29].

The repair-based approach to topological configuration starts with an initial configuration (which typically does not meet the required specifications) and then makes repairs to transform the configuration until it meets the specifications. For example, an initial configuration might just connect all computers, printers and other devices to a single ethernet switch, but this switch might be very large and expensive. A repair rule might replace the switch with a tree of smaller, cheaper switches. Repair-based approaches make sense when the mismatch between specifications and the current configuration can be traced to *local* constraint violations. A repair rule can be written that 'knows how' to repair each kind of violation. Repair-based methods have been very successful in solving scheduling problems [48].

Machine learning approaches to repair-based configuration seek to learn a heuristic function $h(x)$ that estimates the quality of the best solution reachable from configuration x by applying repair operators. If h has been learned correctly, then a hill-climbing search that chooses the repair giving the biggest improvement in h will lead us to the global optimum. One method for learning h is to apply reinforcement learning techniques. Zhang and Dietterich [47] describe a method for learning heuristics for optimizing space shuttle payload scheduling; Boyan and Moore [5] present algorithms that learn heuristics for configuring the functional blocks on integrated circuit chips.

In both refinement and repair-based methods, constraint satisfaction methods are typically applied to determine good parameter values for the current proposed configuration. If no satisfactory parameter values can be found, then a proposed refinement or repair cannot be applied, and some other refinement or repair operator must be tried. It is possible for the process to reach a dead end, which requires backtracking to some previous point or restarting the search.

5.3.3.6 Component Selection and Configuration

The refinement and repair-based methods described above can also be extended to handle component selection and configuration. Indeed, our local network configuration example shows how refinement rules can propose components to include in the configuration. Similar effects can be produced by repair operators.

5.3.3.7 Changing Operating Conditions

The methods discussed so far only deal with the problem of optimizing a configuration under fixed operating conditions. However, in many applications, including networking, the optimal configuration may need to change as a result of changes in the mix of traffic and the set of components in the network. This raises the issue of how data points collected under one operating condition (e.g., one traffic mix) may be used to help optimize performance under a different operating condition. To our knowledge, there is no research on this question.

5.4 Open Issues and Research Challenges

Most research in the field of machine learning has been motivated by problems in pattern recognition, robotics, medical diagnosis, marketing and related commercial areas. This accounts for the predominance of supervised classification and reinforcement learning in current research. The networking domain requires several shifts in focus and raises several exciting new research challenges, which we discuss in this section.

5.4.1 From Supervised to Autonomous Learning

As we have seen above, the dominant problem formulation in machine learning is supervised learning, where a 'teacher' labels the training data to indicate the desired response. While there are some potential applications of supervised learning in Knowledge Plane applications (e.g., for recognizing known networking misconfigurations and intrusions), there are many more applications for autonomous learning methods that do not require a teacher. In particular, many of the networking applications involve looking for anomalies in real-time data streams, which can be formulated as a combination of unsupervised learning and learning for interpretation.

Anomaly detection has been studied in machine learning, but usually it has considered only a fixed level of abstraction. For networking, there can be anomalies at the level of individual packets, but also at the level of connections, protocols, traffic flows, and network-wide disturbances. A very interesting challenge for machine learning is to develop methods that can carry out simultaneous unsupervised learning at all of these levels of abstraction. At very fine levels of detail, network traffic is continually changing, and therefore, is continually novel. The purpose of introducing levels of abstraction is to hide unimportant variation while exposing important variation.

Anomaly detection at multiple levels of abstraction can exploit regularities at these multiple levels to ensure that the anomaly is real. A similar idea – multi-scale analysis – has been exploited in computer vision, where it is reasonable to assume that a real pattern will be observable at multiple levels of abstraction. This helps reduce false alarms.

5.4.2 From Offline to Online Learning

Most applications of machine learning involve offline approaches, where data is collected, labeled, and then provided to the learning algorithm in a batch process. Knowledge Plane applications involve the analysis of real-time data streams, and this poses new challenges and opportunities for learning algorithms.

In the batch framework, the central constraint is usually the limited amount of training data. In contrast, in the data-stream setting, new data is available at every time instant, so this problem is less critical. (Nonetheless, even in a large data stream, there may be relatively few examples of a particular phenomenon of interest, so the problem of sparse training data is not completely eliminated.)

Moreover, the batch framework assumes that the learning algorithm has essentially unlimited amounts of computing time to search through the space of possible knowledge structures. In the online setting, the algorithm can afford only a fixed and limited amount of time to analyze each data point.

Finally, in the batch framework, the criterion to be minimized is the probability of error on new data points. In the online framework, it makes more sense to consider the response time of the system. How many data points does it need to observe before it detects the relevant patterns? This can be reformulated as a mistake-bounded criterion: how many mistakes does the system make before it learns to recognize the pattern?

5.4.3 From Fixed to Changing Environments

Virtually all machine learning research assumes that the training sample is drawn from a stationary data source – the distribution of data points and the phenomena to be learned are not changing with time. This is not true in the networking case. Indeed, the amount of traffic and the structure of the network are changing continuously. The amount of traffic continues to rise exponentially, and new autonomous systems are added to the Internet almost every day. New networking applications (including worms and viruses) are introduced frequently.

An additional challenge is that while some of the changes in the networking environment result simply from new applications and traffic growth, other changes are driven by adversaries who are trying to elude existing intrusion detection mechanisms. This calls for new approaches to machine learning that explicitly consider the game-theoretic aspects of the problem.

Research in machine learning needs to formalize new criteria for evaluating learning systems in order to measure success in these changing environments. A major challenge is to evaluate anomaly detection systems, because by definition they are looking for events that have never been seen before. Hence, they cannot be evaluated on a fixed set of data points, and measures are needed to quantify the degree of novelty of new observations.

5.4.4 From Centralized to Distributed Learning

Another important way in which Knowledge Plane applications differ from traditional machine learning problems is that, in the latter, it has usually been possible to collect all of the training data on a single machine and run the learning algorithm over that data collection. In contrast, a central aspect of the Knowledge Plane is that it is a distributed system of sensors, anomaly detectors, diagnostic engines and self-configuring components.

This raises a whole host of research issues. First, individual anomaly detectors can form models of their local traffic, but they would benefit from traffic models learned elsewhere in the Knowledge Plane. This would help them detect a new event the first time they see it, rather than having to be exposed multiple times before the event pattern emerges.

Second, some events are inherently distributed patterns of activity that cannot be detected at an individual network node. The research challenge here is to determine what kinds of statistics can be collected at the local level and pooled at the regional or global level to detect these patterns. This may involve a bidirectional process of information exchange in which local components report summary statistics to larger scale 'think points'. These think points detect a possible pattern that requires additional data to verify.

So they need to request the local components to gather additional statistics. Managing this bidirectional statistical reasoning is an entirely new topic for machine learning research.

5.4.5 From Engineered to Constructed Representations

An important ingredient in the success of existing learning systems is the careful engineering of the attributes describing the training data. This 'feature engineering' process is not well understood, but it involves combining background knowledge of the application domain with knowledge about learning algorithms. To illustrate this, consider a very simple example that arises in intrusion detection: rather than describing network traffic using absolute IP addresses, it is better to describe packets according to whether they share the same or different IP addresses. This ensures that the learned intrusion detector is not specific to a single IP address but instead looks for patterns among a set of packets sharing a common address, regardless of the absolute value of the address.

A critical challenge for machine learning is to develop more automatic ways of constructing the representations given to the learning algorithms. This requires making explicit the design principles currently used by human data analysts.

5.4.6 From Knowledge-Lean to Knowledge-Rich Learning

An important factor influencing the development of machine learning has been the relative cost of gathering training data versus building knowledge bases. The constructing and debugging of knowledge bases is a difficult and time-consuming process, and the resulting knowledge bases are expensive to maintain. In contrast, there are many applications where training data can be gathered fairly cheaply. This is why speech recognition and optical character recognition systems have been constructed primarily from training data. Any literate adult human is an expert in speech recognition and optical character recognition, so it is easy for them to label data points to train a learning system.

There are other domains (including networking), where there are very few experts available, and their time is perhaps better employed in developing formal representations of the knowledge they possess about network architectures and configurations. This is particularly true in the area of network diagnosis and configuration, where experts can help construct models of network components and prescribe rules for correct configuration. This raises the challenge of how to combine training data with human-provided models and rules. This should become an important goal for future machine learning research.

5.4.7 From Direct to Declarative Models

Most machine learning systems seek to induce a function that maps directly from inputs to outputs and therefore requires little inference at run time. In an optical character recognition system, for example, the learned recognizer takes a character image as input and produces the character name as output without any run-time inference. We will call this 'direct knowledge', because the learned knowledge performs the task directly.

However, as applications become more complex, a simple view of the performance element as a classifier (or direct decision maker) is no longer adequate. Diagnosis and configuration tasks require a more complex performance element that makes a sequence of interacting decisions at run time. These performance elements typically require declarative knowledge such as 'the probability that a misconfigured gateway will exhibit symptom X is P' or 'it is illegal to simultaneously select configuration options Y and Z.' An important goal for machine learning is to learn these forms of declarative knowledge (i.e., knowledge that makes minimal assumptions about how it will be used by the performance element).

Declarative knowledge is easier for people to understand, and it can be more easily combined with human-provided knowledge as well. Hence, acquiring declarative knowledge is an important challenge for machine learning in the context of the Knowledge Plane.

5.5 Challenges in Methodology and Evaluation

Machine learning research has a long history of experimental evaluation, with some examples dating back to the 1960s, well before the field was a recognized entity. However, the modern experimental movement began in the late 1980s, when researchers realized the need for systematic comparisons (e.g., [21]) and launched the first data repository. Other approaches to evaluation, including formal analysis and comparison to human behavior, are still practiced, but, over the past decade, experimentation has come to dominate the literature on machine learning, and we will focus on that approach in our discussions of cognitive networking.

Experimentation involves the systematic variation of independent factors to understand their impact on dependent variables that describe behavior. Naturally, which dependent measures are most appropriate depends on the problem being studied. For fault diagnosis, these might involve the system's ability to infer the correct qualitative diagnosis, its ability to explain future network behaviors, and the time taken to detect and diagnose problems. Similar measures seem appropriate for responding to intruders and worms, though these might also include the speed and effectiveness of response. For studies of configuration, the dependent variables might concern the time taken to configure a new system and the resulting quality, which may itself require additional metrics. Similarly, routing studies would focus on the efficiency and effectiveness of the selected routes.

Note that these behavioral measures have nothing directly to do with learning; they are same measures one would use to evaluate a nonlearning system and even the abilities of a human network manager. Because learning is defined as improvement in performance, we can only measure the effectiveness of learning in terms of the performance it aims to improve. Note also that the metrics mentioned above are quite vague, and they must be made operational before they can be used in experimental evaluations. In doing so, it may seem natural to use variables associated with one's selected formulation of the learning problem, such as predictive accuracy for classification or received reward for action selection. However, we should resist this temptation and instead utilize variables that measure directly what is desired from a networking perspective.

An experimental study also requires the variation of one or more independent factors to determine their effects on behavior. In general, these can deal with

- the effects of experience, such as the number of observations available to the learning system;
- the effects of data characteristics, such as the degree of noise or percentage of features missing;

- the effects of task characteristics, such as the complexity of a configuration problem or the number of simultaneous faults;
- the effects of system characteristics, such as the inclusion of specific learning modules or sensitivity to parameter settings; and
- the effects of background knowledge, such as information about network structure and bandwidth.

Again, which variables are appropriate will depend largely on the networking problem at hand and the specific learning methods being used. However, a full understanding of how machine learning can assist cognitive networking will require studies that examine each of the dimensions above.

Of course, one cannot carry out experiments in the abstract. They require specific domains and problems that arise within them. To study the role of learning in network management, we need a number of testbeds that can foster the experimental evaluation of alternative approaches to learning. At least some of these should involve actual networks, to ensure the collection of realistic data for training and testing the learning methods. However, these should be complemented with simulated networks, which have the advantage of letting one systematically vary characteristics of the performance task, the learning task and the available data. Langley [24] has argued that experiments with both natural and synthetic data are essential, since the former ensures relevance and the latter lets one infer source of power and underlying causes.

Much of the success of the last 15 years of machine learning research can be traced to the establishment of a collection of data sets at the University of California, Irvine [3] [27] [32]. The UCI data sets provided a common set of problems on which to evaluate learning algorithms and greatly encouraged comparative studies. The data sets span a wide range of application problems ranging from basic science and medicine to optical character recognition and speech recognition.

Ideally, we want an analog of this repository to enable the careful evaluation of machine learning in networking domains. However, because the Knowledge Plane envisions an adaptive network that learns about itself over time, it is important that this resource not be limited to static data sets, but also include simulated networks that allow learning methods, and their associated performance elements, to interact with the network environment in an online manner.

5.6 Summary

In this chapter, we have reviewed the current state of research in machine learning and highlighted those aspects that are relevant to the emerging vision of the Knowledge Plane [9] [10] [36]. We began by discussing the major problem formulations in machine learning: learning for classification and regression, learning for acting and planning to act, and learning for interpretation and understanding. We then discussed the various tasks that are raised by the Knowledge Plane and examined how each of these tasks could be mapped to existing problem formulations. In some cases, there is a direct mapping, but in other cases, such as complex configuration and diagnosis, there is hardly any existing work in machine learning.

Furthermore, even when networking problems fit well into existing formulations, this does not mean that existing algorithms can be applied directly. This is because many aspects of computer networking impose new kinds of requirements that are not met by existing algorithms. We reviewed these requirements in the third section of the chapter, where we discussed the need for learning that is autonomous, online, distributed, knowledge-rich, and able to deal with changing environments and to represent learned knowledge declaratively. Finally, we discussed the need for clear articulation of performance criteria for evaluating learning systems and the importance of creating simulation environments to support controlled experimentation.

Computer networking in general, and the Knowledge Plane vision in particular, pose enormous challenges for machine learning research. Conversely, existing algorithms and tools from machine learning have an important role to play in making the Knowledge Plane vision a reality. We can look forward to many exciting research advances in the coming years.

Acknowledgements

The authors thank Christopher Ramming for asking the hard questions that led us to write this chapter and Francisco Martin for providing material on existing practices in intrusion prevention, detection and repair. The authors gratefully acknowledge the support of the Defense Advanced Research Projects Agency under AFRL Contract F30602-03-2-0191 (to Oregon State University) and Grant F30602-03-2-0198 (to the Institute for the Study of Learning and Expertise). Views and conclusions contained in this document are those of the authors and do not necessarily represent the official opinion or policies, either expressed or implied of the US government, of AFRL, or of DARPA.

References

[1] Aha, D.W., Kibler, D. and Albert, M.K. (1991) Instance-based learning algorithms, *Machine Learning*, **6**, 37.

[2] Baluja, S. and Caruana, R. (1995) Removing the genetics from the standard genetic algorithm. *Proceedings of the 12th Annual Conference on Machine Learning*, p. 38–46, San Francisco.

[3] Blake, C.L. and Merz, C.J. (1998) UCI repository of machine learning databases, http://www.ics.uci.edu/~mlearn/MLRepository.html.

[4] Blum, A. and Mitchell, T. (1998) Combining labeled and unlabeled data with co-training. *Proceedings of the 11th Annual Conference on Computing Learning Theory*, pp. 92–100, New York.

[5] Boyan J. and Moore, A. (2000) Learning evaluation functions to improve optimization by local search. *Journal of Machine Learning Research*, **1**, 77–112.

[6] Boyan, J.A. and Littman, M.L. (1994) Packet routing in dynamically changing networks: a reinforcement learning approach. In J.D. Cowan, G. Tesauro and J. Alspector (eds.), *Advances in Neural Information Processing Systems*, volume 6, pp. 671–8, Morgan Kaufmann.

[7] Buntine, W. (1996) A guide to the literature on learning probabilistic networks from data. *IEEE Transactions on Knowledge and Data Engineering*, **8**, 195–210.

[8] Cheeseman, P., Self, M., Kelly, J., Taylor, W., Freeman, D. and Stutz, J. (1988) Bayesian classification, *Proceedings of the Seventh National Conference on Artificial Intelligence*, pp. 607–11, St.Paul, MN.

[9] Clark, D. (2002) A new vision for network architecture, http://www.isi.edu/~braden/know-plane/DOCS/DDC_knowledgePlane_3.ps.

[10] Clark, D.D., Partridge, C., Ramming, J.C. and Wroclawski, J.T. (2003) A knowledge plane for the internet, *Proceedings of the 2003 Conference on Applications, Technologies, Architectures, and Protocols for Computer Communications*, pp. 3–10, Karlsruhe, Germany.

[11] Clark, P. and Niblett, T. (1988) The CN2 induction algorithm. *Machine Learning*, **3**, 261.

[12] Cristianini, N. and Shawe-Taylor, J. (2000) An Introduction to Support Vector Machines (and other kernel-based learning methods), Cambridge University Press.

[13] Cypher, A. (1993) *Watch What I Do: Programming by Demonstration*, MIT Press, Cambridge, MA.

[14] Dechter, R. (2003) *Constraint Processing*, Morgan Kaufmann, San Francisco.

[15] DeJong, G. (2006) Toward robust real-world inference: a new perspective on explanation-based learning. In J. Fürnkranz, T. Scheffer and M. Spiliopoulou (eds.), *Machine Learning: ECML 2006; Lecture Notes in Computer Science*, pp. 102–13, Springer Verlag, Berlin.

[16] Drastal, G., Meunier, R. and Raatz, S. (1989) Error correction in constructive induction. In A.Maria Segre (ed.), *Proceedings of the Sixth International Workshop on Machine Learning*, pp. 81–3, Ithaca, New York.

[17] Fisher, D.H. (1987) Knowledge acquisition via incremental conceptual clustering. *Machine Learning*, **2**, 139.

[18] Flann, N.S. and Dietterich, T.G. (1989) A study of explanation-based methods for inductive learning. *Machine Learning*, **4**, 187–226.

[19] Jaffar, J. and Maher, M.J. (1994) Constraint logic programming: a survey. *Journal of Logic Programming*, **19/20**, 503–81.

[20] Kibler, D. and Langley, P. (1988) Machine learning as an experimental science. *Proceedings of the 3rd European Working Session on Learning*, pp. 81–92, Glasgow.

[21] Kumar, V. (1992) Algorithms for constraints satisfaction problems: a survey. *AI Magazine*, **13**, 32–44.

[22] Langley, P. (1995) *Elements of Machine Learning*, Morgan Kaufmann, San Francisco.

[23] Langley, P. (1996) Relevance and insight in experimental studies. *IEEE Expert*, 11–12.

[24] Langley, P. (1999) User modeling in adaptive interfaces. *Proceedings of the 7th International Conference on User Modeling*, pp. 357–70, Banff, Alberta.

[25] Langley, P. and Simon, H.A. (1995) Applications of machine learning and rule induction. *Communications of the ACM*, **38**, 55–64.

[26] Merz, C.J. and Murphy, P.M. (1996) UCI repository of machine learning databases, http://www.ics.uci.edu/~mlearn/MLRepository.html.

[27] Mitchell, T.M. (1997) *Machine Learning*, McGraw-Hill, New York.

[28] Mitchell, T.M., Mahadevan, S. and Steinberg, L.I. (1985) LEAP: a learning apprentice for VLSI design. *Proceedings of the 9th International Joint Conference on Artificial Intelligence*, pp. 573–80, Los Angeles, CA.

[29] Moore, A., Schneider, J., Boyan, J. and Soon Lee, M. (1998) Q2: Memory-based active learning for optimizing noisy continuous functions. *Proceedings of the 15th International Conference of Machine Learning*, pp. 386–94.

[30] Moriarty, D.E., Schultz, A.C. and Grefenstette, J.J. (1999) Evolutionary algorithms for reinforcement learning. *Journal of Artificial Intelligence Research*, **11**, 241–76.

[31] Murphy, P.M. and Aha, D.W., (1994) UCI repository of machine learning databases, http://www.ics.uci.edu/~mlearn/MLRepository.html.

[32] Myers, R.H. and Montgomery, D.C. (1995) *Response Surface Methodology: Process and Product Optimization Using Designed Experiments*, John Wiley & Sons, Inc., New York.

[33] Oblinger, D., Castelli, V. and Bergman, L. (2006) Augmentation-based learning. *IUI2006: 2006 International Conference on Intelligent User Interfaces*, pp. 202–9.

[34] Ourston, D. and Mooney, R. (1990) Changing the rules: a comprehensive approach to theory refinement. *Proceedings of the 8th National Conference on Artificial Intelligence*, pp. 815–20.

[35] Pack Kaelbling, L., Littman, M.L. and Moore, A.W. (1996) Reinforcement learning: a survey. *Journal of Artificial Intelligence Research*, **4**, 237–85.

[36] Partridge, C., (2003) Thoughts on the structure of the knowledge plane, http://www.isi.edu/~braden/know-plane/DOCS/craig.knowplane.pdf.

[37] Priebe, C.E. and Marchette, D.J. (1993) Adaptive mixture density estimation. *Pattern Recognition*, **26**(5), 771–85.

[38] Quinlan, J.R. (1993) *C4.5: Programs for Empirical Learning*, Morgan Kaufmann, San Francisco.

[39] Rumelhart, D.E., Hinton, G.E. and Williams, R.J. (1986) Learning internal representations by error propagation. In D.E. Rumelhart, J.L. McClelland and the PDP research group (eds.), *Parallel Distributed Processing: Explorations in the Microstructure of Cognition, Volume 1: Foundations*, MIT Press, Cambridge, MA.

[40] Sammut, C., Hurst, S., Kedzier, D. and Michie, D. (1992) Learning to Fly. *Proceedings of the 9th International Conference on Machine Learning*, pp. 385–93, Aberdeen.

[41] Selman, B., Kautz, H.A. and Cohen, B. (1993) Local search strategies for satisfiability testing. *Proceedings of the Second DIMACS Challange on Cliques, Coloring, and Satisfiability*, Providence RI.

[42] Sleeman, D., Langley, P. and Mitchell, T. (1982) Learning from solution paths: an approach to the credit assignment problem, *AI Magazine*, **3**, 48–52.

[43] Stolcke, A. and Omohundro, S. (1994) Inducing probabilistic grammars by Bayesian model merging. In R.C. Carrasco and J. Oncina (eds.), *Grammatical Inference and Applications: Proceedings of the Second International Colloquium on Grammatical Inference*, pp. 106–18, Springer Verlag.

[44] Sutton, R. and Barto, A.G. (1998) *Introduction to Reinforcement Learning*, MIT Press, Cambridge, MA.

[45] Tonge, F.M. (1963) Summary of a heuristic line balancing procedure. In E.A. Feigenbaum and J. Feldman (eds.), *Computers and Thought*, pp. 168–90, AAAI Press/MIT Press, Menlo Park, CA.

[46] Williams, R.J. (1992) Simple statistical gradient-following algorithms for connectionist reinforcement learning. *Machine Learning*, **8**, 229.

[47] Zhang, W. and Dietterich, T.G. (1995) A reinforcement learning approach to job-shop scheduling. *International Joint Conference on Artificial Intelligence*, pp. 1114–20, Montreal, Canada.

[48] Zweben, M., Daun, B. and Deale, M. (1994) Scheduling and rescheduling with iterative repair. In M. Zweben and M.S. Fox (eds.), *Intelligent Scheduling*, pp. 241–55, Morgan Kaufmann, San Francisco.

6

Cross-Layer Design and Optimization in Wireless Networks

Vineet Srivastava
HelloSoft, Inc., Hyderabad, Andhra Pradesh, India

Mehul Motani
National University of Singapore, Singapore

6.1 Introduction

Communication networks have traditionally been engineered following the principle of protocol layering. This means designing specific network functionalities (such as flow control, routing, medium access) in isolation from each other, and putting together the complete system through limited interfaces between the *layers* performing these specific tasks. The layers, which are in fact distributed systems with collaborating entities distributed through the network [6, p. 20], are arranged in a vertical hierarchy. Each layer makes use of the services provided by the layers below itself and, in turn, makes its services available to the layers above itself. Inter-layer communication happens only between adjacent layers and is limited to procedure calls and responses. Familiar examples of layered communication architectures are the seven-layer Open Systems Interconnect (OSI) model [6, p. 20] and the four-layer TCP/IP model [39].

Even though layered communication architectures have been instrumental in the development of many communication networks, most notably the Internet, in recent times, the suitability of protocol layering is being questioned in the research community. This is largely due to wireless networks becoming an integral part of the communications

Cognitive Networks: Towards Self-Aware Networks Edited by Qusay H. Mahmoud

infrastructure, and hence occupying the center stage of research and development activity. Researchers argue that while designing protocols at the different layers in isolation and running them without much interaction between the functionalities may have served well in the case of wired networks, doing so is not suitable for wireless networks due to the peculiarities of the wireless medium. To illustrate their point, researchers usually present a *cross-layer design* proposal. Broadly speaking, cross-layer design refers to protocol design done by actively exploiting the dependence between the protocol layers to obtain performance gains. This is unlike layering, where the protocols at the different layers are designed independently. There have been a large number of cross-layer design proposals in the literature; references [11], [48] and [54], among others, consolidate some of these proposals; reference [6] highlights the potential problems that cross-layer design, if done without appropriate care, can create.

While the presence of wireless links in a network makes strict layering of protocols look like an inefficient solution, the pursuit of cognitive networks makes it look grossly inadequate. This is because a cognitive network not only needs to provide network functionalities to the applications (and hence the users), it also needs to adapt to the application needs, and to the prevailing network conditions (for example to the traffic in the network, to the conditions of the underlying channel, or to the interference in the spectrum bands of operation) [7]. Such adaptability requires rich coordination between the traditional protocol layers, which is not feasible in the constraints of the layered architectures. Thus, when it comes to cognitive networks, cross-layer design is not just a matter of efficiency, but in fact it is central to the very concept of the network. The question thus is not 'whether', but rather 'how', cross-layer design and optimization should be done. For example, which layers should be coupled and how? What should be the roles of the different layers? On a broader note, should there be layers at all or should cognitive networks be built around fundamentally different architectures? As a matter of fact, such questions may not have a single answer given the multiplicity of the communication networks that we have around us, each with their own specific needs and purposes. (See also the concluding remarks in [19].)

The case for cross-layer design from a performance viewpoint, as articulated above, is easy to understand. Unfortunately, performance improvement brought about by cross-layer design is only one part of the picture. To be successful, any cross-layer design idea needs to be implemented and deployed in real systems. This means that the architectural considerations are equally, if not more, important. The success of layering is really due to its architectural merit: that different portions of the network can be developed and innovated upon by different vendors, and still be stitched together to yield a working system. With couplings introduced between the layers, as done with cross-layer design, the desirable architectural qualities of layering may be lost. Thus, a big challenge facing the proponents of cross-layer design is to evaluate the architectural consequences of their work, and to ascertain that their cross-layer design ideas are not eroding away the architectural merits of the layered architectures [34]. Architectural considerations involve questions like the coexistence of different cross-layer design ideas; determining which cross-layer design ideas, if deployed, may hamper further innovation; and so on. Admittedly, such questions are easier stated than answered or even addressed. Nonetheless, these questions have to be tackled head on for the promise of cross-layer design to be realized. Facilitating a healthy debate on such questions is our goal in this chapter. We do so by consolidating the different results, ideas and perspectives pertinent to the area of cross-layer design.

We start by looking at the different definitions of cross-layer design and capture them in a concise and general definition. Next, we look at the broad motivations for cross-layer design purely from a performance viewpoint and, in doing so, we present a quick survey of the research literature in this area. From a performance viewpoint, the motivations for cross-layer design are the unique problems created by the wireless networks, the avenue for opportunistic communications, and the new modalities of communications created by wireless links. Moving on, we categorize the different cross-layer design proposals from an architectural viewpoint, and look at the initial ideas for implementing cross-layer design proposals. Next, we take a brief look at some cross-layer design ideas that have found their way into commercial systems or ongoing standards. Having done the consolidation, we raise what we call the open challenges, which are some important problems researchers may want to address as they move forward. Next, we place the ongoing cross-layer design activity in a brief historic context after which we present our conclusions.

Even though the questions surrounding cross-layer design are multifaceted, we believe understanding cross-layer design at a broad level and not seeing it like a quick fix for specific problems created by wireless networks will help. Our aim in this chapter is to facilitate such holistic thinking incorporating the performance viewpoint, the architectural considerations as well as the future potential modalities of wireless communications about this extremely important and promising design paradigm.

6.2 Understanding Cross-Layer Design

6.2.1 The Different Interpretations

Cross-layer design is described in the research literature in different ways [52], [27]: as a protocol design methodology that exploits the synergy between the layers; as a joint design and optimization across more than one protocol layer; as a protocol design methodology that relies on, and creates, adaptability across the different layers by sharing information between them; and so on. An interpretation of cross-layer design, usually discussed in the context of multimedia transmissions, is that of a design methodology that leverages joint source and channel coding [67] which creates coupling between the application and the physical layers. Yet another interpretation of cross-layer design, usually discussed in the context of communications over fading channels, is that of a design methodology that takes into account the channel characteristics at the higher layers, and the stochastic arrival of traffic at the lower layers [5].

6.2.2 A Definition for Cross-Layer Design

In what follows, we present a definition for cross-layer design by viewing it as a violation of a reference layered architecture. Our definition is concise and yet it encompasses the aforementioned notions about cross-layer design. Our definition also draws a clearer contrast between layered protocol design and cross-layer design, regardless of the layered architecture in question.

As noted earlier a layered architecture, like the seven-layer Open Systems Interconnect (OSI) model [6, p. 20], divides the overall networking task into layers and defines a hierarchy of services to be provided by the individual layers. The services at the layers are realized by designing protocols for the different layers. The architecture forbids direct communication between non-adjacent layers; communication between adjacent layers is limited to procedure calls and responses.

In the framework of a reference layered architecture, the designer has two choices at the time of protocol design. Protocols can be designed by respecting the rules of the reference architecture. In the case of a layered architecture, this would mean designing protocols such that a higher layer protocol only makes use of the services at the lower layers and not be concerned about the details of how the service is being provided. Similarly, the services provided by a lower layer would not depend on the details of the higher layer (e.g., the application) that is requesting or using them. Following the architecture also implies that the protocols would not need any interfaces that are not present in the reference architecture.

Alternatively, protocols can be designed by violating the reference architecture, for example, by allowing direct communication between protocols at non-adjacent layers or by sharing variables between layers. Such violation of a layered architecture is cross-layer design with respect to the reference architecture.

- **Definition 6.1:** Protocol design by the violation of a reference layered architecture is cross-layer design with respect to the particular layered architecture.
- **Comment 6.1:** Examples of violation of a layered architecture include creating new interfaces between layers, redefining the layer boundaries, designing protocol at a layer based on the details of how another layer is designed, joint tuning of parameters across layers, and so on.
- **Comment 6.2:** Violation of a layered architecture involves giving up the luxury of designing protocols at the different layers independently. Protocols so designed may impose some conditions on the processing at the other layer(s).
- **Comment 6.3:** Cross-layer design is defined as a protocol design methodology. However, a protocol designed with this methodology is also termed as cross-layer design.

For exposition, consider a hypothetical three-layer model with the layers denoted by L_1, L_2 and L_3.L_1 is the lowest layer and L_3 the highest. Note that in such an architecture, there is no interface between L_3 and L_1. One could, however, design an L_3 protocol that needs L_1 to pass a parameter to L_3 at run-time. This calls for a new interface, and hence violates the architecture. Alternatively, one could view L_2 and L_1 as a single layer, and design a joint protocol for this 'super-layer.' Or, one could design the protocol at L_3, keeping in mind the processing being done at L_1, again giving up the luxury of designing the protocols at the different layers independently. All these are examples of cross-layer design with respect to the three-layer architecture in question.

Architecture violations, like those introduced by cross-layer design, clearly undermine the significance of the architecture since the architecture no longer represents the actual system. If many architecture violations accumulate over time, the original architecture can completely lose its meaning. Architecture violations can have detrimental impact on the system longevity, as has been argued for the case of cross-layer design in [34].

6.3 General Motivations for Cross-Layer Design

What motivates designers of wireless networks to violate the layered communication architectures? There are a three broad motivations. Several cross-layer design proposals aim to solve some unique problems created by wireless links. An example of such a problem is the classic case of a TCP sender mistaking a wireless error to be an indication of network congestion [52]. Another category of cross-layer design ideas aim to exploit the fundamental characteristics of the wireless medium opportunistically, for example by utilizing channel variations from fading at the higher layers. This is in line with the general goal of an adaptive protocol stack that responds dynamically to the changes in the network conditions. Yet another category of cross-layer design ideas make use of the new modalities of communications that wireless medium creates and that cannot be accomodated within the constraints of layering. An example here is node cooperation, as we discuss later. Cross-layer design with all three motivations are important in the context of cognitive networks of the future.

In this section, we elaborate further on the motivations for cross-layer design. We draw examples from the published research literature. Apart from clarifying the main motivations for cross-layer design, taking relevant examples from the literature also gives a good measure of the wide range of scenarios in which cross-layer design has been applied.

Since there is no overarching layered architecture that is followed in *all* communication systems, the reference layered architecture we assume is a five-layer hybrid reference model presented in [59, p, 44]. This model has the application layer, the transport layer, the network layer, the link layer which comprises the data link control (DLC) and medium access control (MAC) sublayers [6, p. 24], and the physical layer; we assume that all the layers perform their generally understood functionalities.

6.3.1 Problems Created by Wireless Links

6.3.1.1 TCP on Wireless Links

The first motivation for cross-layer design is that wireless links create some new problems for protocol design that cannot be handled well in the framework of the layered architectures. A transmission control protocol (TCP) sender erroneously mistaking a packet error on a wireless link to be an indicator of network congestion is an example [52]. This problem is often resolved by direct communication between the link layer and the transport layer, which is thus an example of cross-layer design motivated by a unique problem created by a wireless link.

6.3.1.2 Real-Time Multimedia on Wireless Links

Likewise, when an error-prone and shared wireless link is used for real-time multimedia communications, the error correction and resource management techniques employed at the lower layers of the stack may need to be adapted jointly with the compression and streaming algorithms applied at the higher layers. In fact, even on wired networks, rich coordination between the layers is needed to handle the real-time nature of multimedia transmission [69]. The time-varying nature of the wireless links, their relatively higher error probabilities and the delays caused in negotiating multiple access over the shared wireless medium further complicate matters and require a cross-layer design approach [51]. Here is thus another example where the problems created by wireless links necessitate violating the layered architectures. A good survey of cross-layer design ideas in this context is presented in [11]. We also refer the readers to [67] which explores the topic of joint source and channel coding as a special case of cross-layer design for the real-time multimedia communications problem, as we discussed earlier.

6.3.1.3 Power Control in Ad Hoc Networks

On a broader note, the wireless medium often creates problems in wireless networks that affect all layers of the traditional layered architectures at once. Some examples are power control in wireless ad hoc networks, energy management in wireless devices and security in wireless networks.

The problem of power control comes up because in a wireless ad hoc network, the transmissions of the different nodes interfere with each other unless they have been assigned orthogonal resources. Thus, even if all the nodes employ a large transmission power, they may not be able to communicate with one another because large transmission powers also lead to higher interference on the other nodes. Thus power control is needed.

Power control clearly influences the network topology, which is a concern of the network layer. It also impacts how much spatial reuse can be achieved, that is, how far apart can two ongoing communication sessions be without interfering with each other, which is a concern of the MAC layer. Power control is also linked to the processing at the physical layer, because the signal processing at the physical layer determines how stringent the requirements on the power control need to be. All these factors determine the end-to-end throughput. Furthermore, the transmitted power(s) determines the lifetime of the nodes (and the network) which one would want to maximize. Hence, the problem of power control cannot possibly be handled at any one layer in isolation, as is done while designing protocols in the framework of the layered architectures. It is thus a problem that, by its very nature, requires cross-layer design. It is no surprise then that a number of cross-layer design proposals in the literature have looked at power control in a cross-layer design framework. We shall discuss some of these ideas in Section 6.4.1.3.

6.3.1.4 Energy Efficiency and Security

Much like power control, energy efficiency of a communication device and network security (see [47] for example) are multifaceted issues that cannot be meaningfully addressed at any one layer of the protocol stack in isolation. The problem of energy-efficient design of wireless ad hoc networks is discussed thoroughly in [27] which also makes a case for cross-layer design. Reference [23] looks at the problem of energy management in wireless communications systems, and presents a thorough discussion on the cross-layer interfaces required for a rich coordination between the layers.

6.3.1.5 Vertical Handovers

Next, consider the problem of vertical handover in wireless networks. Vertical handover refers to a seamless transition of a multimode device from one network interface to another. For example, a person accessing the Internet on the road using the cellular mobile phone network might move into a building served by a wireless LAN. Ideally, from a user's perspective, such a change should be seamless and ongoing connections and data transfers should not be interrupted. To be fair, this problem of guaranteeing seamlessness as described above, though extremely important for mobile wireless networks, is not unique to wireless networks. As [31, p. 14] points out, a person who is accessing the Internet through a cable might also request for the same kind of seamlessness when changing the mode of access. However, we mention it here because the presence of wireless links creates avenues for taking into account cross-layer interactions when conducting vertical handovers, as discussed in detail in [57]. Thus, though not unique to wireless networks, this is another problem that touches all the layers of the protocol stack at once. Vertical handover is an integral component of the cognitive network because through effective vertical handovers the user can be kept connected to the network in the best possible mode at all times. In fact, vertical handovers are likely to have an immediate practical and commercial relevance as telecom service providers and equipment manufacturers worldwide are gearing towards converged networks and services.

6.3.2 The Optimistic Side

6.3.2.1 Fading Channels: Single User

Let us now look at a more optimistic side. It is well known that due to fading and multipath effects, wireless links can show tremendous time and/or frequency selectivity.

The variation in the channel quality creates new opportunities at the higher layers. As [4] puts it, fading allows the physical channel to be viewed as a 'packet pipe' whose delay, rate and/or error probability characteristics vary with time. Contrast this with a wired communication channel whose characteristics remain largely time-invariant. Reference [4] considers a buffered single user point-to-point communication system and proposes a rate and power adaptation policy based on the fluctuations of the channel and the buffer occupancy. Reference [63] considers a similar situation and also comes up with an optimal adaptive policy that minimizes a linear combination of the transmission power and the buffer overflow probability. Such adaptations are examples of cross-layer design motivated by the time-varying nature of the wireless channels.

6.3.2.2 Fading Channels: Multiple Users

In fact, the time variation in the quality of the wireless medium creates even more interesting opportunities in the case of multiuser networks. Consider the problem of downlink scheduling on a cellular network [52]. The situation is as follows: there are a number of users whose data is arriving in a stochastic fashion at the base station. The base station needs to schedule the transmissions of the different users. Since the network traffic is bursty, a fixed assignment of slots/frequency bands to the users is not efficient. Instead, the allocation of the channel to the different users should be done dynamically. In doing the scheduling, significant performance gains can be made if the base station takes into account the state of the downlink channel, and allows transmission to a particular user only if the channel between the base station and that particular user is in a 'good' state. This makes intuitive sense, since, when the channel for a particular user is bad, there is no point in scheduling the transmission for that user. For theoretical bases of such channel-dependent scheduling algorithms, we refer the readers to the results in [61], [62] and the references therein. The problem of resource allocation over single-user and multiuser (e.g., the multiple access and the broadcast) fading channels is also discussed in [5]. Also noteworthy in this context is the work presented in [68] that deals with the problem of transmitting bursty traffic over fading channels, and illustrates how channel variation can be 'exploited (rather than combatted)' to improve system performance. In an ad hoc network setting too, the time variation in the channels between the different users can be taken into account at the MAC layer to make throughput gains. (See [32] and the references therein.)

In short, fading, being unique to wireless links, does create new challenges and problems. But on the optimistic side, it also creates new opportunities. Generally speaking, layered architectures, with their fixed interfaces, appear too stiff to meaningfully address the time variations in the channel caused by fading. Making use of these opportunities for protocol design motivates cross-layer design, as evident from the several works that we have cited above.

6.3.3 The New Modalities

6.3.3.1 Multi-Packet Reception

Apart from the pessimistic and the optimistic views presented above, the wireless medium offers some new modalities of communication that the layered architectures do not accommodate. For instance, the physical layer can be made capable of receiving multiple packets [60] at the same time. This clearly changes the traditional balance of roles between the MAC layer and the physical layer, and the functionalities of the two layers can then be designed jointly, as illustrated in [60].

6.3.3.2 Node Cooperation

As another example, consider node cooperation. Nodes in a wireless network can cooperate with each other in involved ways. Node cooperation is actually a fairly broad concept. It covers a lot of ground, ranging from the nodes with single antenna each cooperating to create distributed coding systems [45] or to obtain the diversity gains provided by multiple-input multiple-output (MIMO) channels [37]; to cooperative beam forming where once again nodes equipped with single antennas cooperate to form antenna arrays; to relaying and mesh networking; and so on. Broadly, the motivation for node cooperation comes from the fundamental problems of wireless communications, such as the scarcity of spectrum, the energy limitations at the wireless devices and inability to accommodate multiple antennas at the same terminal. Cooperation may be imperative to increase the system capacity, the spectrum usage and energy efficiency for wireless communications devices [26]. Node cooperation is a very active field of research, with interest being shown from both the academia and the industry. We refer the readers to [19] and the several other articles in the same volume for a thorough and excellent introduction to the ongoing research in the area of node cooperation for wireless networks.

As far as this chapter is concerned, our interest in the actively researched area of node cooperation comes from the fact that it is a new modality of communications that wireless networks create. And because cooperation can fundamentally change the network model of a collection of point-to-point links that the traditional layered architectures have assumed (and that made sense for the wired networks), it is clear that incorporating node cooperation inevitably requires violating the layered architectures, and hence cross-layer design. See for example the discussion in [19] and also the discussion in [54], where the latter discusses in detail on what kind of architecture violations may be needed for accommodating coded cooperation in wireless ad hoc networks. Reference [70] presents a node cooperation scheme that generalizes the idea of hybrid automatic repeat request (ARQ) to multiple users; the exposition in [70] vividly illustrates the cross-layer interactions that cooperative communications necessitate.

6.3.3.3 Cognitive Radios

Another new modality that the wireless medium is ushering in is the idea of cognitive radios [29]. Cognitive radios can be understood as radios that gain awareness about their surroundings and adapt their behavior accordingly. For instance, a cognitive radio may determine an unused frequency band and use that for a transmission, before jumping to another unused band. Cognitive radios hold tremendous potential to increase the efficiency of wireless spectrum usage and create devices and systems that truly interact with the users. We refer the readers to [42] for a thorough discussion on the potential of the cognitive radio architecture.

Cognitive radios, at the very least, require enhanced sharing of information between the physical and the MAC layers of the protocol stack. Basically, unlike a conventional radio, the frequency band in which a cognitive radio may be operating at any given time depends on the channel occupancy measured at the physical layer, and conveyed to the MAC layer through an appropriate interface. This clearly requires cross-layer design.

6.3.4 Cognitive Networks and Cross-Layer Design

Cross-layer design done with all the three aforementioned motivations is integral to the cognitive networks of the future. Indeed, an efficient protocol stack that responds to the environment, network conditions and user demands is central to the very idea of cognitive networks. As we have seen above, the pursuit of achieving such efficiency and flexibility in wireless networks is not feasible by maintaining the strict boundaries between the different network functionalities, as done by layering. The way forward, thus, is inevitably cross-layer design.

6.4 A Taxonomy of Cross-Layer Design Proposals

In the previous section, we took a quick look at the broad motivations for the different cross-layer design ideas in the literature. Having seen *why* the presence of wireless links motivates designers to violate layered architectures, we now turn our attention to *how* the layered architectures are violated in the different cross-layer design proposals.

Based on the published research literature, we find that the layered architectures have been violated in the following basic ways:

1. Creation of new interfaces (Figure 6.1(a), (b), (c)).
2. Merging of adjacent layers (Figure 6.1(d)).
3. Design coupling without new interfaces (Figure 6.1(e)).
4. Vertical calibration across layers (Figure 6.1(f)).

We shall now discuss the aforementioned four categories in more detail and point out some relevant examples for the different categories. A few points are worth mentioning here. First, the examples that we point out are meant to be representative and not exhaustive. Secondly, the architectural violations that we identify can be combined to yield more

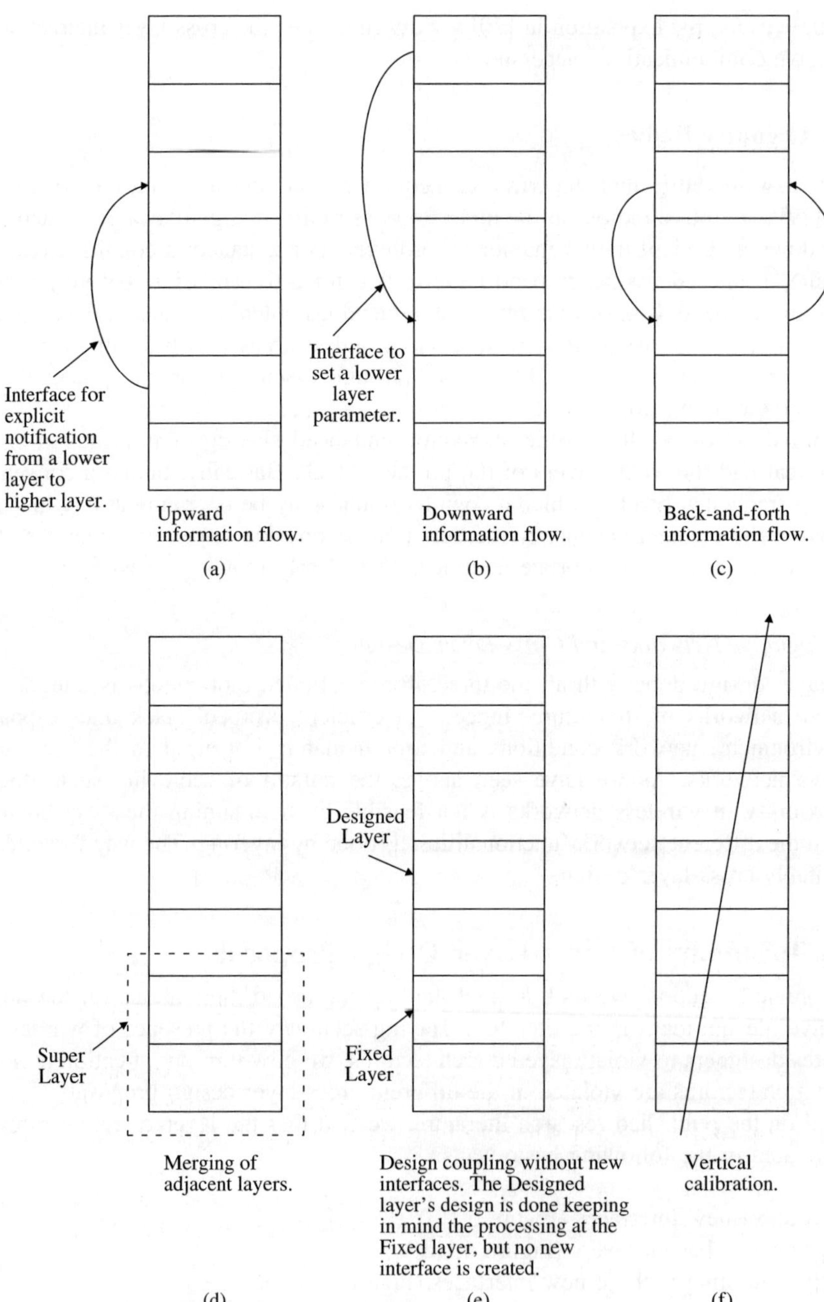

Figure 6.1 Illustrating the different kinds of cross-layer design proposals. The rectangular boxes represent the protocol layers

complex cross-layer designs. Finally, the reference layered architecture we assume is a five-layer hybrid reference model that we discussed in Section 6.3.

6.4.1 Creation of New Interfaces

Several cross-layer designs require creation of new interfaces between the layers. The new interfaces are used for information sharing between the layers at run-time. The architecture violation here is obviously the creation of a new interface not available in the layered architecture. We further divide this category into three categories depending on the direction of information flow along the new interfaces:

1. Upwards: from lower layer(s) to a higher layer.
2. Downwards: from higher layer(s) to a lower layer.
3. Back and forth: iterative flow between the higher and lower layer.

We now discuss the three subcategories in more detail and point out the relevant examples.

6.4.1.1 Upward Information Flow

A higher layer protocol that requires some information from the lower layer(s) at run-time results in the creation of a new interface from the lower layer(s) to the higher layer, as shown in Figure 6.1(a). For instance, if the end-to-end TCP path contains a wireless link, errors on the wireless link can trick the TCP sender into making erroneous inferences about the congestion in the network and as a result the performance deteriorates. Creating interfaces from the lower layers to the transport layer to enable explicit notifications alleviates such situations. For example, the explicit congestion notification (ECN) from the router to the transport layer at the TCP sender can explicitly tell the TCP sender if there is congestion in the network to enable it to differentiate between errors on the wireless link and network congestion [52]. Similarly [66], addressing the problems created for TCP by the disruption of links in the end-to-end route due to mobility in wireless ad hoc networks, also discusses the addition of more such explicit notifications between the link layer and the transport layer.

Examples of upward information flow are also seen in the literature at the MAC layer (link layer in general) in the form of channel-adaptive modulation or link-adaptation schemes [30], [32], [46], [50]. The idea is to adapt the power and data rates in response to the channel condition, which is made known to the MAC layer (link layer) by an interface from the physical layer. Reference [63] extends such a cross-layer adaptation loop by deciding the link layer parameters by considering both the channel condition as well as the instantaneous buffer occupancy at the transmitter, as we discussed in Section 6.3.2.

It is interesting to compare and contrast cross-layer design proposals that rely on upward flow of information to what can be called as self-adaptation loops at a layer. By a self-adaptation loop, we mean an adaptive higher layer protocol that respond to events that, within the constraints of layering, are directly observable at the layer itself. Hence, self-adaptation loops do not require new interfaces to be created from the lower layer(s) to the higher layer and cannot be classified as cross-layer designs. For example, consider the auto-rate fallback mechanism for rate selection [33] in wireless devices with multirate physical layers. The idea is that if some number of packets sent at a particular rate are successfully delivered, the data rate is increased, whereas, if a packet failure is experienced, data rate is dropped. In this case, the rate selection mechanism responds to the acknowledgements, which are directly observable at the MAC layer. Hence we do not treat auto-rate fallback as a cross-layer design. Similar examples of self-adaptations are provided in [7] and [28]. In fact, TCP provides another example of a self-adaptation loop since TCP changes its window size in response to the acknowledgements observed at the transport layer itself.

6.4.1.2 Downward Information Flow

Some cross-layer design proposals rely on setting parameters on the lower layer of the stack at run-time using a direct interface from some higher layer, as illustrated in Figure 6.1(b). As an example, the applications can inform the link layer about their delay requirement, and the link layer can then treat packets from the delay-sensitive applications with priority [65].

A good way to look at the upward and downward information flow is to treat them as notifications and hints, respectively, as proposed in [38]. Upward information flow serves the purpose of notifying the higher layers about the underlying network conditions; downward information flow is meant to provide hints to the lower layers about how the application data should be processed.

6.4.1.3 Back and Forth Information Flow

Two layers, performing different tasks, can collaborate with each other at run-time. Often, this manifests in an iterative loop between the two layers, with information flowing back and forth between the layers, as highlighted in Figure 6.1(c). Clearly, the architecture violation here are the two complimentary new interfaces.

As an example, we refer to the network-assisted diversity multiple access (NDMA) proposal [22], whereby the physical and the MAC layers collaborate in collision resolution in the uplink of a wireless local area network (LAN) system. Basically, with improvements in the signal processing at the physical layer (PHY), it becomes capable of recovering packets from collisions. Thus, upon detecting a collision, the base station first estimates the number of users that have collided, and then requests a suitable number of retransmissions from the set of colliding users. Then PHY signal processing lets the base station separate the signals from all the colliding users.

As another example, consider the problem of joint scheduling and power control in wireless ad hoc networks. Examples include [12], [24], [25], [36]. Basically, power control determines the effective topology of the network by determining which nodes can communicate with one another in a single hop. If the transmitted power is too large, then many nodes may be connected by a single hop, but the interference also would be large. On the other hand, keeping the power too small can make the network fragmented or create too many hops and hence added MAC contention. Protocols resulting from considering the joint problem of power control and scheduling often result in an iterative solution: Trying to keep the power level at an optimal level by responding to the changes in averaged throughput (e.g. see [25]). Reference [24] considers the joint scheduling and power control problem in time-division multiple access (TDMA) based wireless ad hoc networks. A scheduling algorithm chooses the users that will transmit, and then a power control algorithm determines if the transmissions of all the chosen users can simultaneously go on. If no, the scheduling algorithm is repeated. This iteration between scheduling and power control is repeated until a valid transmission schedule has been found. While there are differing views on which layer should power control belong to,[1] the collaborative nature of the cross-layer design mentioned above and the back and forth information flow that they require should be clear.

6.4.2 Merging of Adjacent Layers

Another way to do cross-layer design is to design two or more adjacent layers together such that the service provided by the new 'super-layer' is the union of the services provided by the constituent layers. This does not require any new interfaces to be created in the stack. Architecturally speaking, the super-layer can be interfaced with the rest of the stack using the interfaces that already exist in the original architecture.

Although we have not come across any cross-layer design proposal that explicitly creates a super-layer, it is interesting to note that the collaborative design between the PHY and the MAC layers that we discussed in Section 6.4.1.3 while discussing the NDMA idea tends to blur the boundary between these two adjacent layers.

6.4.3 Design Coupling Without New Interfaces

Another category of cross-layer design involves coupling between two or more layers at design time without creating any extra interfaces for information-sharing at run-time. We illustrate this in Figure 6.1(e). While no new interfaces are created, the architectural cost here is that it may not be possible to replace one layer without making corresponding changes to another layer.

For instance, [60] considers the design of MAC layer for the uplink of a wireless LAN when the PHY layer is capable of providing multipacket reception capability. Multipacket reception capability implies that the physical layer is capable of receiving more than one packet at the same time. Notice that this capability at the physical layer considerably changes the role of the MAC layer; thus, it needs to be redesigned. Similarly [55] considers the design of MAC layer in ad hoc networks with smart antennas at the physical layer.

6.4.4 Vertical Calibration Across Layers

The final category in which cross-layer design proposals in the literature fit into is what we call vertical calibration across layers. As the name suggests, this refers to adjusting parameters that span across layers, as illustrated in Figure 6.1(f). The motivation is easy to understand. Basically, the performance seen at the level of the application is a function of the parameters at all the layers below it. Hence, it is conceivable that joint tuning

[1] Power control is mentioned as a physical layer task in [12]. A different view is taken in [43] and [24] where power control is placed at the MAC layer and in [35] where power control is placed at the network layer.

can help to achieve better performance than individual settings of parameters – as would happen had the protocols been designed independently – can achieve.

As an example, [2] looks at optimizing the throughput performance of the TCP by jointly tuning power management, forward error correction (FEC) and ARQ settings. Similarly, [40] presents an example of vertical calibration where the delay requirement dictates the persistence of the link-layer ARQ, which in turn becomes an input for the deciding the rate selection through a channel-adaptive modulation scheme.

Vertical calibration can be done in a static manner, which means setting parameters across the layers at design time with the optimization of some metric in mind. It can also be done dynamically at run-time, which emulates a flexible protocol stack that responds to the variations in the channel, traffic and overall network conditions. Examples of dynamic vertical calibrations can be found in [49] and [1]. Both these references describe prototype systems that employ adaptations across several layers of the protocol stack.

Static vertical calibration does not create significant consideration for implementations since the parameters can be adjusted once at design time and left untouched thereafter. Dynamic vertical calibration, on the other hand, requires mechanisms to retrieve and update the values of the parameters being optimized from the different layers. This may incur significant cost in terms of overheads and also impose strict requirements on the parameter retrieval and update process to make sure that the knowledge of state of the stack is current and accurate. We again refer the readers to [49] and [1] for a thorough discussion of the challenges involved in implementing dynamic vertical calibrations.

6.5 Proposals for Implementing Cross-Layer Interactions

Alongside the cross-layer design proposals that we discussed in Section 6.4, initial proposals on how cross-layer interactions can be implemented are also being made in the literature. These can be put into three categories:

1. Direct communication between layers (Figure 6.2(a)).
2. A shared database across the layers (Figure 6.2(b)).
3. Completely new abstractions (Figure 6.2(c)).

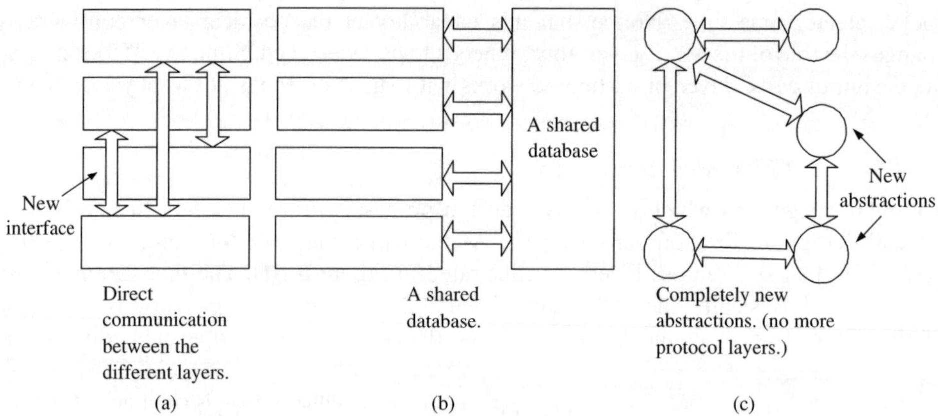

Figure 6.2 Proposals for architectural blueprints

6.5.1 Direct Communication Between Layers

A straightforward way to allow run-time information sharing between the layers is to allow them to communicate with each other, as depicted schematically in Figure 6.2(a). Note that this is applicable when there has to be run-time information sharing between layers (for example, in cross-layer designs that rely on new interfaces or in dynamic vertical calibrations). Practically speaking direct communication between the layers means making the variables at one layer visible to the other layers at run-time. By contrast, under a strictly layered architecture, every layer manages its own variables and its variables are of no concern to other layers.

There are many ways in which the layers can communicate with one another. For instance, protocol headers may be used to allow flow of information between the layers. Alternatively, the extra 'inter-layer' information could be treated as internal packets. The work in [64] presents a comparative study of several such proposals and goes on to present another such proposal, namely, the cross-layer signaling shortcuts (CLASS). CLASS allows any two layers to communicate directly with one another.

These proposals are appealing in the case where just a few cross-layer information exchanges are to be implemented in systems that were originally designed in conformance with the layered architectures. In that case, one can conceivably 'punch' a few holes in the stack while still keeping it tractable. However, in general, when variables and internal states from the different layers are to be shared as prescribed by such proposals, a number of implementation issues relating to managing shared memory spaces between the layers may need to be resolved.

6.5.2 A Shared Database Across the Layers

The other class of ideas propose using a common database that can be accessed by all the layers, as illustrated in Figure 6.2(b). See for instance [15], [41], [48] and [57]. In one sense, the common database is like a new layer, providing the service of storage/retrieval of information to all the layers.

The shared database approach is particularly well suited to vertical calibrations across layers. An optimization program can interface with the different layers at once through the shared database. Similarly, new interfaces between the layers can also be realized through the shared database. The main challenge here is the design of the interactions between the different layers and the shared database.

6.5.3 Completely New Abstractions

The third set of proposals present completely new abstractions, which we depict schematically in Figure 6.2(c). Consider, for example, the proposal in [10] which presents a new way to organize the protocols: in heaps and not in stacks as done by layering.

Such novel organizations of protocols are appealing as they allow rich interactions between the building blocks of the protocols. Hence, potentially they offer great flexibility, both during the design as well as at run-time. However, they change the very way protocols have been organized, and hence may require completely new system-level implementations.

6.6 Cross-Layer Design Activity in the Industry and Standards

So far we have been talking about the cross-layer design activity largely from the research literature. Interestingly, cross-layer design ideas have been making their way into several ongoing standardization activities and commercial products too. In this section we take a brief look at some of this industry activity.

6.6.1 3G Cellular Networks

The channel-dependent scheduling that we discussed in Section 6.3.2 has been incorporated into the high data rate (HDR) version of CDMA2000 (1xEV-DO) and enhanced general packet radio service (EGPRS) in the EDGE extension of GSM [52]. In these systems, the users periodically update the base station about the state of the downlink channel. This information is used by the base station for scheduling the transmissions to the different users.

6.6.2 Vertical Handovers

The problem of vertical handovers that we discussed in Section 6.3.3 is being addressed by the IEEE 802.21 standards group. The IEEE 802.21 group is defining standardized interfaces for sharing information from the lower layers (e.g. link layer and physical layer) with the higher layers (e.g. network layer). The IEEE 802.21 effort thus is an interesting example of convergence between the performance viewpoint, which dictates the need for cross-layer information sharing, and the architectural viewpoint, which requires standardized and well-defined interfaces between the modules.

6.6.3 Wireless Regional Area Networks

The IEEE 802.22 is working to standardize the operation of a wireless regional area network (WRAN). An excellent overview of this standardization effort can be found in [17]. The distinctive feature of the IEEE 802.22 is that it defines a wireless network that is to operate within the spectrum that has been assigned to the television broadcast systems. In order to protect the 'primary' users (the TV broadcasters) the 802.22 standard makes use of the cognitive radio techniques, thus representing another example of cross-layer design (between the MAC and the PHY layers).

6.6.4 Wireless Local Area Networks

The IEEE 802.11 standards cover the wireless local area networks that are primarily used within homes or small offices for wireless Internet access. The IEEE 802.11 a/b/g suite of technologies has found its way into many homes and offices worldwide. These technologies, however, do not perform well for multimedia traffic, primarily because the IEEE 802.11 MAC does not provide any kind of QoS guarantees. This problem has been addressed in an extension to the 802.11 MAC under the effort called IEEE 802.11e [44]. One of the features of IEEE 802.11e MAC that has a strongly cross-layer design flavor is in the realm of what is called the enhanced distributed channel access (EDCA). The idea is to categorize the incoming data from the higher layer into four priority classes at the MAC layer. This allows different kinds of data (for instance voice, video, time-insensitive data) to be treated differently at the MAC layer, such that their individual QoS requirements may be satisfied. In architectural terms, this is equivalent to having a direct communication interface between the higher layers and the MAC layer. The IEEE 802.11e has strong support from the industry. As an example, several of the IEEE 802.11e features have been incorporated into the 'Super-G' chipsets from the wireless LAN chipset vendor Atheros (see http://www.atheros.com).

6.6.5 Multimedia Over Wireless

In the context of multimedia over wireless networks, we point out Ruckus Wireless (http://www.ruckuswireless.com), a privately held company, which has also taken a cross-layer design approach remarkably similar to the 802.11e idea above in solving the problem of maintaining the QoS requirements of the different kinds of traffic types. The company calls its technology 'Smart-Cast'; the key idea is to apply a traffic-dependent scheduling. That is, at the MAC layer, the packets are classified according to their latency requirements, and a scheduling algorithm then determines the ordering in which the packets are transmitted over the wireless medium. The Ruckus Wireless system also features a proprietary technology called 'Beam Flex'. The idea in Beam Flex is to continuously monitor the interference in the wireless medium, and using the appropriate combination of the smart antenna system, route the traffic around the interference. This, thus, once again is an example of cross-layer design involving the physical layer and the MAC layer.

Digital Fountain (http://www.digitalfountain.com) provides error protection by employing a type of erasure correction code known as a fountain code (termed DF Raptor FEC by the company). This technology has been standardized for use in multimedia broadcast and multicast services (MBMS) in 3G. Basically, the use of error correction at the higher layers provides an enhanced line of defense against impairments caused by the wireless channel and network congestion. Fountain codes allow the receiver to recover lost packets without requiring the sender to keep track of and retransmit missing packets. This is done by cleverly introducing redundancy via the generation of a potentially large amount of encoded data for a given piece of content (say an mp3 file). The receiver then just has to collect a sufficient number of coded packets to decode the multimedia content, without keeping track of exactly which packets it has. When it has decoded the file, it can inform the sender to stop. Digital Fountain's DF Raptor FEC efficiently mitigates the need for end-to-end reliability at the packet level, and pushes it to the file level. Furthermore, since the codes in question are fountain codes, they are highly scalable and provide adaptive error protection in face of variations in the wireless transmission environment and the network quality. That the DF Raptor FEC technology takes a cross-layer design approach is evident in the fact that it can be employed at the transport layer or at the application layer and it uses forward error correction to provide end-to-end reliability.

6.6.6 Mesh Networks

In recent times there has been a considerable interest in the deployments of mesh networks. MIT Roofnet [8] is an example of an experimental mesh network deployed at Cambridge, Massachusetts, in the United States. The MIT Roofnet makes use of a routing protocol called ExOR which is in fact an integrated routing and link layer protocol [9]. ExOR relies on the broadcast nature of the wireless medium. (MIT Roofnet uses omnidirectional antennas.) Whenever the source has a packet to send, it broadcasts the packet which is received by a number of potential forwarders. One out of the set of potential forwarders, chosen based on the reliability of the link between the potential forwarders and the destination, forwards the packet further. This process continues until the packet is delivered at the destination. We refer the reader to [9] for more details. As far as we are concerned, we just notice that the cross-layer design involving the network and the link layers results in throughput improvements compared with the traditional approaches. Tropos Networks (http://www.tropos.com) deploys commercial wireless mesh networks. Though the details of Tropos' proprietary routing algorithm, the Predictive Wireless Routing Protocol (PWRP), are not known, we believe, based on the cursory descriptions that we have read, that PWRP also makes use of throughput information from the MAC layer in making the routing decisions at the network layer. This represents another example of cross-layer design between the network and the MAC layers.

This look at the standardization and the industry activity that we have taken is no doubt an extremely brief sampling of the activity in the industry. It, nonetheless, clearly illustrates that cross-layer design ideas are finding their way into commercial products and standards, which should be an encouragement for researchers in this field.

6.7 The Open Challenges

In Sections 6.3 through 6.6, we looked at the ongoing work in the area of cross-layer design. In doing so, we came face to face with the general motivations behind cross-layer design, the several different interpretations of cross-layer design, many examples of cross-layer design, some initial ideas on how cross-layer interactions may be implemented, and a quick snapshot on how cross-layer design ideas are getting into ongoing standardizations and commercial deployments. Having taken stock of the ongoing work, we now raise and discuss what we call the 'open challenges' for cross-layer design. Broadly speaking, we include two kinds of issues in the open challenges: questions about cross-layer design that, in our opinion, are important but are not getting sufficient attention in the research literature; and some questions regarding the fundamental nature of the wireless medium, questions whose answers will influence how communication architectures for the wireless networks should look like, and hence are pertinent to the cross-layer design effort. We note that some of these issues have been raised elsewhere in the literature too. As before, our purpose here is to consolidate the different issues and discuss their significance with respect to the cross-layer design activity.

The following are some open challenges for the designers proposing cross-layer design ideas:

1. How do the different cross-layer design proposals coexist with one another?
2. Will a given cross-layer design idea possibly stifle innovation in the future?
3. What are the cross-layer designs that will have the most significant impact on network performance, and hence should be most closely focused on?
4. Has a given design proposal been made with a thorough knowledge of the effect of the interactions between the parameters at different layers on network performance?
5. Under what network and environmental conditions be a particular cross-layer design proposal be invoked?
6. Can the mechanisms/interfaces used to share information between the layers be standardized?

7. What should be the role of the physical layer in wireless networks?
8. Is the conventional view of the network – that of a collection of point-to-point links – appropriate for wireless networks?

We now look at some of these issues in greater detail.

6.7.1 The Important Cross-Layer Couplings

While there are a number of cross-layer design proposals in the literature today, it is not clear which are the most important ones. A thorough cost-benefit analysis of the different cross-layer design proposals in terms of the implementation complexity versus the performance improvement is needed. To be fair, the important cross-layer design ideas depend on the specific network scenario considered and the metric of interest. For example, in the case of wireless ad-hoc networks, one can make the following inferences from the literature today: cross-layer design is needed between network layer and the MAC layer for wireless ad hoc networks, as we discussed in Section 6.6.6, since the functionalities of the two layers interact [3]; explicit notifications by new interfaces to the transport layer improve end-to-end performance [52]; making use of the channel knowledge at the MAC layer allows opportunistic usage of the channel and improves performance ([32] and the references therein); and energy, delay and security-related issues need to be handled across the layers in a holistic manner. It is time to move ahead from these general insights and harmonize the different cross-layer design ideas. This requires comparative quantitative study of the different cross-layer design proposals, and is an open challenge for the community. Also relevant in this context is the question of coexistence of cross-layer design proposals, which we discuss next.

6.7.2 Co-Existence of Cross-Layer Design Proposals

An important question to be answered is how different cross-layer design proposals can coexist with one another. To clarify by example, say the MAC layer in a stack responds to the variation in the channel by adjusting the data rate. The question is, will additionally adjusting the frame length at the link layer help further? How will an overriding control from, say, the transport layer, trying to control the link layer parameters, interact with these adaptation loops?

The question of coexistence of cross-layer design ideas is pertinent when it comes to determining whether some cross-layer design proposals can stifle further innovation. Let us say the physical layer and the link layer are optimized for a certain performance metric in a cross-layer design scheme. If this scheme is deployed first, can other schemes that also rely on some (other) cross-layer couplings, or those that assume no coupling between the link layer and the physical layer be deployed too at a later time?

Apart from presenting new cross-layer design proposals, designers need to start establishing which other cross-layer design interaction may or may not be employed together with their proposal. This has also been stressed in [34].

6.7.3 When to Invoke a Particular Cross-Layer Design?

The network conditions in a wireless network are usually time varying. In such a situation, one of the stated motivations behind cross-layer design is to achieve the network equivalent of impedance matching [52]. The idea is to make the protocol stack responsive to the variations in the underlying network conditions so that an optimal operating point is always maintained.

The pursuit of achieving such optimal operation throws up two complimentary challenges. First, designers need to establish the network conditions under which the proposed cross-layer designs would result in performance improvements. Reference [34] presents an example to illustrate how a cross-layer design involving an iterative optimization of throughput and power leads to a loss in performance under a certain pathological network condition. The example in [34] underscores the need for designers to establish the network conditions under which their design proposals should and should not be used. Secondly, efficient mechanisms to make a timely and accurate assessment of the state of the network need to be built into the stack, and the corresponding overheads need to be taken into account [1], [49]. This is also related to the question of interfaces between the modules, discussed next.

6.7.4 Standardization of Interfaces

The one thing that layering achieved was to present standardized boundaries and interfaces between the modules of the system, namely, the protocol layers. Now that the layered architecture is being violated in different ways, finding the new reference architecture becomes a challenge. What should be the boundaries between the modules? Should we stick to the traditional layer boundaries as in Figure 6.2(a) and (b) and determine the new interfaces from there, or should we look at completely new boundaries, as in Figure 6.2(c)? Or a combination? What should the interfaces between the modules look like?

Addressing this challenge requires greater synergy between the performance viewpoint and the implementation concerns than what is seen in the literature today. Basically, the organization of the modules (layers or otherwise) and the interfaces between them determine how efficiently can information be shared between them, at what kinds of overheads and delays. This, in turn, determines how effective cross-layer design proposals that rely on sharing dynamic information between the modules can be. Hence, proposers of cross-layer design relying on back and forth information flow between layers or dynamic vertical calibrations need to start considering the impact of delays in the retrieval/updating of information on the protocol performance. They also need to quantify the overheads associated with their cross-layer design proposals.

One point that should be mentioned here is that the interfaces get standardized during the process of drafting technical standards. A good example is the IEEE 802.21 standardization which is codifying the interfaces to be followed for the transfer of information between the different layers of the stack.

6.7.5 The Role of the Physical Layer

In wired networks, the role of the physical layer has been rather small: sending and receiving packets when required to do so from the higher layers. This is also the case in the current generation of wireless technologies. As we have seen in Section 6.4.3, advances in the signal processing at the physical layer can allow it to play a bigger role in wireless networks. This begs a question as to how much of a role should the physical layer play? This question is relevant to the cross-layer design effort because first, layered architectures like the OSI reference model do not allow much role for the physical layer besides providing a bit-pipe; and second, enhancements in the physical layer will have to be balanced by corresponding changes to the higher layers. Hence, figuring out the role to be played by the physical layer is an important question. Cross-layer designs relying on advanced signal processing at the physical layer can be an interesting research ground for the future.

6.7.6 The Right Communication Model

Wired networks, by their very nature, are essentially a collection of well-defined point-to-point communication links. The same cannot be said about wireless networks because the wireless medium is inherently broadcast and there is no clear-cut concept of a communication link in wireless networks. This gives rise to a fundamental question of whether it still makes sense to 'create' links in a wireless network, which is what has been done in all the networks of the past. As we discussed in Section 6.3.3, cooperative communications, when done at the link/physical layer, fundamentally changes the abstraction that the higher layers have of the network since the network no longer looks like a collection of point-to-point links [37]. By coupling physical, link and network layers in intricate ways, cooperative communications inevitably invite a cross-layer design approach. Hence, the open question facing the research community is to explore such modalities of communications in context of end-to-end network performance, and to design suitable communications architectures to support such modalities.

6.8 Discussion

We have been looking at the ongoing work in the area of cross-layer design for wireless communications and networks throughout this chapter. Implicit in this entire discussion, as it is in most of the published research, is the existence and acceptance of a reference layered architecture. A good number of communication systems broadly follow the five-layer model, which is a hybrid between the two most commonly taught models: the OSI seven-layer model and the four-layer TCP/IP model. The question is, how did these models come about in the first place?

The four-layer TCP/IP model came into being as part of an initiative undertaken by the United States (US) Department of Defense to connect different kinds of packet-switched data networks. The idea was to enable a communication device connected to a packet-switched network to communicate with any other communication device connected to any other packet-switched network. Reference [13] discusses the key considerations that motivated the development of some specific features of the TCP/IP protocols (for example a connectionless mode of communications with end-to-end flow and error control provided by the transport layer). In effect, the motivations elaborated in [13] highlight the design principles of the Internet since, in time, the TCP/IP model became the model that shaped the Internet.

On the other hand, the OSI seven-layer model came into being as a result of an initiative taken up by the International Standards Organization (ISO) to come up with a set of standards that would allow disparate systems anywhere in the world to communicate with each other. The idea was to allow interoperability between communication equipments developed by the different vendors. The basic seven-layer reference model was published in early 1980s and it became an international standard (ISO 7498) in 1983 [20]. The OSI model had split the lowest layer of the TCP/IP model into three separate layers. It had also done the same for the highest layer.

Interestingly, in the case of the TCP/IP model, the protocols at the different layers were developed before formal layer definitions had been made [56], [59, p. 39]. Basically, as the TCP/IP was getting bundled with the UNIX operating system, the popularity of these protocols in the research community shot up [56]. In due course, layer definitions were attached to the model [59, p. 39] and thus came the four-layer TCP/IP model. By contrast, in the case of the OSI seven-layer model, the layers were defined *before* the development of protocols [59]. As mentioned earlier, the idea behind the OSI model was to standardize the protocol development effort, and to do so, a flexible architecture was provided in the seven-layer model. Subsequent to the publication of the seven-layer model, protocols at the different layers were developed and published as separate international standards.

Reference [59] provides an insightful comparison and critique of both the models. On a broader note, it discusses the relative merits of the contrasting approaches: architecture before protocols, or protocols before architecture. Interestingly, engineers actually implementing the protocols, or considering implementation issues, have long raised questions about the suitability of strict protocol layering, whether defined (as in OSI) or adopted (as in TCP/IP). For instance, [16] advocated the concept of 'soft-layers' that allow more flexible information sharing between adjacent layers, [58] discussed in detail the inefficiencies and problems that resulted in implementing layered OSI systems, and [14] and [18] talked of integrated layer processing – none because of wireless links in the network, or because of a desire to create cognitive networks! In fact based on the feedback from the designers, the original OSI model was revised and a new reference was published. An interesting summary of the major changes to the original reference model and their motivation, peppered with palpable political overtones, can be found in [21].

Seen in this historical light, there is really no surprise that the increasing proliferation of wireless networking and the pursuit of cognitive networks is ushering in a large number of cross-layer design proposals. After all, the layered architectures themselves have been created in response to specific technical and industrial needs, and are not in any way all encompassing. What we are witnessing today are, by and large, minor tweaking to these well-understood layered architectures. It is entirely conceivable that in future the architectures of today will evolve and may or may not look similar to the ones today. Every cross-layer design proposal serves to highlight a specific shortcoming of traditional protocol layering for the scenario in question. Over time these ideas, if significant, will be incorporated into the standards or proprietary systems, and hence the architectures will evolve.

That said, we hasten to add that while the protocol layers as we have been taught in networking courses may not survive, the core principle of modularity inevitably will, as that is the very principle that underpins the modern economy. Possibly a new kind of layering will evolve for wireless networks. As researchers in the field, we will do well to see our cross-layer design ideas in the bigger context of specific architectures that we are looking at, and hence evaluate their coexistence with other ideas, their impact on the deployability of other ideas, and so on. In other words, even though cross-layer design does inevitably look like the way forward, researchers should look not just at the performance viewpoint, but also at the larger architectural consequences of their work. In stating this, we are unanimous with [34].

6.9 Conclusions

In this chapter, we consolidated the several different ideas and results falling under the broad purview of cross-layer design. We saw that cross-layer design can be defined as a violation of a reference layered architecture. Cross-layer design is done to either solve a unique problem or exploit a new opportunity or new modality created by wireless links. We taxonomized the different cross-layer design proposals according to the kind of architecture violations they represent, looked at the initial ideas for implementing cross-layer interactions and briefly described some cross-layer design ideas that have made their way into commercial products and industry standards. We then raised some open questions related to cross-layer design that researchers may want to address as they move forward. Finally we placed the cross-layer design of today in a brief historic perspective.

Through the broad-based discussion that we undertook in this chapter, we have seen both the wide-applicability of cross-layer design ideas, as well as the inevitability of cross-layer design in the pursuit of cognitive wireless networks of the future. There is little doubt that cross-layer design holds tremendous potential to unleash the true potential of wireless communications. However, for that potential to be fulfilled, it is imperative for the designers to view cross-layer design holistically by considering both performance and architectural viewpoints. We hope that by consolidating the varied ideas, categorizing the initial thoughts and by raising some open questions, this chapter will facilitate such holistic thinking about cross-layer design.

References

[1] Akyildiz, I., Altunbasak, Y., Fekri, F. and Sivakumar, R. (2004) AdaptNet: an adaptive protocol suite for the next-generation wireless Internet. *IEEE Communications Magazine*, **42**(3), 128–36.

[2] Barman, D., Matta, I., Altman, E. and Azuozi, R.E. (2004) TCP optimization through FEC, ARQ and transmission power tradeoffs. *Proceedings of the International Conference on Wired/Wireless Internet Communications*, Frankfurt.

[3] Barrett, C., Drozda, M., Marathe, A. and Marathe, M.V. (2002) Characterizing the interaction between routing and MAC protocols in ad-hoc networks. Proceedings of the ACM Annual International Symposium on Mobile Ad-Hoc Networking and Computing (MobiHoc'02), pp. 92–103, June, Lausanne.

[4] Berry, R.A. and Gallager, R.G. (2002) Communication over fading channels with delay constraints. *IEEE Transactions on Information Theory*, May, 1135–49.

[5] Berry, R.A. and Yeh, E. (2004) Cross-layer wireless resource allocation. *IEEE Signal Processing Magazine*, September, 59–68.

[6] Bertsekas, D. and Gallager, R. (1992) *Data Networks*, 2nd edn, Prentice Hall, New Jersey.

[7] Bhagwat, P., Bhattacharya, P., Krishna, A. and Tripathi, S.K. (1996) Enhancing throughput over wireless LANs using channel state dependent packet scheduling. *Proceedings of the IEEE Infocom'96*, pp. 1133–40.

[8] Bicket, J., Aguayo, D., Biswas, S. and Morris, R. (2005) Architecture and evaluation of an unplanned 802.11b mesh network. *Proceedings of the ACM Annual International Conference on Mobile Computing and Networking (MobiCom'05)*, pp. 31–42, August.

[9] Biswas, S. and Morris, R. (2005) Opportunistic routing in multi-hop wireless networks. Proceedings of the ACM (SIGCOMM'05), pp. 69–74, August.

[10] Braden,, R., Faber, T. and Handley, M. (2002) From protocol stack to protocol heap – role-based architecture. *Proceedings of the Hot Topics in Networking (HOTNETS I)*, March, Princeton, NJ.

[11] Burbank, J.L. and Kasch, W.T. (2005) Cross-layer design for military networks. *Proceedings of the IEEE Military Communications Conference (MILCOM'05)*, pp. 1912–18, October.

[12] Chiang, M. (2004) To layer or not to layer: balancing transport and physical layers in wireless multihop networks. *Proceedings of the IEEE Infocom'04*, pp. 2525–36, March, Hong Kong.
[13] Clark, D.D. (1988) The design philosophy of the DARPA Internet protocols. *Proceedings of theACM Symposium on Communications Architectures and Protocols (SIGCOMM'88)*, pp. 106–14.
[14] Clark, D.D. and Tennenhouse, D.L. (1990) Architectural considerations for a new generation of protocols. *Proceedings of the ACM Symposium on Communications Architectures and Protocols (SIGCOMM'90)*, Philadelphia.
[15] Conti, M., Maselli, G., Turi, G. and Giordano, S. (2004) Cross-layering in mobile ad hoc network design. *IEEE Computer Magazine*, February, 48–51.
[16] Cooper, G.H. (1983) *An argument for soft layering of protocols*, Master's thesis, Masachussets Institute of Technology.
[17] Cordeiro, C., Challapali, K., Birru, D. and Shankar, N.S. (2005) IEEE 802.22: the first worldwide wireless standard based on cognitive radios. *Proceedings of the IEEE International Symposium on Dynamic Spectrum Access Networks (DySPAN'05)*, pp. 328–37, November.
[18] Crowcroft, J., Wakeman, I., Wang, Z. and Sirovica, D. (1992) Is layering harmful? *IEEE Network*, **6**(1), 20–4.
[19] Cui, S. and Goldsmith, A.J. (2006) Cooperation techniques in cross-layer design. In F.H.P. Fitzek and M.D. Katz (eds.), Cooperation in *Wireless Networks: Principles and Applications*, Springer.
[20] Day, J. (1989) The reference model for open systems interconnection. In C.A. Sunshine (ed.), *Computer Network Architectures and Protocols*, 2nd edn, Plenum Press, New York.
[21] Day, J.D. (1995) The (un)revised OSI reference model. *ACM Computer Communication Review*, **25**(10), 39–55.
[22] Dimić, G., Sidiropoulos, N. and Zhang, R. (2004) Medium access control – physical cross-layer design. *IEEE Signal Processing Magazine*, **21**(9), 40–50.
[23] Eberle, W., Bougard, B., Pollin, S. and Catthoor, F. (2005) From myth to methodology: cross-layer design for energy-efficient wireless communication. *Proceedings of the ACM Design Automation Conference (DAC'05)*, pp. 303–8, June, Anaheim, CA.
[24] Elbatt, T. and Ephremides, A. (2004) Joint scheduling and power control for wireless ad hoc networks. *IEEE Transactions on Wireless Communications*, **3**(1), 74–85.
[25] Elbatt, T., Krishnamurthy, S., Connors, D. and Dao, S. (2000) Power management for throughput enhancement in wireless ad-hoc networks. *Proceedings of the IEEE International Conference on Communications (ICC'2000)*, pp. 1506–13, New Orleans.
[26] Fitzek, F.H.P. and Katz, M.D. (2006) Cooperation in nature and wireless communications. In F.H.P. Fitzek and M.D. Katz (eds.), Cooperation in *Wireless Networks: Principles and Applications*, Springer.
[27] Goldsmith, A. and Varaiya, P. (1995) Capacity of fading channel with channel side information. *IEEE Transactions on Information Theory*, **43**, 1986–92.
[28] Hara, S., Ogino, A., Araki, M., Okada, M. and Morinaga, M. (1996) Throughput performance of SAW-ARQ protocol with adaptive packet length in mobile packet data transmission. *IEEE Transactions on Vehicular Technology*, **45**, 561–9.
[29] Haykin, S. (2005) Cognitive radio: brain-empowered wireless communications. *IEEE Journal on Selected Areas in Communications*, **23**, 201–20.
[30] Holland, G., Vaidya, N. and Bahl, P. (2001) A rate-adaptive MAC protocol for multi-hop wireless networks. *Proceedings of the ACM Annual International Conference on Mobile Computing and Networking (MobiCom'01)*, pp. 236–51, July, Rome.
[31] Jamalipour, A. (2003) *The Wireless Mobile Internet*, John Wiley & Sons, Co., Chichester, UK.
[32] Ji, Z., Yang, Y., Zhou, J., Takai, M. and Bagrodia, R. (2004) Exploiting medium access diversity in rate adaptive wireless LANs. *Proceedings of the ACM Annual International Symposium on Mobile Computing and Networking (MobiCom'04)*, pp. 345–59, October, Philadelphia.
[33] Kamerman, A. and Monteban, L. (1997) WaveLAN II: a high-performance wireless LAN for the unlicensed band. *Bell Labs Technical Journal*, 118–33.
[34] Kawadia, V. and Kumar, P. (2005) A cautionary perspective on cross layer design. *IEEE Wireless Communications Magazine*, **12**, 3–11.
[35] Kawadia, V. and Kumar, P.R. (2005) Principles and protocols for power control in ad hoc networks. *IEEE Journal on Selected Areas in Communications*, **5**, 76–88.

[36] Kozat, U., Koutsopoulos, I. and Tassiulas, L. (2004) A framework for cross-layer design of energy-efficient communication with QoS provisioning in multi-hop wireless networks. *Proceedings of the IEEE Infocom'04*, pp. 1446–56, March, Hong Kong.
[37] Laneman, J.N. (2006) Cooperative diversity. In F.H.P. Fitzek and M.D. Katz (eds.), Cooperation in *Wireless Networks: Principles and Applications*, Springer.
[38] Larzon, L.-A., Bodin, U. and Schelén, O. (2002) Hints and notifications. *Proceedings of the IEEE Wireless Communications and Networking Conference (WCNC'02)*, pp. 635–41, March, Orlando, FL.
[39] Leiner, B.M., Cole, R., Postel, J. and Mills, D. (1985) The DARPA Internet protocol suite. *IEEE Communications Magazine*, **23**, 29–34.
[40] Liu, Q., Zhou, S. and Giannakis, G.B. (2004) Cross-layer combining of adaptive modulation and coding with truncated ARQ over wireless links. *IEEE Transactions on Wireless Communications*, **3**(9), 1746–55.
[41] Mahonen, P., Petrova, M., Riihijarvi, J. and Wellens, M. (2006) Cognitive wireless networks: your network just became a teenager. *Proceedings of the IEEE Infocom'06.*
[42] Mitola, J. (2006) Cognitive radio architecture. In F.H.P. Fitzek and M.D. Katz (eds.), Cooperation in *Wireless Networks: Principles and Applications*, Springer.
[43] Muqattash, A. and Krunz, M. (2004) A single-channel solution for transmission power control in wireless ad hoc networks. *Proceedings of the ACM Annual International Symposium on Mobile Ad-Hoc Networking and Computing (MobiHoc'04)*, pp. 210–21, Tokyo.
[44] Ni, Q. (2005) Performance analysis and enhancements for IEEE 802.11e wireless networks. *IEEE Network*, **19**(7), 21–7.
[45] Nosratinia, A., Hunter, T.E. and Hedayat, A. (2004) Cooperative communications in wireless networks. *IEEE Communications Magazine*, **42**(10), 74–80.
[46] Qiao, D. and Choi, S. (2001) Goodput enhancement of 802.11a wireless LAN via link adaptation. *Proceedings of the IEEE International Conference on Communications (ICC'01)*, pp. 1995–2000, Helsinki.
[47] Radosavac, S., Moustakides, G.V. and Baras, J.S. (2006) Impact of the optimal MAC layer attacks on the network layer. *Proceedings of the 4th ACM Workshop on Security of Ad-hoc and Sensor Networks (SASN'06)*, pp. 135–46, October.
[48] Raisinghani, V.T. and Iyer, S. (2004) Cross-layer design optimizations in wireless protocol stacks. *Computer Communications*, **27**, 720–4.
[49] Sachs, D.G., Yuan, W., Hughes, C.J. *et al.* (2004) *GRACE: a hierarchical adaptation framework for saving energy*, Technical report, University of Illinois, Urbana Champagne.
[50] Sadeghi, B., Kanodia, V., Sabharwal, A. and Knightly, E. (2002) Opportunistic media access for multirate ad-hoc networks. *Proceedings of the ACM Annual International Conference on Mobile Computing and Networking (MobiCom'02)*, September, Atlanta.
[51] Schaar, M.V.D. and Shankar, N.S. (2005) Cross-layer wireless multimedia transmission: challenges, principles, and new paradigms. *IEEE Wireless Communications Magazine*, August, 50–8.
[52] Shakkottai, S., Rappaport, T.S. and Karlsson, P.C. (2003) Cross-layer design for wireless networks. *IEEE Communications Magazine*, **41**, 74–80.
[53] Srivastava, V. and Mehul, M. (2005) The road ahead for cross-layer design. *Proceedings of the IEEE Broadnets'05*, pp. 551–60, October.
[54] Srivastava, V. and Motani, M. (2005) Cross-layer design: a survey and the road ahead. *IEEE Communications Magazine*, **43**(12), 112–19.
[55] Sundaresan, K. and Sivakumar, R. (2004) A unified MAC layer framework for ad-hoc networks with smart antennas. *Proceedings of the ACM Annual International Symposium on Mobile Ad-Hoc Networking and Computing (MobiHoc'04)*, pp. 244–55, Tokyo.
[56] Sunshine, C.A. (1989) A brief history of computer networking. In C.A. Sunshine (ed.), *Computer Network Architectures and Protocols*, 2nd edn, Plenum Press, New York.
[57] Sur, A. and Sicker, D.C. (2005) Multi layer rules based framework for vertical handoff. *Proceedings of the IEEE Broadnets'05*, pp. 571–80, October.
[58] Svobodova, L. (1989) Implementing OSI systems. *IEEE Journal on Selected Areas in Communications*, **7**, 1115–30.
[59] Tanenbaum, A.S. (1996) *Computer Networks*, 3rd edn, Prentice-Hall, New Jersey.
[60] Tong, L., Naware, V. and Venkitasubramaniam, P. (2004) Signal processing in random access, *IEEE Signal Processing Magazine*, **21**, 29–39.

[61] Tse, D. and Hanly, S. (1998) Multi-access fading channels: part I: polymatroid structure, optimal resource allocation and throughput capacities. *IEEE Transactions on Information Theory*, **44**, 2796–815.

[62] Tse, D. and Hanly, S. (1998) Multi-access fading channels: part II: Delay-limited capacities. *IEEE Transactions on Information Theory*, **44**, 2816–31.

[63] Tuan, H.A. and Mehul, M. (2003) Buffer and channel adaptive modulation for transmission over fading channels. *Proceedings of the IEEE International Conference on Communications ((ICC'03)*, pp. 2748–52, July, Anchorage.

[64] Wang, Q. and Abu-Rgheff, M.A. (2003) Cross-layer signalling for next-generation wireless systems. *Proceedings of the IEEE Wireless Communications and Networking Conference (WCNC'03)*, pp. 1084–9, March, New Orleans.

[65] Xylomenos, G. and Polyzos, G.C. (2001) Quality of service support over multi-service wireless Internet links. *Computer Networks*, **37**, 601–5.

[66] Yu, X. (2004) Improving TCP performance over mobile ad hoc networks by exploiting cross-layer information awareness. *Proceedings of the ACM Annual International Conference on Mobile Computing and Networking (MobiCom'04)*, pp. 231–244, October, Philadelphia.

[67] Zhai, F., Eisenberg, Y. and Katsaggelos, A.K. (2005) Joint source-channel coding for video communications. In A. Bovik (ed.), *Hand of Image and Video Processing*, 2nd edn, Elsevier Academic Press.

[68] Zhang, J. (2002) Bursty traffic meets fading: a cross-layer design perspective. *Proceedings of the Allerton Conference on Communications, Control and Computing'02*.

[69] Zhang, Q., Zhu, W. and Zhang, Y.-Q. (2001) Resource allocation for multimedia streaming over the Internet. *IEEE Transactions on Multimedia*, **3**, 339–55.

[70] Zhao, B. and Valenti, M.C. (2005) Practical relay networks: a generalization of hybrid-ARQ. *IEEE Journal on Selected Areas in Communications*, **23**, 7–18.

7

Cognitive Radio Architecture[1]

Joseph Mitola III
The MITRE Corporation[2]

7.1 Introduction

Cognitive radio (CR) today includes a relatively wide range of technologies for making wireless systems computationally intelligent. This has resulted from an interdisciplinary integration of complementary but somewhat isolated technologies: perception, planning and machine learning technologies from artificial intelligence on the one hand, and on the other hand software radio technologies that had come to include self-description in the extensible markup language, XML [16] [19] [20]. The first significant application proposed for such smarter radios was the autonomous sharing of pooled spectrum [18], which the US Federal Communications Commission (FCC) endorsed relatively soon thereafter to encourage the development of secondary spectrum markets [25]. The original visionary formulation of the ideal cognitive radio (iCR) remains underdeveloped: an autonomous agent that perceives the user's situation (shopping or in distress) to proactively assist the user (kiosk or ambulance), particularly if the user is too busy or otherwise occupied to go through the tedium of using the cell phone, such as when in distress. At the 2004 Dagstuhl workshop [30], the notion was extended to cognitive wireless networks (CWN).

This chapter summarizes important aspects of the architecture of the iCR that is more fully developed in the foundation text [22], particularly with respect to the critical machine learning technologies. The iCR vision includes isolated radio devices and CWNs with

[1] Adapted from *Cognitive Radio Architecture: The Engineering Foundations of Radio XML*, John Wiley & Sons, Inc, New York, 2006. With Permission, All Rights Reserved

[2] The author's affiliation with The MITRE Corporation is provided for identification purposes only and should not be interpreted as the endorsement of the material by The MITRE Corporation or any of its sponsors.

Cognitive Networks: Towards Self-Aware Networks Edited by Qusay H. Mahmoud

machine perception – vision, speech and other language skills – to ground the user continuously in a <Scene/>[3] that includes significant physical (space-time), social and radio dimensions. Thus grounded, the iCR's embedded intelligent agent can respond more accurately to the user's current situation, interpreting location awareness signals (e.g., GPS) more accurately and focusing radio resources on the user's specific needs. Modern radio resources include not just dynamic radio spectrum [37] and air interface channels, but also the four-dimensional space-time resource created by directional antennas, multi-input multi-output (MIMO) processing [7], and high band hot spot technologies. The result is a complex adaptive system [44] of cooperating radio devices and networks that manage the RF environment to the user's needs for quality of information (QoI) given the fine structure of the RF, physical, and social setting in which the user and device happen to be situated.

7.1.1 Ideal CRs Know Radio Like TellMe® Knows 800 Numbers

When you dial 1-800-555-1212, an algorithm may say 'Toll Free Directory Assistance powered by TellMe® [33]. Please say the name of the listing you want.' If you travel like I do, it may say 'OK, United Airlines. If that is not what you wanted press 9, otherwise wait while I look up the number.' Ninety-nine point nine percent of the time TellMe gets it right, replacing thousands of directory assistance operators of yore. TellMe, a speech-understanding system, achieves such a high degree of success in part by its focus on just one task: finding a toll-free telephone number. Narrow task focus is one of the keys to a successful speech recognition interface in this application.

The cognitive radio architecture (CRA) of this chapter frames the functions, components and design rules of CWNs, in many ways as the conceptual offspring of TellMe. CWNs are emerging in research settings as real-time, focused applications of radio with computational intelligence technologies. CWNs differ from the more general artificial intelligence (AI) based services like intelligent agents, computer speech and computer vision in degree of focus. Like TellMe, CWNs focus on very narrow tasks. Broader than TellMe, the task is to adapt radio-enabled information services to the specific needs of a specific user to achieve high QoI. TellMe, a network service, requires substantial network computing resources to serve thousands of users at once. CWNs, on the other hand, may start with a CR in your purse or on your belt, a cell phone on steroids, focused on the narrow task of creating from the myriad available wireless information networks and resources just what is needed by just one user, you. TellMe interacts with anybody, but each ideal cognitive radio (iCR) is self-aware and owner-aware via sensory perception and autonomous machine learning (AML) technologies, earning the term 'cognitive.' Each iCR fanatically serves the needs and protects the personal information of just one owner via the CRA using its audio and visual sensory perception and AML.

TellMe is here and now, while CWNs are emerging in global wireless research centers and industry forums like the Software-Defined Radio (SDR) Forum and Wireless World Research Forum (WWRF). This chapter summarizes CRA systems architecture challenges and approaches, emphasizing iCR as a technology enabler for rapidly emerging commercial CWN services and generation-after-next military communications, based on the foundation technologies of computer vision, computer speech, AML and SDR.

[3] Such closed XML tags highlight concepts with ontological roles in organizing cognitive radio architecture.

7.1.2 CRs See What You See, Discovering RF Uses, Needs and Preferences

In 2002, GRACE (Graduate Robot Attending Conference) [35], an autonomous mobile robot with a CRT for a face, entered the International Joint Conference on Artificial Intelligence (IJCAI). It completed the mobile robot challenge by finding the registration desk; registering by talking to the receptionist; following the signs that said 'ROBOTS' this way and 'HUMANS' the other way, when called on giving a five-minute talk about herself; and then answering questions. She was the first to complete this challenge first articulated in the 1980s. There were no joysticks and no man behind the curtain: just a robot that can autonomously see, hear and interact with the people and the environment to accomplish a specific task.

Compared to GRACE, the standard cell phone is not too bright. Although the common cell phone has a camera, it lacks GRACE's vision algorithms so it does not know what it is seeing. It can send a video clip, but it has no perception of the visual scene in the clip. If it had GRACE-like vision algorithms, it could perceive the visual scene. It could tell if it were at home, in the car, at work, shopping or driving up the driveway on the way home. If GRACE-like vision algorithms show that you are entering your driveway in your car, a cognitive SDR could learn to open the garage door for you wirelessly. Thus, you would not need to fish for the garage door opener, yet another wireless gadget. In fact, you would no longer need a garage door opener, as iCRs enter the market. To open the car door, you will not need a key fob either. As you approach your car, your personal CR perceives the common scene and, as trained, synthesizes the fob RF transmission and opens the car door for you.

Your CR may perceive visual scenes continuously via a pair of glasses enhanced with cell phone cameras and iris display to serve as a visual information appliance, continuously searching for visual–RF correlations, cues to your needs for information and thus for wireless services. An iCR learns to open your garage door when you arrive home from your use of the garage door opener. When first you open the garage door with the wireless garage door opener, your iCR correlates the visual and RF scenes: owner's hand on device, then RF signal in the ISM band, and then the garage door opens. The next time, your iCR verifies through reinforcement learning that your hand on the button, the RF signal and the opening of the garage door form a sequential script, a use-case. The third time, your cognitive radio detects the approach to the garage door and offers to complete the RF use-case for you, saying, 'I see we are approaching the garage. Would you like me to open the door for us?' Thereafter, it will open the garage door when you drive up the driveway unless you tell it not to. It has transformed one of your patterns of RF usage – opening the garage door – into a cognitive (self-user perceptive) service, offloading one of your daily tasks. Since the CR has learned to open the garage door, you may unclutter your car by just one widget, that door opener.

Since your CR learned to open the garage door by observing your use of the radio via AML, you did not pay the cell phone company, and you did not endure pop-up advertising to get this personalized wireless service. As you enter the house with arms full of packages, your CR closes the garage door and locks it for you, having learned that from you as well. For the iCR vision system to see what you see, today's Bluetooth earpieces evolve to iCR Bluetooth glasses, complete with GRACE-like vision as a far term embodiment of iCR.

CRs do not attempt everything. They learn about your radio use patterns because they know a lot about radio and about generic users and legitimate uses of radio. CRs have the *a priori* knowledge needed to detect opportunities to assist you with your use of the radio spectrum accurately, delivering that assistance with minimum intrusion. TellMe is not a generic speech understanding system and CR is not a generic AI service in a radio.

Products realizing the visual perception of this vignette are realizable on laptop computers today. Reinforcement learning (RL) and case-based reasoning (CBR) are mature AML technologies with radio network applications now being demonstrated in academic and industrial research settings as technology pathfinders for iCR [20] and CWN [11]. Two or three Moore's law cycles or three to five years from now, these vision and learning algorithms will fit in your cell phone. In the interim, CWNs will begin to offer such services via evolutionary market-driven pathways, offering consumers new trade-offs between privacy and ultra-personalized convenience.

7.1.3 Cognitive Radios Hear What You Hear, Augmenting Your Personal Skills

Compared to GRACE, the cell phone on your waist is deaf. Although your cell phone has a microphone, it lacks GRACE's speech understanding technology, so it does not perceive what it hears. It can let you talk to your daughter, but it has no perception of your daughter, nor of the content of your conversation. If it had GRACE's speech understanding technology, it could perceive your speech dialog. It could detect that you and your daughter are talking about common subjects like homework, or your favorite song. With iCR, GRACE-like speech algorithms would detect your daughter saying that your favorite song is now playing on WDUV. As an SDR, not just a cell phone, your iCR tunes to FM 105.5 so that you can hear 'The Rose.' With your iCR, you no longer need a transistor radio. Your iCR eliminates from your pocket, purse or backpack yet another RF gadget. In fact, you may not need iPOD®, GameBoy® and similar products as high-end CRs enter the market (such as an evolved iPhone®, of course). Your iCR will learn your radio listening and information use patterns, accessing the songs, downloading games, snipping broadcast news, sports, stock quotes as you like, as the iCR reprograms its internal SDR to better serve your needs and preferences.

Combining vision and speech perception, as you approach your car your iCR perceives this common scene and, as you had the morning before, tunes your car radio to WTOP to your favorite 'Traffic and weather together on the eights.' With GRACE's speech understanding algorithms, your iCR recognizes such regularly repeated catch phrases, turning up the volume for the traffic report and then turning it down or off after the weather report, avoiding annoying commercials and selecting relevant ones, having learned that from you rather than having been preprogrammed like a TIVO®. If you actually need a tax service, it could remember (record and acquire web pointers to) *those* radio commercials for your listening pleasure at tax time when you actually need them, accessed for you when you are actually thinking about that (at your convenience), instead of blasted in your direction when you are thinking about something else, synthesizing via AML your personal yellow pages.

For AML of this sophistication, iCRs need to save speech, RF and visual cues, all of which may be accessed by the user, an information prosthetic that expands the user's ability to remember details of conversations and snapshots of scenes, augmenting the skills of the <Owner/>.[4] Because of the brittleness of speech and vision technologies, iCRs would try to 'remember everything' like a continuously running camcorder. Since CRs detect content such as speakers' names, and keywords like 'radio' and 'song,' they can retrieve some content asked for by the user, expanding the user's memory in a sense. CRs thus could enhance the personal skills of their users such as memory for detail.

High-performance dialog and audio-video retrieval technologies are cutting edge but not out of reach for suitably narrow domains like TellMe and CR evolution of customized wireless services. Casual dialog typically contains anaphora and ellipsis, using words like 'this' and 'that' to refer to anonymous events like playing a favorite song. Although innovative, speech research systems already achieve similar dialogs in limited domains [43]. When the user says, 'How did you do that?' the domain of discourse may be the <Self/> and its contemporaneous actions. Since CR can do only one or two things at once, the question, 'How did you do that?' may have only one primary semantic referent, e.g. playing the song. Reasoning using analogy, also cutting edge, is no longer beyond the pale for tightly limited domains like CR and thus is envisioned in the CRA.

7.1.4 CRs Learn to Differentiate Speakers to Reduce Confusion

To further limit combinatorial explosion in speech, iCRs may form speaker models, statistical summaries of the speech patterns of speakers, particularly of the <Owner/>. Speaker modeling is particularly reliable when the <Owner/> uses the CR as a cell phone to place a phone call. Contemporary speaker recognition algorithms differentiate male from female speakers with high (>95%) probability. With a few different speakers to be recognized (e.g., fewer than 10 in a family) and with reliable side information like the speaker's telephone number, today's algorithms recognize individual speakers with 80 to 90% probability. Speaker models can become contaminated, such as erroneously including both <Owner/> and <Daughter/> speech in the <Owner/> model. Insightful product engineering could circumvent such problems, eventually rendering <Owner/> interactions as reliable as TellMe®.

Over time, each iCR learns the speech patterns of its <Owner/> in order to learn from the <Owner/> and not be confused by other speakers. iCR thus leverages experience incrementally to achieve increasingly sophisticated dialog. Directional microphones are rapidly improving to service video teleconference (VTC) markets. Embedding VTC-class microphones into iCR 'glasses' would enable iCR to differentiate user speech from backgrounds like radio and TV. Today, a 3 GHz laptop supports this level of speech understanding and dialog synthesis in real time, making it likely available in a cell phone in three to five years.

[4] Semantic web: researchers may formulate iCR as sufficiently speech-capable to answer questions about <Self/> and the <Self/> use of <Radio/> in support of its <Owner/>. When an ordinary concept like 'owner' has been translated into a comprehensive ontological structure of computational primitives, e.g., via semantic web technology [18], the concept becomes a computational primitive for autonomous reasoning and information exchange. Radio XML, an emerging CR derivative of the eXtensible Markup Language, XML, offers to standardize such radio-scene perception primitives. They are highlighted in this chapter by <Angle-brackets/>. All iCR know of a <Self/>, a <Name/>, and an <Owner/>. The <Self/> has capabilities like <GSM/> and <SDR/>, a self-referential computing architecture, which is guaranteed to crash unless its computing ability is limited to real-time response tasks [18]; this is appropriate for CR but may be inappropriate for general-purpose computing.

At present, few consumers train the speech recognition systems embedded in most laptop computers. It's too much work and many speech recognition algorithms of 2006 did not take dictation well enough for regular consumer use. Thus, although speech recognition technology is growing, many of today's products are not as effective at the general task of converting speech to text as TellMe is in finding an 800 number. The iCR value proposition overcomes this limit by embedding machine learning, so your iCR continually learns about you by analyzing your voice, speech patterns, visual scene and related use of the RF spectrum from garage door openers to weather, from cell phone and walkie-talkie to wireless home computer network. Do you want to know if your child's plane is in the air? Ask your iCR and it could find 'NiftyAir 122 Heavy cleared for takeoff by Dulles Tower.' Again, in order to customize services for you, the <Owner/>, the CR must both know a lot about radio and learn a lot about you, the <Owner/>, recording and analyzing personal information. The CRA therefore incorporates speech recognition. However, this mix places a premium on trustable privacy technologies.

7.1.5 More Flexible Secondary Use of Radio Spectrum

Consider a vignette with Lynne, the <Owner/> and Barb, the <Daughter/>. Barb drives to Lynne's house in her car. Coincidentally, Lynne asks Genie, the iCR <Self/> 'Can you call Barb for me?'

Genie: 'Sure. She is nearby so I can use the TV band for a free video call if you like.'

Lynne: 'Is that why your phone icon has a blue TV behind it?'

Genie: 'Yes. I can connect you to her using unused TV channel 43 instead of spending your cell phone minutes. The TV icon shows that you are using free airtime as a secondary user of TV spectrum. I sent a probe to her cognitive radio to be sure it could do this. Your NiftyWireless monthly subscription includes this capability for you free of charge.'

Lynne: 'OK, thanks for saving cell time for me. Let me talk to her. [Barb's face appears on the screen.]

Barb: 'Wow, where did you come from?' [Barb had never seen her cell phone display her Mom in a small TV picture in real time before, only in video clips.]

Lynne: 'Isn't this groovy. Genie, my new cognitive radio, hooked us up on a TV channel. It says you are nearby. Oh, I see you are out front and need help with the groceries. Here I come.'

Since a VTC can be realized via a sequence of JPEG images, such palmtop VTCs may not be far off. In 2004, the US FCC issued a Report and Order that radio spectrum allocated to TV, but unused in a particular broadcast market, could be used by CR as secondary spectrum users under Part 15 rules for low power devices, e.g., to create ad hoc networks. SDR Forum member companies have demonstrated CR products with these elementary spectrum-perception and use capabilities. Wireless products – military and commercial – realizing the FCC vignettes already exist. Complete visual and speech perception capabilities are not many years distant. Productization is underway. Thus, the CRA emphasizes CR spectrum agility, but in a context of enhanced sensory-perception technologies, an architecture with a long-term growth path.

7.1.6 SDR Technology Underlies Cognitive Radio

To conclude the overview, take a closer look at the enabling radio technology, SDR. Samuel F.B. Morse's code revolutionized telegraphy in the late 1830s, becoming the standard for 'telegraph' by the late 1800s. Thus when Marconi and Tesla brought forward wireless technology in 1902, Morse code was already a standard language for communications. Today, as then, a radio includes an antenna, an RF power amplifier to transmit, and RF conversion to receive; along with a modulator/demodulator (modem) to impart the code to and from the RF channel; and a coder-decoder (codec) to translate information from human-readable form to a form coded for efficient radio transmission. Today as then, RF conversion depends on capacitive and inductive circuits to set the radio frequency, but then some such circuits were the size of a refrigerator, while today they can be micro electromechanical systems (MEMS) on a chip. Then, the modulator consisted of the proverbial telegraph key, a switch to open and close the transmission circuit for on-off-keyed (OOK) data encoding. Morse code, a series of short (dots) or long marks (dashes) and spaces – sounds and silence – is still the simplest, cheapest way to communicate across a continent, and Morse code over HF radio is still used today in remote regions from the Australian outback to Africa and Siberia. Then and now, the 'coder' was the person who had memorized the Morse code, manually converting dot-dot-dash-dot from and to the letters of the alphabet. Radio engineers almost never abandon an RF band (HF) or mode (Morse code). Instead, the use morphs from mainstream to a niche market like sports, amateur radio, remote regions or developing economies. Today there are nice user interfaces and digital networking, but radio engineering has not taken anything away from legacy modes like HF manual Morse code. At the relatively low data rates of mobile radio (<1 Mbps), networking (routing and switching) is readily accomplished in software, unlike wired networks where data rates reach gigabits per second and dedicated hardware is needed for high-speed switching.

The essential functional blocks of radio have not changed for a century and are not likely to change either because the laws of physics and information define them: antenna, RF conversion, power amplification, modem, and codec. Today, however, microelectronics technologies enable one to pack low power RF, modem and codecs into single-chip packages while even relatively high gain antennas may fit neatly into the palm of your hand. Today, there are a myriad of modems evolved from the single RF of Morse to the sharing of RF bands in frequency, time and code-space. The manual codec has evolved to include communications security (COMSEC) coding, authentication and multilayered digital protocol stacks. Cognitive radio embraces all the broad classes of modulation, each with unique modems, codecs and most importantly content, the reason people use the radio, after all.

The SDR Forum, IEEE, and Object Management Group (OMG) have standardized software architecture for wireless plug and play of the myriad band-mode combinations: the software communications architecture (SCA) and software radio architecture (SRA), respectively. But, the next enabler for SDR may be the increasingly programmable analog RF of SDR: antennas, RF conversion and amplifiers. Historically, the analog RF had fixed frequency and bandwidth, optimized for a small RF band such as 88 to 108 MHz for FM broadcast, 850 to 950 MHz for cell phones and 1.7 to 1.8 GHz for personal communications systems (PCS), or a third-generation cellular band. Today's cellular radios typically include four chipsets, one for Bluetooth hands free headset, one optimized for first-generation 'roaming' where infrastructure is not well built out; one for second-generation digital service such as GSM; and one for PCS or NexTel®.

Each of these chipsets accesses only the narrow band needed for the service, so today's cell phones can't open the garage door, not without another chipset. In 1990–95, DARPA demonstrated SPEAKeasy II, the first SDR with continuous RF from 2 MHz to 2 GHz in just three analog RF bands: HF (2–30 MHz), mid-band (30–500 MHz) and high band (0.5–2 GHz).

MEMS technology is makes it increasingly practical to reprogram analog RF components digitally, so a cell phone could some day synthesize the garage door opener as the new digitally controlled analog RF MEMS technology emerges. RF MEMS digitally controls analog RF devices [29]. In some RF MEMS devices, a controller commands a micro-scale motor to move the interdigitated fingers of a capacitor, changing its analog value and hence changing the RF center frequency of the analog radio circuit. As the fingers move in and out by a few microns, the RF resonant frequency changes up and down by MHz. As this technology matures and enters service, RF chipsets will be reconfigurable across radio bands and modes, realizing affordable nearly ideal SDRs. FM and TV broadcasts inform large markets with news, sports, weather, music and the like. From boom box to weather radio, people around the world still depend on AM, FM and TV broadcasts for such information. In the past, you had to buy a specialized radio receiver and tune it manually to the station you like. With RF MEMS SDR, you tell the CR what you want to hear and it finds it for you. Your approval or disapproval constitutes training of the AML algorithms that tuned the MEMS SDR for your user-specific content preferences.

RF MEMS have been demonstrated to reduce the size, weight and power of analog RF subsystems by two to three orders of magnitude, and by over 1000:1 in some cases, but they have been slow to enter markets because of lower than necessary reliability, a focus of both academic and commercial RF MEMS research and development. To facilitate the insertion of RF MEMS and other enabling technologies, the CRA embraces hardware abstraction.

7.1.7 Privacy is Paramount

A CR that remembers all your conversations for several years needs only a few hundred gigabytes of data memory, readily achieved in wearable CR-PDA even today. Many such conversations will be private, and some will include credit card numbers, social security numbers, bank account information and the like. When my laptop was stolen with five years of tax returns, the process of dealing with identity theft was daunting and not foolproof. How can one trust a CR with all that personal information? Why would it need to remember all that stuff anyway?

One value proposition of CWNs is the reduction of tedium. Thus, asking the new owner to program the CR or to train it for an hour in the way that one is supposed to train the speech recognition system in a new laptop would be to increase tedium, not to decrease it. CR therefore aggregates experience, reprocessing the raw speech, vision and RF data during sleep cycles so that it learns from experience with minimum tedious training interactions with the user. Although based solidly on contemporary RL and CBR technology, task-focused introspective learning for nearly unsupervised dialog acquisition, e.g., via text mining tools [13] is on the cutting edge of autonomous product development while the more general problem of minimally supervised dialog acquisition in general is at the cutting edge of language research. Thus, CR products will always 'cheat' in the same way that TellMe cheats; CR products pick a small, workable set of tasks that consumers will pay for and use, mini-killer apps. The resulting revenue streams build technology for increasingly capable tasks, evolving towards the vision-RF-dialog skills of the previous vignettes. However, to learn this way, CR really must remember all the raw data – all your keystrokes, emails and conversations, to learn your use patterns and preferences autonomously, thus capturing private personal data.

If CR must remember your private personal data, then it must protect that information. Fingerprint readers are not perfect, as is any single information assurance (IA) measure, so CR may use a mix of IA measures. Candidate IA measures include soft biometrics like face and voice recognition along with more obtrusive measures like iris recognition. Layers of public key infrastructure (PKI), GSM-like randomized challenges and signed responses with network validation of identity, and battery backup of IA protection skills, e.g., that erase all user data when the CR detects that it is being physically compromised, e.g., by the unexplained removal of screws of its case. Privacy is paramount, and practical products must protect personal information, identity, medical information and the like with high reliability. Thus, a mix of soft biometrics like face and voice recognition coupled with selective hard biometrics like a fingerprint reader, PKI and other encryption methods. Given the limits of speech and visual perception technologies, CRs employ a large fraction of their sensory perception resources recognizing the face, voice and daily habits of the <Owner/>. Some robots accumulate stimuli in a way that simulate human emotion, e.g., happiness or distress. If the robot detects its <Owner's/> voice and face, then it knows what to expect based on having learned the owner's patterns. If the voice and face are not recognized, then the CR might become defensive, protecting the owner's data and potentially erasing it rather than divulge personal data to someone the <Owner/> has not previously authorized. Embedding a backup battery deep within the motherboard and embedding sensors in screws in the motherboard might dissuade all but the most sophisticated criminals from stealing such CRs. Therefore, CRA explicitly includes hardware and software facilities to implement trustable protection of privacy.

7.1.8 Military Applications Abound

Military applications of CR in CWNs abound. It is easy to imagine realistic vignettes where radios relay the commander's change to an operations order in his own words, 'Coalition partners are now located at grid square 76-11, so hold your fire. Rendezvous at Checkpoint Charlie at 1700.' There might be little doubt about the authenticity of an order if it can be recalled and distributed digitally, authenticated and suitably protected to military standards, of course. Tactical military radio communications are notoriously noisy. Thus, a radio that conveys such critical information error-free and in the voice of the commander could reduce the fog of war, potentially saving lives.

With autonomous machine learning skills, military iCRs would learn coalition RF use patterns. Autonomously reprogramming their SDR transceivers, coalition CRs could learn to connect commanders directly with each other, avoiding the need for dedicated military radio operators per se and either reducing the size of a squad from 10 to nine or enhancing the squad's capabilities by the 10% no longer needed to just operate the radio. Although one can never completely replace the flexibility and insight of skilled people, as iCRs offload mundane radio operation tasks from the radio operator, the team's effectiveness will increase, beneficial in the short run even if it takes decades to realize 'Radar O'Reilly' in software.

Although the Phraselator [1] experiment showed the promise of real-time language translation in a handheld device for coalition operations, a Phraselator is yet another widget like the garage door opener. Envisioned iCR offer a flexible hardware platform in which to embed Phraselator algorithms invoked by language identification algorithms that detect non-native language and hence the need for real-time translation. Since iCR is about enhancing the effectiveness of communications, language translation embedded in CR to translate when and where needed certainly has the potential to enhance communications among coalition partners who speak different languages, again reducing the fog of war and improving the likelihood of success.

The CRA is not specifically designed for military applications, but its open and evolutionary nature enable a wide range of commercial and military applications.

7.1.9 Quality of Information (QoI) Metric

QoI concerns the information that meets a specific user's need at a specific time, place, physical location and social setting. If information is available, then the quality, quantity, timeliness and suitability may be measured. One expression for QoI [22] is given in Equation (7.1):

$$\text{QoI} = \text{Quantity} * \text{Timeliness} * \text{Validity} * \text{Relevance} * \text{Accuracy} * \text{Detail}$$

If there is no information, then $Quantity = 0$ as does QoI. If all the required information is present, then Quantity = 1.0. Since different users require different information to be fully satisfied, this user-dependent parameter is at best difficult to measure.

Timeliness must be defined in terms of the iCR user's timeline along which the information would be used. If the information is needed immediately, then the quality may be characterized as inversely proportional to excessive time delay. To avoid division by zero, one may consider timeliness to be 1.0 if the information is available before a minimum delivery time:

$$\text{Tmin(time, place, social setting, topic)}$$

For simplicity, let's adopt the convention that a situation is a specifiable subspace of time, place and social setting. The concept of a social setting must be defined in terms that the user accepts, such as 'shopping' or 'getting mugged'. Suppose the shortest time delay in such a setting is ε so the maximum contribution of timeliness to QoI would be $1/\varepsilon$. If timeliness is normalized by ε, then maximum timeliness would be 1.0.

If *validity* is +1 if true and – 1.0 if false, with the possibility of fuzzy set membership, then the validity value is an element of [−1, 1], and QoI may be positive or negative. Information that is known to be or that winds up being false has a qualitatively different kind of value than information the validity of which is unknown (validity = 0). Information of unknown validity may safely be ignored, so the QoI value of zero seems appropriate. Information that turns out to be false may in fact be misleading, yielding negative results because the user behaved in accordance with the falsehood. This is the sense of negative QoI.

Relevance is the degree to which the information corresponds to the need, measured in terms of precision and recall. In information retrieval, recall is the fraction of relevant documents retrieved from a corpus by a query and precision is the fraction of documents actually retrieved that turn out to be relevant. Recall of 1.0 indicates that all relevant documents are retrieved, while precision of 1.0 indicates that no irrelevant documents have been retrieved. Adapting this well-known metric to QoI, one may define relevance as the product of precision and recall. This metric may not be ideal for information retrieval purposes, but it suffices in its role as a QoI metric that can be used to give an iCR feedback from its user by observing user behavior (e.g., asking for more or apparently not using items retrieved).

Accuracy refers to the quantitative aspects of the information. Quantitative errors include factual correctness (e.g., spell the president's name right) and numerical errors. Numerical accuracy reflects numerical error of the information represented with arbitrary precision, while QoI precision reflects the least error that it is possible to represent in a given numerical string. These could be differentiated, but for simplicity, one may measure whether the precision in which the number is expressed supports the required accuracy. If the accuracy required by the user is met, the value of the accuracy metric is 1.0. The rate of degradation of the accuracy metric from 1.0 may be linear, quadratic, exponential, fractal or defined by table look-up, provided it falls in the range (0,1].

Finally if sufficient detail is provided to justify the information delivered, then detail = 1.0, gradually dropping to zero if no elaborating detail is provided.

Consider the following example:

<Query> Name of the largest state in the USA</Query> <Quantity> name</Quantity>
<Timeliness> in the next few seconds</Timeliness><Validity> Must be true</Validity>
<Accuracy> Name must be spelled correctly</Accuracy>< Detail/> </Query>

Example 7.1 Illustrative information query

Since <Detail/> is null, the user is not asking for any special supporting information. In response to this query, the name 'Texas' was valid until 'Alaska' became a state. The user didn't specify a time frame so <Present/> may be assumed, but if the <User/> happens to be interested in history, such an assumption would not adequately reflect the user's QoI needs. In addition, Texas remains the largest state in the contiguous lower 48 states, so the geospatial scope might render Texas as accurate. A high QoI response from a CWN might provide both Alaska and Texas with the associated validity. If such a complete answer were provided quickly and were spelled right, then QoI = 1.0. If the query were met an hour later because the iCR couldn't reach the cell phone network or WLAN for that length of time, then the QoI is less than 1.0. The amount of degradation from 1.0 depends on the urgency of the need. If the user were playing Trivial Pursuit with a few friends, then the penalty for time delay might not be great. If the user were playing Who Wants to be a Millionaire? on TV and asked the iCR for help as a phone-in, then even a few minutes of delay could yield unacceptably low QoI.

Given a working definition of QoI, the iCR could automatically manipulate the parameters of the air interface (s) as a function of the user's specific needs for QoI.

7.1.10 Architecture

Architecture is a comprehensive, consistent set of *design rules* by which a specified set of *components* achieves a specified set of *functions* in products and services that evolve through multiple design points over time. This section introduces the fundamental design rules by which SDR, sensors, perception and AML may be integrated to create aware, adaptive and cognitive radios (iCRs) with better QoI through capabilities to Observe (sense, perceive), Orient, Plan, Decide, Act and Learn (the OOPDAL loop) in RF and user domains, transitioning from merely adaptive to demonstrably cognitive radio.

This section develops five complementary perspectives of architecture called CRA I through CRA V. CRA I defines six functional components, black boxes to which are ascribed a first-level decomposition of iCR functions and among which important interfaces are defined. One of these boxes is SDR, a proper subset of iCR. One of these boxes performs cognition via the <Self/>, a self-referential subsystem that strictly embodies finite computing (e.g., no while or until loops) avoiding the Gödel–Turing paradox.

CRA II examines the flow of inference through a cognition cycle that arranges the core capabilities of *ideal* CR (iCR) in temporal sequence for a logical flow and circadian rhythm for the CRA. CRA III examines the related levels of abstraction for iCR to sense elementary sensory stimuli and to perceive QoS-relevant aspects of a <Scene/> consisting of the <User/> in an <Environment/> that includes <RF/>. CRA IV of the foundation text [22] examines the mathematical structure of this architecture, identifying mappings among topological spaces represented and manipulated to preserve set-theoretic properties. Finally, CRA V of the foundation text [22] reviews SDR architecture, sketching an evolutionary path from the SCA/SRA to the CRA. The CRA <Self/> provided in CRA Self. xml of that text expresses in radio XML (RXML) the CRA introduced in this chapter along with *a priori* knowledge for autonomous machine learning.

7.2 CRA I: Functions, Components and Design Rules

The *functions* of iCR exceed those of SDR. Reformulating the iCR <Self/> as a *peer* of its own <User/> establishes the need for added functions by which the <Self/> accurately perceives the local scene including the <User/> and autonomously learns to tailor the information services to the specific <User/> in the current RF and physical <Scene/>.

7.2.1 ICR Functional Component Architecture

The SDR components and the related cognitive components of iCR appear in Figure 7.1. The cognition components describe the SDR in Radio XML so that the <Self/> can know that it is a radio and that its goal is to achieve high QoI tailored to its own users. RXML intelligence includes *a priori* radio background and user stereotypes as well as knowledge of RF and space-time <Scenes/> perceived and experienced. This includes both structured reasoning with iCR peers and CWNs, and ad hoc reasoning with users, all the while learning from experience.

The detailed allocation of functions to components with interfaces among the components requires closer consideration of the SDR component as the foundation of CRA.

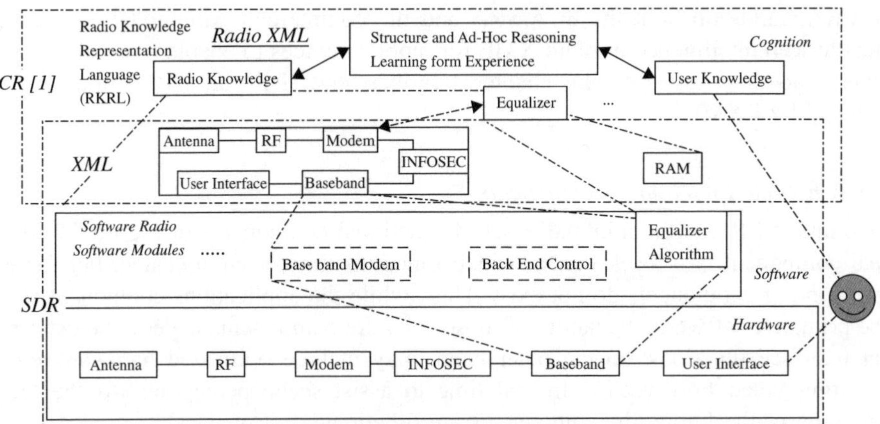

Figure 7.1 The cognitive radio architecture augments software defined radio with computational intelligence and learning capacity. © Dr. Joseph Mitola III, used with permission

7.2.2 SDR Components

SDRs include a hardware platform with RF access and computational resources, plus at least one software-defined personality. The SDR Forum has defined its software communications architecture (SCA) and the Object Management Group (OMG) has defined its software radio architecture (SRA), similar fine-grain architecture constructs enabling reduced cost wireless connectivity with next-generation plug and play. These SDR architectures are defined in unified modeling language (UML) object models [4], CORBA interface design language (IDL) [24], and XML descriptions of the UML models. The SDR Forum's SCA [41] and OMG SRA [39] standards describe the technical details of SDR both for radio engineering and for an initial level of wireless air interface ('waveform') plug and play. The SCA/SRA was sketched in 1996 at the first DoD-inspired MMITS Forum, developed by the US DoD in the 1990s and the architecture is now in production for the US military [38]. This architecture emphasizes plug-and-play wireless personalities on computationally capable mobile nodes where network connectivity is often intermittent at best.

The commercial wireless community [36], on the other hand, led by cell phone giants Motorola, Ericsson and Nokia envisions a much simpler architecture for mobile wireless devices, consisting of two APIs, one for the service provider and another for the network operator. They define a knowledge plane in the future intelligent wireless networks that is not dissimilar from a distributed CWN. That forum promotes the business model of the user → service provider → network operator → large manufacturer → device, where the user buys mobile devices consistent with services from a service provider, and the technical emphasis is on *intelligence in the network*. This perspective no doubt will yield computationally intelligent networks in the near- to mid-term.

The CRA developed in this text, however, envisions the computational intelligence to create ad hoc and flexible networks with the *intelligence in the mobile device*. This technical perspective enables the business model of user → device → heterogeneous networks, typical of the Internet model where the user buys a device (e.g., a wireless laptop) that can connect to the Internet via any available Internet service provider (ISP).

The CRA builds on both the SCA/SRA and the commercial API model but integrates semantic web intelligence in radio XML for mobile devices to enable more of an Internet business model to advance. This chapter describes how SDR and iCR form a continuum facilitated by RXML.

7.2.3 iCR Node Functional Components

The simplest CRA is the minimalist set of functional components of Figure 7.2. A functional component is a black box to which functions have been allocated, but for which implementing components do not exist. Thus, while the applications component is likely to be primarily software, the nature of those software components is yet to be determined. User interface functions, on the other hand, may include optimized hardware, e.g., for computing video flow vectors in real time to assist scene perception. At the level of abstraction of the figure, the components are functional, not physical.

These functional components are:

1. The *user sensory perception* (User SP) interface includes haptic, acoustic, and video sensing and perception functions.
2. The local *environment* sensors (location, temperature, accelerometer, compass, etc.).
3. The *system applications* (media independent services like playing a network game).
4. The *SDR* functions (which include RF sensing and SDR radio applications).
5. The *cognition* functions (symbol grounding for system control, planning, learning).
6. The local effector functions (speech synthesis, text, graphics and multimedia displays).

These functional components are embodied on an iCR platform, a hardware realization of the six functions. In order to support the capabilities described in the previous sections, these components go beyond SDR in critical ways. First, the user interface goes well beyond buttons and displays. The traditional user interface has been partitioned into a substantial user sensory subsystem and a set of local effectors. The user sensory interface includes buttons (the haptic interface) and microphones (the audio interface) to include acoustic sensing that is directional, capable of handling multiple speakers simultaneously and to include full motion video with visual scene perception. In addition, the audio subsystem does not just encode audio for (possible) transmission; it also parses and interprets the audio from designated speakers such as the <User/> for a high performance spoken natural language interface. Similarly, the text subsystem parses and interprets the language

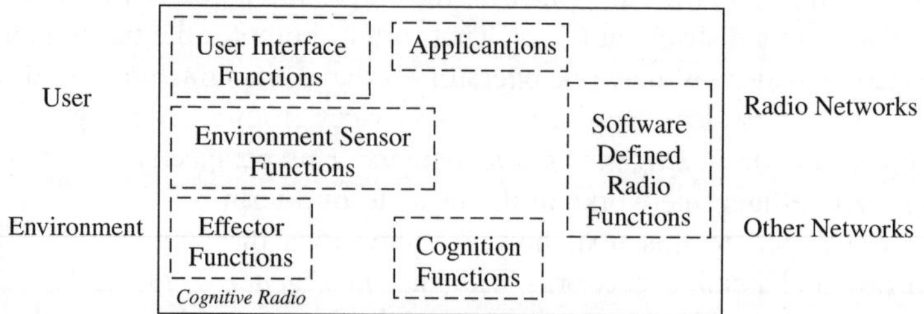

Figure 7.2 Minimal adaptive, aware, cognitive radio (iCR) node architecture. © Dr. Joseph Mitola III, used with permission

to track the user's information states, detecting plans and potential communications and information needs unobtrusively as the user conducts normal activities. The local effectors synthesize speech along with traditional text, graphics and multimedia displays.

Systems applications are those *information services* that define value for the user. Historically, voice communications with a phone book, text messaging and the exchange of images or video clips comprised the core value proposition of systems applications for SDR. These applications were generally integral to the SDR application, such as data services via the GSM Packet Radio Service (GPRS), which is really a wireless SDR personality more than an information service. iCR systems applications break the service out of the SDR waveform so that the user need not deal with details of wireless connectivity unless that is of particular interest. Should the user care whether he plays the distributed video game via 802.11 or Bluetooth over the last three meters? Probably not. The typical user might care if the iCR wants to switch to 3G at $5 per minute, but a particularly affluent user might not care and would leave all that up to the iCR.

The cognition component provides all the cognition functions from the semantic grounding of entities from the perception system to controlling the overall system through planning and initiating actions, learning user preferences and RF situations in the process.

Each of these subsystems contains its own processing, local memory, integral power conversion, built-in test (BIT) and related technical features.

7.2.4 The Ontological <Self/>

iCR may consist of the six functional components: user SP, environment, effectors, SDR, system applications and cognition. Example 7.2 describes those components of the <Self/>, enables external communications and internal reasoning about the <Self/>, using the RXML syntax.

```
<Self/>
<iCR-Platform/>
<Functional-Components/>
    <User SP/><Environment/><Effectors/><SDR/><Sys Apps/><Cognition/>
</Functional-Components>
</Self>
```

Example 7.2 The iCR<Self/>is defined to be an iCR platform, consisting of six functional components using the RXML syntax

Given the top-level outline of these functional components along with the requirement that they be embodied in physical hardware and software (the platform), the six functional components are defined ontologically in Example 7.2. In part, this example states that the hardware–software platform and the functional components of the iCR are independent. Platform-independent computer languages like Java are well understood. This ontological perspective envisions platform independence as an architecture design principle for iCR. In other words, the burden is on the (software) functional components to adapt to whatever RF-hardware-OS platform might be available.

7.2.5 Design Rules Include Functional Component Interfaces

The functional components of Table 7.1 imply associated functional interfaces. In architecture, design rules may include a list of the quantities and types of components as well

Table 7.1 iCR N-Squared diagram characterizes internal interfaces between functional processes. © Dr. Joseph Mitola III, used with permission

From\To	User SP	Environment	Sys Apps	SDR	Cognition	Effectors
User SP	1	7	13 PA	19	25 PA	31
Environment	2	8	14 SA	20	26 PA	32
Sys Apps	3	9	15 SCM	21 SD	27 PDC	33 PEM
SDR	4	10	16 PD	22 SD	28 PC	34 SD
Cognition	5 PEC	11 PEC	17 PC	23 PAE	29 SC	35 PE
Effectors	6 SC	12	18	24	30 PCD	36

P – Primary; A – Afferent; E – Efferent; C – Control; M – Multimedia; D – Data; S – Secondary; Others not designated P or S are ancillary.
The information services API (the relevant SysApps interfaces) consists of interfaces 13–18, 21, 27 and 33. This set of interfaces establishes the way in which conventional applications are hosted in the CRA. The cognition API (the complete set of Cognition interfaces) consists of interfaces 25–30, 5, 11, 23 and 35. The cognitive applications layer above this API realizes user adaptation, situation discovery and learning, resource planning, and other high level perception-experience-response facilities that require something like the iCR's OOPDAL architecture. The interfaces between the information services and cognition APIs define the ways in which cognition controls applications (such as suspending a game when cognition detects an emergency) and applications request services from cognition (such as requesting apparent user level of interest and skill level with other interactive games when initializing a new game).

as the interfaces among those components. This section addresses the interfaces among the six functional components.

The iCR N-Squared diagram [8] characterizes iCR interfaces. These constitute an initial set of iCR APIs. In some ways these APIs augment the established SDR APIs. For example, the cognition API brings a planning capability to SDR. This is entirely new and much needed in order for basic adaptive and aware radios to migrate towards dynamic spectrum.

In other ways, these APIs supersede the existing SDR APIs. In particular, the SDR user interface is partitioned into the user sensory-perception (SP) and effector APIs. User SP includes acoustics, voice and video, while the effectors include speech synthesis to give the iCR <Self/> its own voice to the degree that a particular user prefers such interaction (e.g. in case of emergencies). In addition, wireless applications are expanding rapidly. Voice and short message service are being augmented with an ability to acquire images and video clips locally with exchange of content-specific ontological tags among wireless and network users. The distinctions among cell phone, PDA and game box continue to disappear.

These new capabilities shape SP interface changes that enable the iCR to sense the situation represented in the environment with increasing autonomy, to interact with the user as a function of the perceived situation, and to access radio networks on behalf of the user in a situation-tailored way.

Interface notes follow the numbers of the table:

1. **User SP–User SP:** Cross-media correlation interfaces (video–acoustic, haptic–speech, etc.) to limit search and reduce uncertainty (e.g., if video indicates user

is not talking, acoustics may be ignored or processed less aggressively for command inputs than if user is speaking).

2. **Environment–User SP:** Environment sensors parameterize user sensor perception. Temperature below freezing may limit video.
3. **Sys Apps–User SP:** Systems applications may focus scene perception by identifying entities, range, expected sounds for video, audio and spatial perception processing.
4. **SDR–User SP:** SDR applications may provide expectations of user input to the perception system to improve probability of detection and correct classification of perceived inputs.
5. **Cognition–User SP:** This is the *primary control efferent* path from cognition to the control of the user sensory perception subsystem, controlling speech recognition, acoustic signal processing, video processing and related sensory perception. Plans from cognition may set expectations for user scene perception, improving perception.
6. **Effectors–User SP:** Effectors may supply a replica of the effect to user perception so that self-generated effects (e.g., synthesized speech) may be accurately attributed to the <Self/>, validated as having been expressed, and/or cancelled from the scene perception to limit search.
7. **User SP–Environment:** Perception of rain, buildings, indoor/outdoor can set GPS integration parameters.
8. **Environment–Environment:** Environment sensors would consist of location sensing such as GPS or Glonass; temperature of the ambient; light level to detect inside versus outside locations; possibly smell sensors to detect spoiled food, fire, etc. There seems to be little benefit to enabling interfaces among these elements directly.
9. **Sys Apps–Environment:** Data from the systems applications to environment sensors would also be minimal.
10. **SDR–Environment:** Data from the SDR personalities to the environment sensors would be minimal.
11. **Cognition–Environment:** (primary control path) Data from the cognition system to the environment sensors controls those sensors, turning them on and off, setting control parameters, and establishing internal paths from the environment sensors.
12. **Effectors–Environment:** Data from effectors directly to environment sensors would be minimal.
13. **User SP–Sys Apps:** Data from the user sensory perception system to systems applications is a *primary afferent path* for multimedia streams and entity states that effect information services implemented as systems applications. Speech, images and video to be transmitted move along this path for delivery by the relevant systems application or information service to the relevant wired or SDR communications path. Sys Apps overcomes the limitations of individual paths by maintaining continuity of conversations, data integrity and application coherence, e.g., for multimedia games. While the cognition function sets up, tears down and orchestrates the systems applications, the primary API between the user scene and the information service consists of this interface and its companions, the environment afferent path; the effector efferent path; and the SDR afferent and efferent paths.
14. **Environment–Sys Apps:** Data on this path assists systems applications in providing location-awareness to services.

15. **Sys Apps–Sys Apps:** Different information services interoperate by passing control information through the cognition interfaces and by passing domain multimedia flows through this interface. The cognition system sets up and tears down these interfaces.
16. **SDR–Sys Apps:** This is the primary afferent path from external communications to the iCR. It includes control and multimedia information flows for all the information services. Following the SDR Forum's SCA, this path embraces wired as well as wireless interfaces.
17. **Cognition–Sys Apps:** Through this path the iCR <Self/> exerts control over the information services provided to the <User/>.
18. **Effectors–Sys Apps:** Effectors may provide incidental feedback to information services through this afferent path, but the use of this path is deprecated. Information services are supposed to control and obtain feedback through the mediation of the cognition subsystem.
19. **User SP–SDR:** Although the sensory perception system may send data directly to the SDR subsystem, e.g., in order to satisfy security rules that user biometrics must be provided directly to the wireless security subsystem, the use of this path is deprecated. Perception subsystem information is supposed to be interpreted by the cognition system so that accurate information can be conveyed to other subsystems, not raw data.
20. **Environment–SDR:** Environment sensors like GPS historically have accessed SDR waveforms directly, such as providing timing data for air-interface signal generation. The cognition system may establish such paths in cases where cognition provides little or no value added, such as providing a precise timing reference from GPS to an SDR waveform. The use of this path is deprecated because all of the environment sensors including GPS are unreliable. Cognition has the capability to de-glitch GPS, e.g., recognizing from video that the <Self/> is in an urban canyon and therefore not allowing GPS to report directly, but reporting on behalf of GPS to the GPS subscribers location estimates based perhaps on landmark correlation, dead reckoning, etc.
21. **Sys Apps–SDR:** This is the primary efferent path from information services to SDR through the services API.
22. **SDR–SDR:** The linking of different wireless services directly to each other is deprecated. If an incoming voice service needs to be connected to an outgoing voice service, then there should be a bridging service in Sys Apps through which the SDR waveforms communicate with each other. That service should be set up and taken down by the cognition system.
23. **Cognition–SDR:** This is the primary control interface, replacing the control interface of the SDR SCA and the OMG SRA.
24. **Effectors–SDR:** Effectors such as speech synthesis and displays should not need to provide state information directly to SDR waveforms, but if needed, the cognition function should set up and tear down these interfaces.
25. **User SP–Cognition:** This is the primary afferent flow for the results from acoustics, speech, images, video, video flow and other sensory-perception subsystems. The primary results passed across this interface should be the specific states of <Entities/> in the scene, which would include scene characteristics such as the recognition of landmarks, known vehicles, furniture and the like. In other words, this is the interface

by which the presence of <Entities/> in the local scene is established and their characteristics are made known to the cognition system.

26. **Environment–Cognition:** This is the primary afferent flow for environment sensors.
27. **Sys Apps–Cognition:** This is the interface through which information services request services and receive support from the iCR platform. This is also the control interface by which cognition sets up, monitors and tears down information services.
28. **SDR–Cognition:** This is the primary afferent interface by which the state of waveforms, including a distinguished RF-sensor waveform is made known to the cognition system. The cognition system can establish primary and backup waveforms for information services enabling the services to select paths in real time for low latency services. Those paths are set up, monitored for quality and validity (e.g., obeying XG rules) by the cognition system, however.
29. **Cognition–Cognition:** The cognition system as defined in this six-component architecture entails: (i) orienting to information from <RF/> sensors in the SDR subsystem and from scene sensors in the user sensory perception and environment sensors; (ii) planning; (iii) making decisions; and (iv) initiating actions, including the control over all of the resources of the <Self/>. The <User/> may directly control any of the elements of the systems via paths through the cognition system that enable it to monitor what the user is doing in order to learn from a user's direct actions, such as manually tuning in the user's favorite radio station when the <Self/> either failed to do so properly or was not asked.
30. **Effectors–Cognition:** This is the primary afferent flow for status information from the effector subsystem, including speech synthesis, displays and the like.
31. **User SP–Effectors:** In general, the user sensory-perception system should not interface directly to the effectors, but should be routed through the cognition system for observation.
32. **Environment–Effectors:** The environment system should not interface directly to the effectors. This path is deprecated.
33. **Sys Apps–Effectors:** Systems applications may display streams, generate speech, and otherwise directly control any effectors once the paths and constraints have been established by the cognition subsystem.
34. **SDR–Effectors:** This path may be used if the cognition system establishes a path, such as from an SDR's voice track to a speaker. Generally, however, the SDR should provide streams to the information services of the Sys Apps. This path may be necessary for legacy compatibility during the migration from SDR through iCR to iCR but is deprecated.
35. **Cognition–Effectors:** This is the primary efferent path for the control of effectors. Information services provide the streams to the effectors, but cognition sets them up, establishes paths and monitors the information flows for support to the user's <Need/> or intent.
36. **Effectors–Effectors:** These paths are deprecated, but may be needed for legacy compatibility.

The above information flows aggregated into an initial set of iCR APIs define an information services API by which an information service accesses the other five components

(ISAPI consisting of interfaces 13–18, 21, 27 and 33). They would also define a cognition API by which the cognition system obtains status and exerts control over the rest of the system (CAPI consisting of interfaces 25–30, 5, 11, 23 and 35). Although the constituent interfaces of these APIs are suggested in the table, it would be premature to define these APIs without first developing detailed information flows and interdependencies, which are defined and analyzed in the remainder of this chapter. It would also be premature to develop such APIs without a clear idea of the kinds of RF and user domain knowledge and performance that are expected of the iCR architecture over time. These aspects are developed in the balance of the text, enabling one to draw some conclusions about these APIs in the final sections.

A fully defined set of interfaces and APIs would be circumscribed in RXML. For the moment, any of the interfaces of the N-squared diagram may be used as needed.

7.2.6 Near Term Implementations

One way to implement this set of functions is to embed into an SDR a reasoning engine such as a rule base with an associated inference engine as the cognition function. If the effector functions control parts of the radio, then we have the simplest iCR based on the simple six-component architecture of Figure 7.2. Such an approach may be sufficient to expand the control paradigm from today's state machines with limited flexibility to tomorrow's iCR control based on reasoning over more complex RF states and user situations. Such simple approaches may well be the next practical steps in iCR evolution from SDR towards iCR.

This incremental step doesn't suggest how to mediate the interfaces between multisensory perception and situation-sensitive prior experience and *a priori* knowledge to achieve situation-dependent radio control that enables the more sophisticated information services of the use-cases. In addition, such a simple architecture does not proactively allocate machine learning functions to fully understood components. For example, will autonomous machine learning require an embedded radio propagation modeling tool? If so, then what is the division of function between a rule base that knows about radio propagation and a propagation tool that can predict values like RSSI? Similarly, in the user domain, some aspects of user behavior may be modeled in detail based on physics, such as movement by foot and in vehicles. Will movement modeling be a separate subsystem based on physics and GPS? How will that work inside buildings? How is the knowledge and skill in tracking user movements divided between physics-based computational modeling and the symbolic inference of a rule base or set of Horn clauses [15] with a Prolog engine? For that matter, how will the learning architecture accommodate a variety of learning methods like neural networks, PROLOG, forward chaining, SVM if learning occurs entirely in a cognition subsystem?

While hiding such details may be a good thing for iCR in the near term, it may severely limit the mass customization needed for iCRs to learn user patterns and thus to deliver RF services dramatically better than mere SDRs. Thus, we need to go 'inside' the cognition and perception subsystems further to establish more of a fine-grained architecture. This enables one to structure the data sets and functions that mediate multisensory domain perception of complex scenes and related learning technologies that can autonomously adapt to user needs and preferences. The sequel thus proactively addresses the embedding of machine learning (ML) technology into the radio architecture.

Next, consider the networks. Network-independent SDRs retain multiple personalities in local storage, while network-dependent SDRs receive alternate personalities from a supporting network infrastructure – CWNs. High-end SDRs both retain alternate personalities locally and have the ability to validate and accept personalities by download from trusted sources. Whatever architecture emerges must be consistent with the distribution of RXML knowledge aggregated in a variety of networks from a tightly coupled CWN to the Internet, with a degree of <Authority/> and trust reflecting the pragmatics of such different repositories.

The first two sections of this chapter therefore set the stage for the development of CRA. The next three sections address the cognition cycle, the inference hierarchies and the SDR architecture embedded both into the CRA with the knowledge structures of the CRA.

7.2.7 The Cognition Components

Figure 7.1 shows three computational-intelligence aspects of CR:

1. Radio knowledge – RXML:RF
2. User knowledge – RXML:User
3. The capacity to learn

The minimalist architecture of Figure 7.2, and the functional interfaces of Table 7.1 do not assist the radio engineer in structuring knowledge, nor does it assist much in integrating machine learning into the system. The finegrained architecture developed in this chapter, on the other hand, is derived from the functional requirements to fully develop these three core capabilities.

7.2.8 Radio Knowledge in the Architecture

Radio knowledge has to be translated from the classroom and engineering teams into a body of computationally accessible, structured technical knowledge about radio. Radio XML is the primary enabler and product of this foray into formalization of radio knowledge. This text starts a process of RXML definition and development that can only be brought to fruition by industry over time. This process is similar to the evolution of the SCA of the SDR Forum [41]. The SCA structures the technical knowledge of the radio components into UML and XML. RXML will enable the structuring of sufficient RF and user world knowledge to build advanced wireless-enabled or enhanced information services. Thus while the SRA and SCA focus on building radios, RXML focuses on using radios.

The World Wide Web is now sprouting with computational ontologies some of which are non-technical but include radio, such as the open CYC ontology [8]. They bring the radio domain into the semantic web, which helps people know about radio. This informal knowledge lacks the technical scope, precision and accuracy of authoritative radio references such as the ETSI documents defining GSM and ITU definitions, e.g., of 3GPP.

Not only must radio knowledge be precise, it must be stated at a useful level of abstraction, yet with the level of detail appropriate to the use-case. Thus, ETSI GSM in most cases would overkill the level of detail without providing sufficient knowledge of the user-centric functionality of GSM. In addition, iCR is multiband, multimode radio (MBMMR), so the knowledge must be comprehensive, addressing the majority of radio bands and modes available to an MBMMR. Therefore, in the development of CR technology below, this text captures radio knowledge needed for competent CR in the MBMMR bands from HF through millimeter wave. This knowledge is formalized with precision that should be acceptable to ETSI, the ITU and regulatory authorities (RAs) yet at a level of abstraction appropriate to internal reasoning, formal dialog with a CWN or informal dialog with users.

This kind of knowledge is to be captured in RXML:RF.

The capabilities required for an iCR node to be a cognitive entity are to sense, perceive, orient, plan, decide, act and learn. To relate ITU standards to these required capabilities is a process of extracting content from highly formalized knowledge bases that exist in a unique place and that bear substantial authority, encapsulating that knowledge in less complete and therefore somewhat approximate form that can be reasoned with on the iCR node and in real time to support RF-related use-cases. Table 7.2 illustrates this process.

The table is illustrative and not comprehensive, but it characterizes the technical issues that drive an information-oriented iCR node architecture. Where ITU, ETSI, . . . (meaning other regional and local standards bodies) and CWN supply source knowledge, the CWN is the repository for authoritative knowledge derived from the standards bodies and RAs,

Table 7.2 Radio knowledge in the node architecture

Need	Source knowledge	iCR internalization
Sense RF	RF platform	Calibration of RF, noise floor, antennas, direction
Perceive RF	ITU, ETSI, ARIB, RA's	Location-based table of radio spectrum allocation
Observe RF (Sense & Perceive)	Unknown RF	RF sensor measurements and knowledge of basic types (AM, FM, simple digital channel symbols, typical TDMA, FDMA, CDMA signal structures)
Orient	XG-like policy	Receive, parse and interpret policy language
	Known waveform	Measure parameters in RF, space and time
Plan	Known waveform	Enable SDR for which licensing is current
	Restrictive Policy	Optimize transmitted waveform, space-time plan
Decide	Legacy waveform, policy	Defer spectrum use to legacy users per policy
Act	Applications layer	Query for available services (White/Yellow Pages)
	ITU, ETSI, . . . CWN	Obtain new skills encapsulated as download
Learn	Unknown RF	Remember space-time-RF signatures; discover spectrum use norms and exceptions
	ITU, ETSI, . . . CWN	Extract relevant aspects such as new feature

the <Authorities/>. A user-oriented iCR may note differences in the interpretation of source knowledge from <Authorities/> between alternate CWNs, precipitating further knowledge exchanges.

7.2.9 User Knowledge in the Architecture

Next, user knowledge is formalized at the level of abstraction and degree of detail necessary to give the CR the ability to acquire from its owner and other designated users, the user knowledge relevant to information services incrementally. Incremental knowledge acquisition was motivated in the introduction to AML by describing how frequent occurrences with similar activity sequences identifies learning opportunities. AML machines may recognize these opportunities for learning through joint probability statistics <Histogram/>. Effective use-cases clearly identify the classes of user and the specific knowledge learned to customize envisioned services. Use-cases may also supply sufficient initial knowledge to render incremental AML not only effective, but also – if possible – enjoyable to the user.

This knowledge is defined in RXML:User. As with RF knowledge, the capabilities required for an iCR node to be a cognitive entity are to observe (sense, perceive), orient, plan, decide, act and learn. To relate a use-case to these capabilities, one extracts specific and easily recognizable <Anchors/> for stereotypical situations observable in diverse times, places and situations. One expresses the anchor knowledge in using RXML for use on the iCR node.

7.2.10 Cross-Domain Grounding for Flexible Information Services

The knowledge about radio and about user needs for wireless services must be expressed internally in a consistent form so that information services relationships may be autonomously discovered and maintained by the <Self/> on behalf of the <User/>. Relationships among user and RF domains are shown in Figure 7.3.

Staying better connected requires the normalization of knowledge between <User/> and <RF/> domains. If, for example, the <User/> says ‘What’s on one oh seven – seven,’ near the Washington, DC area, then the dynamic <User/> ontology should enable the CR

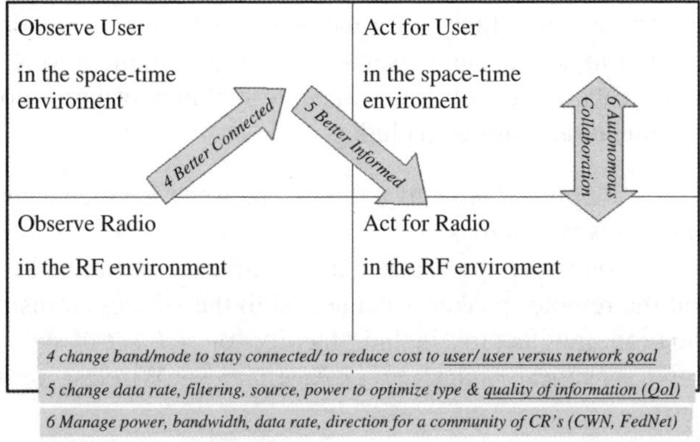

Figure 7.3 Discovering and maintaining services. © Dr. Joseph Mitola III, used with permission

to infer that the user is talking about the current FM radio broadcast, the units are in MHz and the user wants to know what is on WTOP. If it can't infer this, then it should ask the user or discover by first dialing a reasonable default, such as 107.7 FM, a broadcast radio station and asking 'Is this the radio station you want?' Steps 4, 5 and 6 in Figure 7.3 all benefit from agreement across domains on how to refer to radio services. Optimizing behavior to best support the user requires continually adapting the <User/> ontology with repeated regrounding of terms in the <User/> domain to conceptual primitives and actions in the <RF/> domain.

The CRA facilitates this by seeding the speech recognition subsystem with the expressions a particular <User/> employs when referring to information services. These would be acquired a priori by class of use, and from current users via text and speech recognition, with dialogs oriented towards continual grounding e.g. by posing simple questions, either verbally or in displays or both, obtaining reinforcement verbally or via haptic interaction or both. The required degree of mutual grounding would benefit from specific grounding-oriented features in the iCR information architecture, developed below.

The process of linking user expressions of interest to the appropriate radio technical operations sometimes may be extremely difficult. Military radios, for example, have many technical parameters. For example, a 'channel' in SINCGARS consists of de-hopped digital voice in one context (voice communications) or a 25 kHz band of spectrum in another context (and that may be either an FM channel into which its frequency hop waveform has hopped or an FDMA channel when in single channel mode). If the user says 'Give me the Commander's Channel' the SINCGARS user is talking about a 'de-hopped CVSD voice stream'. If the same user a few seconds later says 'This sounds awful. Who else is in this channel?' the user is referring to interference with a collection of hop sets. If the CR observes 'There is strong interference in almost half of your assigned channels,' then the CR is referring to a related set of 25 kHz channels. If the user then says 'OK, notch the strongest three interference channels' he is talking about a different subset of the channels. If in the next breath the user says 'Is anything on our emergency channel?' then the user has switched from SINCGARS context to <Self/> context, asking about one of the cognitive military radio's physical RF access channels. The complexity of such exchanges demands cross-domain grounding; and the necessity of communicating accurately under stress motivates a structured natural language (NL) and rich radio ontology aspects of the architecture developed further below.

Thus, both commercial and military information services entail cross-domain grounding with ontology oriented to NL in the <User> domain and oriented to RXML formalized *a priori* knowledge in the <RF> domain. Specific methods of cross-domain grounding with associated architectural features include:

1. *<RF> to <User> shaping dialog* to express precise <RF> concepts to non-expert users in an intuitive way, such as
   - grounding: 'If you move the speaker box a little bit, it can make a big difference in how well the remote speaker is connected to the wireless transmitter on the TV'
   - iCR information architecture: include facility for *rich set of synonyms* to mediate cognition–NL–synthesis interface (<Antenna>≅<Wireless remote speaker>≅ 'Speaker box')
2. *<RF> to <User> Learning jargon* to express <RF> connectivity opportunities in <User> terms.

- Grounding: 'tee oh pee' for 'WTOP', 'Hot ninety two' for FM 97.7, 'Guppy' for 'E2C Echo Grand on 422.1 MHz'
- iCR information architecture: *NL-visual facility for single-instance update of user jargon*

3. *<User> to <RF> Relating values to actions:* Relate <User> expression of values ('low cost') to features of situations ('normal') that are computable (<NOT> (<CONTAINS> <Situation> <Unusual/>)) and that relate directly to <RF> domain decisions
   - Grounding: normally wait for free WLAN for big attachment; if situation is <unusual>, ask if user wants to pay for 3G
   - iCR information architecture: *associative inference hierarchy* that relates observable features of a <Scene> to user sensitivities, such as <Late-for-work>=> <Unusual>; 'The President of the company needs this' =><Unusual> because 'President'=><VIP> and <VIP> is not in most scenes.

7.2.12 Self-Referential Components

The cognition component must assess, manage and control all of its own resources, including validating downloads. Thus, in addition to <RF> and <User> domains, RXML must describe the <Self/>, defining the iCR architecture to the iCR itself in RXML.

7.2.13 Self-Referential Inconsistency

This class of self-referential reasoning is well known in the theory of computing to be a potential black hole for computational resources. Specifically, any Turing–Capable (TC) computational entity that reasons about itself can encounter unexpected Gödel–Turing situations from which it cannot recover. Thus TC systems are known to be 'partial' – only partially defined because the result obtained when attempting to execute certain classes of procedure are not definable because the computing procedure will never terminate.

To avoid this paradox, CR architecture mandates the use of only 'total' functions, typically restricted to bounded-minimalization [17]. Watchdog 'step-counting' functions [6] or timers must be in place in all its self-referential reasoning and radio functions. The timer and related computationally indivisible control construct is equivalent to the computer-theoretic construct of a step-counting function over 'finite minimalization.' It has been proven that computations that are limited with certain classes of reliable watchdog timers on finite computing resources can avoid the Gödel–Turing paradox or at least reduce it to the reliability of the timer. This proof is the fundamental theorem for practical self-modifying systems.

Briefly: If a system can compute in advance the amount of time or the number of instructions that any given computation should take, then if that time or step-count is exceeded, the procedure returns a fixed result such as 'Unreachable in Time T.' As long as the algorithm does not explicitly or implicitly restart itself on the same problem, then with the associated invocation of a tightly time- and computationally-constrained alternative tantamount to giving up, it

(a) is not Turing capable, but
(b) is sufficiently computationally capable to perform real-time communications tasks such as transmitting and receiving data as well as bounded user interface functions, and

(c) is not susceptible to the Turing–Gödel incompleteness dilemma and thus
(d) will not crash because of consuming unbounded or unpredictable resources in unpredictable self-referential loops.

This is not a general result. This is a highly radio domain-specific result that has been established only for isochronous communications domains in which:

(a) Processes are defined in terms of *a priori* tightly bounded time epochs such as CDMA frames and SS7 time-outs; and
(b) For every situation, there is a default action that has been identified in advance that consumes O(1) resources; and
(c) The watchdog timer or step-counting function is reliable.

Since radio air interfaces transmit and receive data, there are always defaults such as 'repeat the last packed' or 'clear the buffer' that may degrade the performance of the overall communications system. A default has O (1) complexity and the layers of the protocol stack can implement the default without using unbounded computing resources.

7.2.13 Watchdog Timer

Without the reliable watchdog timer in the architecture and without this proof to establish the rules for acceptable computing constructs on cognitive radios, engineers and computer programmers would build CRs that would crash in extremely unpredictable ways as their adaptation algorithms get trapped in unpredictable unbounded self-referential loops. Since there are planning problems that can't be solved with algorithms so constrained, either an unbounded community of CRs must cooperatively work on the more general problems or the CN must employ a TC algorithm to solve the harder problems (e.g., NP-hard with large N) off line. There is also the interesting possibility of trading off space and time by remembering partial solutions and restarting NP-hard problems with these subproblems already solved. While it doesn't actually avoid any necessary calculations, with O(N) pattern matching for solved subproblems, it may reduce the total computational burden, somewhat like the FFT which converts $O(N^2)$ steps to O(N log N) by avoiding the recomputation of already computed partial products. This class of approach to parallel problem solving is similar to the use of pheromones by ants to solve the traveling salesman problem in less than (2^N)/M time with M ants. Since this is an engineering text, not a text on the theory of computing, these aspects are not developed further here, but it suffices to show the predictable finiteness and proof that the approach is boundable and hence compatible with the real-time performance needs of cognitive radio.

This timer-based finite computing regime also works for user interfaces since users will not wait forever before changing the situation, e.g., by shutting the radio off or hitting another key; and the CR can always kind of throw up its hands and ask the user to take over.

Thus, with a proof of stability based on the theory of computing, the CRA structures systems that not only can modify themselves, but can do it in such a way that they are not likely to induce non-recoverable crashes from the 'partial' property of self-referential computing.

7.2.14 Flexible Functions of the Component Architecture

Although this chapter develops the six-element component architecture of a particular information architecture and one reference implementation, there are many possible architectures. The purpose is not to try to sell a particular architecture, but to illustrate the architecture principles. The CRA and research implementation, CR1 [22], therefore, offer open-source licensing for non-commercial educational purposes.

Table 7.3 further differentiates architecture features.

These functions of the architecture are not different from those of the six-component architecture, but represent varying degrees of instantiation of the six components. Consider the following degrees of architecture instantiations:

- ***Cognition functions of radio*** entail the monitoring and structuring knowledge of the behavior patterns of the <Self/>, the <User>, and the environment (physical, user situation and radio) in order to provide information services, learning from experience to tailor services to user preferences and differing radio environments.
- ***Adaptation functions of radio*** respond to a changing environment, but can be achieved without learning if the adaptation is preprogrammed.
- ***Awareness functions of radio*** extract usable information from a sensor domain. Awareness stops short of perception. Awareness is required for adaptation, but awareness does not guarantee adaptation. For example, embedding a GPS receiver into a cell phone makes the phone more location-aware, but unless the value of the current location is actually used by the phone to do something that is location-dependent, the phone is not location-adaptive, only location aware. These functions are a subset of the CRA that enable adaptation.
- ***Perception functions of radio*** continuously identify and track knowns, unknowns and backgrounds in a given sensor domain. Backgrounds are subsets of a sensory domain that share common features that entail no particular relevance to the functions of the radio. For a CR that learns initially to be a single-owner radio, in a crowd, the owner is

Table 7.3 Features of iCR to be organized via architecture. © 2005 Dr. Joseph Mitola III, used with permission

Feature	Function	Examples (RF, vision, speech, location, motion)
Cognition	Monitor and learn	Get to know user's daily patterns and model the local RF scene over space, time and situations
Adaptation	Respond to changing environment	Use unused RF, protect owner's data
Awareness	Extract information from sensor domain	Sense or perceive
Perception	Continuously identify knowns, unknowns and backgrounds in the sensor domain	TV channel; depth of visual scene, identity of objects; location of user, movement and speed of <Self/>
Sensing	Continuously sense and preprocess single sensor field in single sensory domain	RF FFT; binary vision; binaural acoustics; GPS; accelerometer

the object that the radio continuously tracks in order to interact when needed. Worn from a belt as a cognitive wireless personal digital assistant (CWPDA), the iCR perception functions may track the entities in the scene. The non-owner entities comprise mostly irrelevant background because no matter what interactions may be offered by these entities, the CR will not obey them, just the perceived owner. These functions are a subset of the CRA that enable cognition.

- ***The sensory functions of radio*** entail those hardware and/or software capabilities that enable a radio to measure features of a sensory domain. Sensory domains include anything that can be sensed, such as audio, video, vibration, temperature, time, power, fuel level, ambient light level, sun angle (e.g., through polarization), barometric pressure, smell and anything else you might imagine. Sensory domains for vehicular radios may be much richer if less personal than those of wearable radios. Sensory domains for fixed infrastructure could include weather features such as ultraviolet sunlight, wind direction and speed, humidity, traffic flow rate, or rain rate. These functions are a subset of the CRA that enable perception.

The platform independent model (PIM) in the UML of SDR [40] provides a convenient, industry-standard computational model that an iCR can use to describe the SDR and computational resource aspects of its own internal structure, as well as describing facilities that enable radio functions. The general structure of hardware and software by which a CR reasons about the <Self/> in its world is also part of its architecture defined in the SDR SCA/SRA as resources.

7.3 CRA II: The Cognition Cycle

The CRA consists of a set of design rules by which the cognitive level of information services may be achieved by a specified set of components in a way that supports the cost-effective evolution of increasingly capable implementations over time [22]. The cognition subsystem of the architecture includes an inference hierarchy and the temporal organization and flow of inferences and control states, the cognition cycle.

7.3.1 The Cognition Cycle

The cognition cycle developed for CR1 [22] is illustrated in Figure 7.4. This cycle implements the capabilities required of iCR in a reactive sequence. Stimuli enter the cognitive radio as sensory interrupts, dispatched to the cognition cycle for a response. Such an iCR continually observes (senses and perceives) the environment, orients itself, creates plans, decides, and then acts. In a single-processor inference system, the CR's flow of control may also move in the cycle from observation to action. In a multiprocessor system, temporal structures of sensing, preprocessing, reasoning and acting may be parallel and complex. Special features synchronize the inferences of each phase. The tutorial code of [22] all works on a single processor in a rigid inference sequence defined in the figure. This process is called the wake epoch because the primary reasoning activities during this large epoch of time are reactive to the environment. We will refer to 'sleep epochs' for power down condition, 'dream epochs' for performing computationally intensive pattern recognition and learning, and 'prayer epochs' for interacting with a higher authority such as network infrastructure.

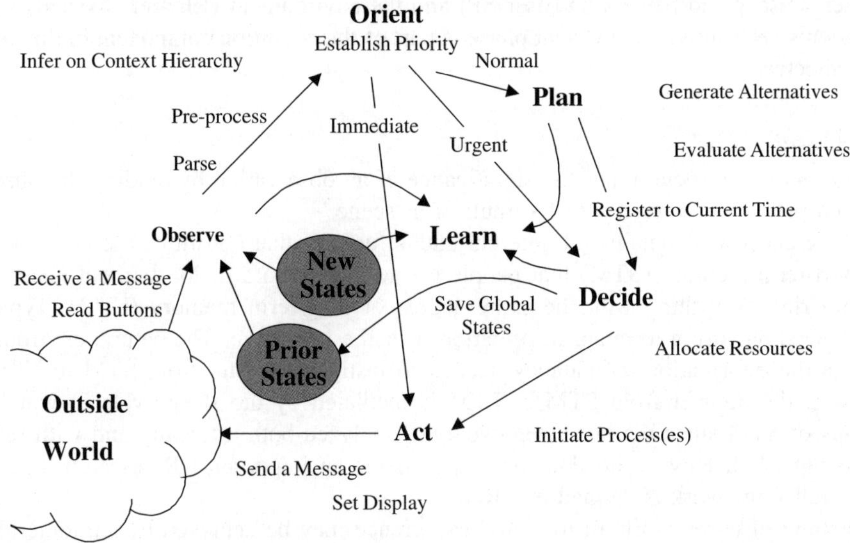

Figure 7.4 Simplified cognition cycle. The Observe-Orient-Decide-Act (OODA) loop is a primary cycle, however, learning, planning and sensing the outside world are crucial phases necessary to be properly prepared for the OODA loop. © Dr. Joseph Mitola III, used with permission

During the wake epoch, the receipt of a new stimulus on any of a CR's sensors or the completion of a prior cognition cycle initiates a new primary cognition cycle. The cognitive radio observes its environment by parsing incoming information streams. These can include monitoring and speech-to-text conversion of radio broadcasts, e.g., the weather channel, stock ticker tapes, etc. Any RF-LAN or other short-range wireless broadcasts that provide environment awareness information may be also parsed. In the observation phase, a CR also reads location, temperature and light level sensors, etc. to infer the user's communications context.

7.3.2 Observe (Sense and Perceive)

The iCR senses and perceives the environment (via 'Observation Phase' code) by accepting multiple stimuli in many dimensions simultaneously and by binding these stimuli – all together or more typically in subsets – to prior experience so that it can subsequently detect time-sensitive stimuli and ultimately generate plans for action.

Thus, iCR continuously aggregates experience and compares prior aggregates to the current situation. A CR may aggregate experience by remembering everything. This may not seem like a very smart thing to do until you calculate that all the audio, unique images and emails the radio might experience in a year only takes up a few hundred gigabytes of memory, depending on image detail. So the computational architecture for remembering and rapidly correlating current experience against everything known previously is a core capability of the CRA. A *novelty* detector identifies new stimuli, using the new aspects of partially familiar stimuli to identify incremental learning primitives.

In the six-component (user SP, environment, effectors, SDR, sys apps and cognition) functional view of the architecture defined above, the Observe phase comprises both the user sensory and perception (user SP) and the environment (RF and physical) sensor subsystems. The subsequent Orient phase is part of the cognition component in this model of architecture.

7.3.3 Orient

The 'Orient phase' determines the significance of an observation by binding the observation to a previously known set of stimuli of a 'scene.'

The Orient phase contains the internal data structures that constitute the equivalent of the short-term memory (STM) that people use to engage in a dialog without necessarily remembering everything with the same degree of long-term memory (LTM). Typically people need repetition to retain information over the long term. The natural environment supplies the information redundancy needed to instigate transfer from STM to LTM. In the CRA, the transfer from STM to LTM is mediated by the sleep cycle in which the contents of STM since the last sleep cycle are analyzed both internally and with respect to existing LTM. How to do this robustly remains an important CR research topic, but the overall framework is defined in CRA.

Matching of current stimuli to stored experience may be achieved by stimulus recognition or by 'binding'. The Orient phase is the first collection of activity in the cognition component.

7.3.3.1 Stimulus Recognition

Stimulus recognition occurs when there is an exact match between a current stimulus and a prior experience. CR1 is continually recognizing exact matches and recording the number of exact matches that occurred along with the time in number of cognition cycles between the last exact match. By default, the response to a given stimulus is to merely repeat that stimulus to the next layer up the inference hierarchy for aggregation of the raw stimuli. But if the system has been trained to respond to a location, a word, an RF condition, a signal on the power bus, etc., then it may either react immediately or plan a task in reaction to the detected stimulus. If that reaction were in error, then it may be trained to ignore the stimulus given the larger context which consists of all the stimuli and relevant internal states, including time.

Sometimes, the Orient phase causes an action to be initiated immediately as a 'reactive' stimulus–response behavior. A power failure, for example, might directly invoke an act that saves the data (the 'Immediate' path to the Act phase in the figure). A non-recoverable loss of signal on a network might invoke reallocation of resources, e.g., from parsing input to searching for alternative RF channels. This may be accomplished via the path labeled 'Urgent' in the figure.

7.3.3.2 Binding

The binding occurs when there is a nearly exact match between a current stimulus and a prior experience and very general criteria for applying the prior experience to the current situation are met. One such criterion is the number of unmatched features of the current scene. If only one feature is unmatched and the scene occurs at a high level, such as the phrase or dialog level of the inference hierarchy, then binding is the first step in generating a plan for behaving similarly in the given state as in the last occurrence of the stimuli. In addition to numbers of features that match exactly, which is a kind of Hamming code, instance-based learning (IBL) supports inexact matching and binding. Binding also determines the priority associated with the stimuli. Better binding yields higher priority for autonomous learning, while less effective binding yields lower priority for the incipient plan.

7.3.4 Plan

Most stimuli are dealt with 'deliberatively' rather than 'reactively.' An incoming network message would normally be dealt with by generating a plan (in the Plan phase, the 'Normal' path). Planning includes plan generation. In research-quality or industrial-strength CRs, formal models of causality must be embedded into planning tools [27]. The Plan phase should also include reasoning about time. Typically, reactive responses are preprogrammed or defined by a network (the CR is 'told' what to do), while other behaviors might be planned. A stimulus may be associated with a simple plan as a function of planning parameters with a simple planning system. Open source planning tools enable the embedding of planning subsystems into the CRA, enhancing the Plan component. Such tools enable the synthesis of RF and information access behaviors in a goal-oriented way based on perceptions from the visual, audio, text and RF domains as well as RA rules and previously learned user preferences.

7.3.5 Decide

The 'Decide' phase' selects among the candidate plans. The radio might have the choice to alert the user to an incoming message (e.g., behaving like a pager) or to defer the interruption until later (e.g., behaving like a secretary who is screening calls during an important meeting).

7.3.6 Act

'Acting' initiates the selected processes using effector modules. Effectors may access the external world or the CR's internal states.

7.3.6.1 Externally Oriented Actions

Access to the external world consists primarily of composing messages to be spoken into the local environment or expressed in text form locally or to another CR or CN using KQML, RKRL, OWL, RXML, or some other appropriate knowledge interchange standard.

7.3.6.2 Internally Oriented Actions

Actions on internal states include controlling machine-controllable resources such as radio channels. The CR can also affect the contents of existing internal models, e.g., by adding a model of stimulus-experience-response (serModel) to an existing internal model structure. The new concept itself may assert related concepts into the scene. Multiple independent sources of the same concept in a scene reinforce that concept for that scene. These models may be asserted by the <Self/> to encapsulate experience. The experience may be reactively integrated into RXML knowledge structures as well, provided the reactive response encodes them properly.

7.3.7 Learning

Learning is a function of perception, observations, decisions and actions. Initial learning is mediated by the Observe-phase perception hierarchy in which all sensory perceptions are continuously matched against all prior stimuli to continually count occurrences and to remember time since last occurrence of the stimuli from primitives to aggregates.

Learning also occurs through the introduction of new internal models in response to existing models and CBR bindings. In general, there are many opportunities to integrate ML into iCR. Each of the phases of the cognition cycle offers multiple opportunities for discovery processes like <Histogram> above, as well as many other ML approaches to be developed below. Since the architecture includes internal reinforcement via counting occurrences and via serModels, ML with uncertainty is also supported in the architecture.

Finally, there is a learning mechanism that occurs when a new type of serModel is created in response to an Action to instantiate an internally generated serModel. For example, prior and current internal states may be compared with expectations to learn about the effectiveness of a communications mode, instantiating a new mode-specific serModel.

7.3.8 Self Monitoring Timing

Each of the prior phases must consist of computational structures for which the execution time may be computed in advance. In addition, each phase must restrict its computations to consume not more resources (time × allocated processing capacity) than the precomputed upper bound. Therefore, the architecture has some prohibitions and some data set requirements needed to obtain an acceptable degree of stability of behavior for CR as self-referential self-modifying systems.

Since first order predicate calculus (FOPC) used in some reasoning systems is not decidable, one cannot in general compute in advance how much time an FOPC expression will take to run to completion. There may be loops that will preclude this, and even with loop detection, the time to resolve an expression may be only loosely approximated as an exponential function of some parameters (such as the number of statements in the FOPC data base of assertions and rules). Therefore unrestricted FOPC is not allowed.

Similarly, unrestricted For, Until and While loops are prohibited. In place of such loops are bounded iterations in which the time required for the loop to execute is computed or supplied independent of the computations that determine the iteration control of the loop. This seemingly unnatural act can be facilitated by next-generation compilers and CASE tools. Since self-referential self-modifying code is prohibited by structured design and programming practices, there are no such tools on the market today. But since CR is inherently self-referential and self-modifying, such tools most likely will emerge, perhaps assisted by the needs of CR and the architecture framework of the cognition cycle.

Finally, the cognition cycle itself can't contain internal loops. Each iteration of the cycle must take a defined amount of time, just as each frame of a 3G air interface takes 10 milliseconds. As CR computational platforms continue to progress, the amount of computational work done within the cycle will increase, but under no conditions should explicit or implicit loops be introduced into the cognition cycle that would extend it beyond a given cycle time.

7.3.9 Retrospection

Since the assimilation of knowledge by machine learning can be computationally intensive, cognitive radio has 'sleep' and 'prayer' epochs that support machine learning. A sleep epoch is a relatively long period of time (e.g., minutes to hours) during which the radio will not be in use, but has sufficient electrical power for processing. During the sleep epoch, the radio can run machine learning algorithms without detracting from its ability to support its user's needs. Machine learning algorithms may integrate experience by aggregating statistical parameters. The sleep epoch may re-run stimulus–response sequences over the experience base with new learning parameters, notionally analogous to the way that people dream. The sleep cycle could be less anthropomorphic, employing a genetic algorithm to explore a rugged fitness landscape, potentially improving the decision parameters from recent experience.

7.3.11 Reaching Out

Learning opportunities not resolved in the sleep epoch can be brought to the attention of the user, the host network, or a designer during a prayer epoch, e.g. to acquire skills downloads from a trusted network. The sleep and prayer epochs are possible modalities for the autonomous enhancement of local knowledge, expertise, and embedded skills.

7.4 CRA III: The Inference Hierarchy

The phases of inference from observation to action show the flow of inference, a top-down view of how cognition is implemented algorithmically. The inference hierarchy is the part of the algorithm architecture that organizes the data structures. Inference hierarchies have been in use since Hearsay II in the 1970s, but the CR hierarchy is unique in its method of integrating machine learning with real-time performance during the wake epochs. An illustrative inference hierarchy includes layers from atomic stimuli at the bottom to information clusters that define action contexts as in Figure 7.5.

Sequence	*Level of Abstraction*
Context Cluster	***Scenes*** in a play, Session
Sequence Clusters	***Dialogs***, Paragraphs, Protocol
Basic Sequences	***Phrases***, video clip, message
Primitive Sequences	***Words***, token, image
Atomic Symbols	***Raw Data,*** Phoneme, pixel
Atomic Stimuli	External Phenomena

Figure 7.5 Standard inference hierarchy. © Dr. Joseph Mitola III, used with permission

The pattern of accumulating elements into sequences begins at the bottom of the hierarchy. Atomic stimuli originate in the external environment including RF, acoustic, image and location domains among others. The atomic symbols extracted from them are the most primitive symbolic units in the domain. In speech, the most primitive elements are the phonemes. In the exchange of textual data (e.g., in email), the symbols are the typed characters. In images, the atomic symbols may be the individual picture elements (pixels) or they may be small groups of pixels with similar hue, intensity, texture, etc.

A related set of atomic symbols forms a primitive sequence. Words in text, tokens from a speech tokenizer and objects in images (or individual image regions in a video flow) are the primitive sequences. Primitive sequences have spatial and/or temporal coincidence, standing out against the background (or noise), but there may be no particular meaning in that pattern of coincidence. Basic sequences, on the other hand, are space–time–spectrum sequences that entail the communication of discrete messages.

These discrete messages (e.g., phrases) are typically defined with respect to an ontology of the primitive sequences (e.g., definitions of words). Sequences cluster together because of shared properties. For examples, phrases that include words like 'hit,' 'pitch,' 'ball' and 'out' may be associated with a discussion of a baseball game. Knowledge discovery and data mining (KDD) and the semantic web offer approaches for defining or inferring the presence of such clusters from primitive and basic sequences.

A scene is a context cluster, a multidimensional space–time–frequency association, such as a discussion of a baseball game in the living room on a Sunday afternoon. Such clusters may be inferred from unsupervised machine learning, e.g., using statistical methods or nonlinear approaches such as support vector machines (SVM).

Although presented above in a bottom-up fashion, there is no reason to limit multidimensional inference to the top layers of the inference hierarchy. The lower levels of the inference hierarchy may include correlated multisensor data. For example, a word may be characterized as a primitive acoustic sequence coupled to a primitive sequence of images of a person speaking that word. In fact, since infants seem to thrive on multisensory stimulation, the key to reliable machine learning may be the use of multiple sensors with multisensor correlation at the lowest levels of abstraction.

Each of these levels of the inference hierarchy is now discussed further.

7.4.1 Atomic Stimuli

Atomic stimuli originate in the external environment and are sensed and preprocessed by the sensory subsystems which include sensors of the RF environment (e.g., radio receiver and related data and information processing) and of the local physical environment including acoustic, video and location sensors. Atomic symbols are the elementary stimuli extracted from the atomic stimuli. Atomic symbols may result from a simple noise-riding threshold algorithm, such as the squelch circuit in RF that differentiates signal from noise. Acoustic signals may be differentiated from simple background noise this way, but generally the result is the detection of a relatively large speech epoch which contains various kinds of speech energy. Thus, further signal processing is typically required in a preprocessing subsystem to isolate atomic symbols.

The transformation from atomic stimuli to atomic symbols is the job of the sensory preprocessing system. Thus, for example, acoustic signals may be transformed into phoneme hypotheses by an acoustic signal preprocessor. However, some speech-to-text software tools may not enable this level of interface via an API. To develop industrial strength CR, contemporary speech-to-text and video processing software tools are needed. Speech-to-text tools yield an errorful transcript in response to a set of atomic stimuli. Thus, the speech-to-text tool is an example of a mapping from atomic stimuli to basic sequences. One of the important contributions of architecture is to identify such maps and to define the role of the level mapping tools.

Image processing software available for the Wintel-Java development environment JBuilder has the ability to extract objects from images and video clips. In addition, research such as that of Goodman *et al.* defines algorithms for what the AAAI calls cognitive vision [2].

But there is nothing about the inference hierarchy that forces data from a preprocessing system to be entered at the lowest level. In order for the more primitive symbolic abstractions such as atomic symbols to be related to more aggregate abstractions, one may either build up the aggregates from the primitive abstractions or derive the primitive abstractions from the aggregates. Since people are exposed to 'the whole thing' by immersion in the full experience of life – touch, sight, sound, taste and balance – all at once, it seems possible – even likely – that the more primitive abstractions are somehow derived through the analysis of aggregates, perhaps by cross-correlation. This can be accomplished in a CRA sleep cycle. The idea is that the wake cycle is optimized for immediate reaction to stimuli, such as our ancestors needed to avoid predation, while the sleep cycle is optimized for introspection, for analyzing the day's stimuli to derive those objects that should be recognized and acted upon in the next cycle.

Stimuli are each counted. When an iCR that conforms to this architecture encounters a stimulus, it both counts how many such stimuli have been encountered and resets a timer to zero that keeps track of the time since the last occurrence of the stimulus.

7.4.2 Primitive Sequences: Words and Dead Time

The accumulation of sequences of atomic symbols forms primitive sequences. The key question at this level of the data structure hierarchy is the sequence boundary. The simplest situation is one in which a distinguished atomic symbol separates primitive sequences, which is exactly the case with white space between words in typed text. In general, one would like a machine learning system to determine on its own that the white space (and a few special symbols: , . ; : etc.) separates the keyboard input stream into primitive sequences.

7.4.3 Basic Sequences

The pattern of aggregation is repeated vertically at the levels corresponding to words, phrases, dialogs and scenes. The data structures generated by PDA nodes create the concept hierarchy of Figure 7.5. These are the reinforced hierarchical sequences. They are reinforced by the inherent counting of the number of times each atomic or aggregated stimulus occurs. The phrase level typically contains or implies a verb (the verb 'to be' is implied if no other verb is implicit).

Unless digested (e.g., by a sleep process), the observation phase hierarchy accumulates all the sensor data, parsed and distributed among PDA nodes for fast parallel retrieval. Since the hierarchy saves everything and compares new data to memories, it is a kind of memory-base learning technique. This is a memory-intensive approach, taking a lot of space. When the stimuli retained are limited to atomic symbols and their aggregates, the total amount of data that needs to be stored is relatively modest. In addition, recent research shows the negative effects of discarding cases in word pronunciation. In word pronunciation, no example can be discarded even if 'disruptive' to a well-developed model. Each exception has to be followed. Thus in CR1, when multiple memories match partially, the most nearly exact match informs the orientation, planning and action.

Basic sequences are each counted. When an iCR that conforms to this architecture encounters a basic sequence, it both counts how many such sequences have been encountered and resets a timer to zero that keeps track of the time since the last occurrence.

7.4.4 Natural Language in the CRA Inference Hierarchy

In speech, words spoken in a phrase may be co-articulated with no distinct boundary between the primitive sequences in a basic sequence. Therefore, speech detection algorithms may reliably extract a basic sequence while the parsing of that sequence into its constituent primitive sequences may be much less reliable. Typically, the correct parse is within the top 10 candidates for contemporary speech-to-text software tools. But the flow of speech signal processing may be something like:

1. Isolate a basic sequence (phrase) from background noise using an acoustic squelch algorithm.
2. Analyze the basic sequence to identify candidate primitive sequence boundaries (words).
3. Analyze the primitive sequences for atomic symbols.
4. Evaluate primitive and basic sequence hypotheses based on a statistical model of language to rank-order alternative interpretations of the basic sequence.

So a practical speech processing algorithm may yield alternative strings of phonemes and candidate parses 'all at once.' NLP tool sets may be embedded into the CRA inference hierarchy as illustrated in Figure 7.6. Speech and/or text channels may be processed via natural language facilities with substantial *a priori* models of language and discourse. The use of those models should entail the use of mappings among the word, phrase, dialog and scene levels of the observation phase hierarchy and the encapsulated component(s).

It is tempting to expect cognitive radio to integrate a commercial natural language processing system such as IBM's ViaVoice or a derivative of an NLP research system such as SNePS [31], AGFL [10], or XTAG [42] perhaps using a morphological analyzer like PCKimmo [26]. These tools both go too far and not far enough in the direction needed for CRA. One might like to employ existing tools using a workable interface between the domain of radio engineering and some of the above natural language tool sets. The definition of such cross-discipline interfaces is in its infancy. At present, one cannot just express a radio ontology in interlingua and plug it neatly into XTAG to get a working cognitive radio. The internal data structures that are used in radio mediate the performance of radio tasks (e.g., 'transmit a waveform'). The data structures of XTAG, AGFL, etc. mediate the conversion of language from one form to another. Thus, XTAG wants to know that 'transmit' is a verb and 'waveform' is a noun. The CR needs to know that if the user says 'transmit' and a message has been defined, then the CR should call the SDR function *transmit()*. NLP systems also need scoping rules for transformations

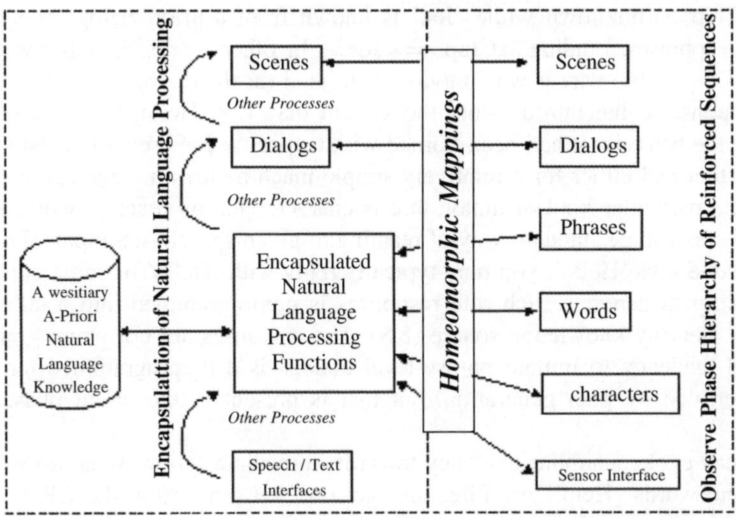

Figure 7.6 Natural language encapsulation in the observation hierarchy. © Dr. Joseph Mitola III, used with permission

on the linguistic data structures. The way in which domain knowledge is integrated in linguistic structures of these tools tends to obscure the radio engineering aspects.

Natural language processing systems work well on well-structured speech and text, such as the prepared text of a news anchor. But they do not work well yet on noisy, nongrammatical data structures encountered when a user is trying to order a cab in a crowded bar. Thus, less-linguistic or meta-linguistic data structures may be needed to integrate core cognitive radio reasoning with speech- and/or text-processing front ends. The CRA has the flexibility illustrated in Figure 7.6 for the subsequent integration of evolved NLP tools. The emphasis of this version of the CRA is a structure of sets and maps required to create a viable cognitive radio architecture. Although introducing the issues required to integrate existing natural language processing tools, the text does not pretend to present a complete solution to this problem.

7.4.5 Observe-Orient Links for Scene Interpretation

CR may use an algorithm-generating language with which one may define self-similar inference processes. In one example, the first process (Proc1) partitions characters into words, detecting novel characters and phrase boundaries as well. Proc2 detects novel words and aggregates known words into phrases. Proc3 detects novel phrases, aggregating known phrases into dialogs. Proc4 aggregates dialogs into scenes, and Proc5 detects known scenes. In each case, a novel entity at level N will be bound in the context of the surrounding known entities at that level to the closest match at the next highest level, N + 1. For example at the word–phrase intersection of Proc2, would map the following phrases:

'Let me introduce Joe'
'Let me introduce <u>Chip</u>'

Since 'Chip' is unknown while 'Joe' is known from a prior dialog, integrated CBR matches the phrases, binding <Chip>=<Joe>. In other words, it will try to act with respect to Chip in the way it was previously trained (at the dialog level) to interact with Joe. In response to the introduction, the system may say 'Hello, Chip, How are you?' mimicking the behavior it had been trained with respect to Joe previously. Not too bright, but not all that bad either for a relatively simple machine learning algorithm.

There is a particular kind of dialog that is characterized by reactive world knowledge in which there is some standard way of reacting to given speech–act inputs. For example, when someone says 'Hello', you may typically reply with 'Hello' or some other greeting. The capability to generate such rote responses is preprogrammed into a lateral component of the Hearsay knowledge source (KS). The responses are not preprogrammed, but the general tendency to imitate phrase-level dialogs is a preprogrammed tendency that can be overruled by plan generation, but that is present in the orient phase, which is Proc6.

Words may evoke a similar tendency towards immediate action. What do you do when you hear the words 'Help!' or 'Fire, fire, get out, get out!' You, the CR programmer, can capture reactive tendencies in your CR by preprogramming an ability to detect these kinds of situations in the word-sense knowledge source, as implied by Figure 7.7. When confronted with them (which is preferred), CR should react appropriately if properly trained, which is one of the key aspects of this text. To cheat, you can preprogram a wider array of stimulus–response pairs so that your CR has more *a priori* knowledge, but some of it may not be appropriate. Some responses are culturally conditioned. Will your CR be too rigid? If it has too much *a priori* knowledge, it will be perceived by its users as too rigid. If it doesn't have enough, it will be perceived as too stupid.

7.4.6 Observe–Orient Links for Radio Skill Sets

Radio knowledge may be embodied in components called radio skills. Radio knowledge is static, requiring interpretation by an algorithm such as an inference engine in order to

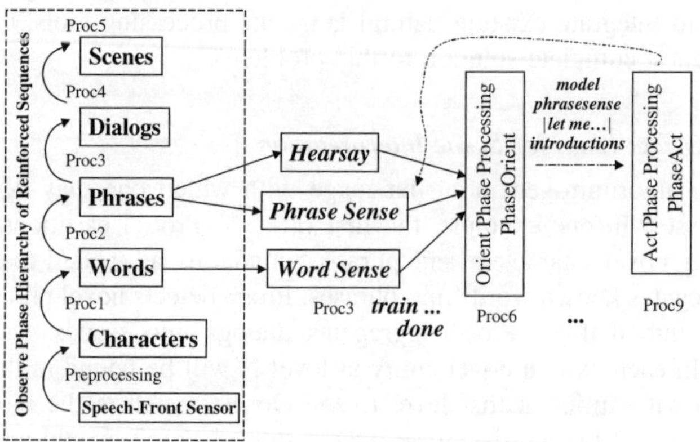

Figure 7.7 The inference hierarchy supports lateral knowledge sources. © Dr. Joseph Mitola III, used with permission

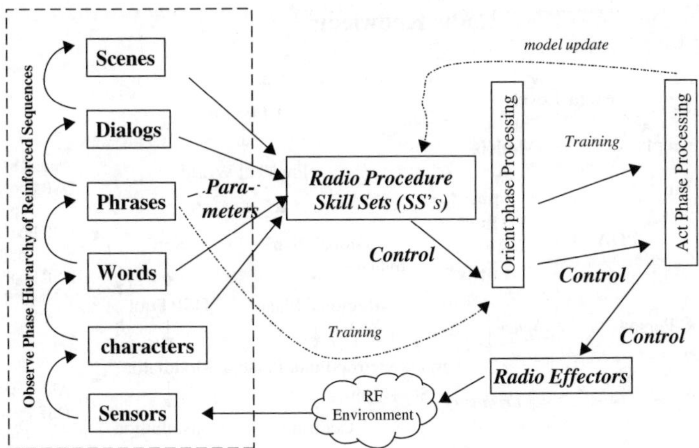

Figure 7.8 Radio skills respond to observations. © 1999 Dr. Joseph Mitola III, used with permission

accomplish anything. Radio skills, on the other hand, are knowledge embedded in serModels through the process of training or sleeping/dreaming. This knowledge is continually pattern-matched against all stimuli in parallel. That is, there are no logical dependencies among knowledge components that mediate the application of the knowledge. With FOPC, the theorem prover must reach a defined state in the resolution of multiple axioms in order to initiate action. In contrast, serModels are continually compared to the level of the hierarchy to which they are attached, so their immediate responses are always cascading towards action. Organized as maps primarily among the wake-cycle phases 'observe' and 'orient', the radio procedure skill sets (SSs) control radio personalities as illustrated in Figure 7.8.

These skill sets may either be reformatted into serModels directly from the *a priori* knowledge of an RKRL frame, or they may be acquired from training or sleep/dreaming. Each skill set may also save the knowledge it learns into an RKRL frame.

7.4.7 General World Knowledge

An iCR needs substantial knowledge embedded in the inference hierarchies. It needs both external RF knowledge and internal radio knowledge. Internal knowledge enables it to reason about itself as a radio. External radio knowledge enables it to reason about the role of the <Self/> in the world, such as respecting rights of other cognitive and not-so-cognitive radios.

Figure 7.9 illustrates the classes of knowledge an iCR needs to employ in the inference hierarchies and cognition cycle. It is one thing to write down that the universe includes a physical world (there could also be a spiritual world, and that might be very important in some cultures). It is quite another thing to express that knowledge in a way that the iCR will be able to use that knowledge effectively. Symbols like 'universe"' take on meaning by their relationships to other symbols and to external stimuli. In this ontology, meta-level

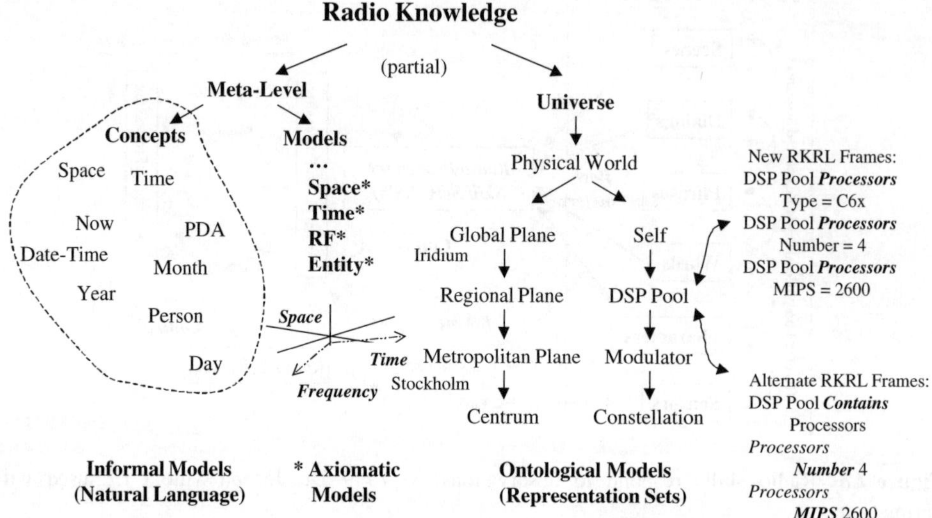

Figure 7.9 External radio knowledge includes concrete and abstract knowledge. © 1999, 2000 Dr. Joseph Mitola III, used with permission

knowledge consists of *abstractions*, distinct from existential knowledge of the physical universe. In RXML, this ontological perspective includes all in a universe of discourse, <Universe> expressed as follows:

<Universe>
<Abstractions> <Time> <Now/> </Time> <Space> <Here/> </Space> ... <RF/> ... <Intelligent-Entities/> ... </Abstractions>
<Physical-universe> ... <Instances/> of Abstractions ... </Physical-universe>
</Universe>

Example 7.3 The universe of discourse of iCR consists of abstractions plus the physical universe

Abstractions include informal and formal meta-level knowledge from unstructured knowledge of concepts to the more mathematically structured models of space, time, RF and entities that exist in space-time. To differentiate 'now' as a temporal concept from 'Now' as the Chinese name of a plant, the CRA includes both the *a priori* knowledge of 'now' as a space-time locus, <Now/>, as well as functions that access and manipulate instances of the concept <Now/>. <Now/> is axiomatic in the CRA, so code refers to 'now' (as n.o.w) in planning actions. The architecture allows an algorithm to return the date-time code from Windows to define instances of <Now/>. Definition-by-algorithm permits an inference system like the cognition subsystem to reason about whether a given event is in the past, present or future. What is the present? The present is some region of time between 'now' and the immediate past and future. If you are a paleontologist, 'now' may consist of the million year epoch in which we all are thought to have evolved from apes. If you are a rock star, 'now' is probably a lot shorter than that to you. How will your CR learn the user's concept of 'now'? The CRA design offers an axiomatic treatment of time, but the axioms were not programmed into Java explicitly. The CRA aggregates knowledge of time by a temporal CBR that illustrates the key principles. The CRA does not fix the definition of <Now/> but enables the <Self/> to define the details in an <Instance> in the physical world about which it can learn from the user, paleontologist or rock star.

Given the complexity of a system that includes both a multi-tiered inference hierarchy and the cognition cycle's observe–orient–plan–decide–act sequence with AML throughout, it is helpful to consider the mathematical structure of these information elements, processes and flows.

The mathematical treatment, CRA IV, is provided elsewhere [20].

7.5 CRA V: Building the CRA on SDR Architectures

A cognitive radio is a software radio (SWR) or software-defined radio (SDR) with flexible formal semantics based entity to entity formal messaging via RXML and integrated machine learning of the self, the user, the environment and the 'situation.' This section reviews SWR, SDR and the software communications architecture (SCA) or software radio architecture (SRA) for those who may be unfamiliar with these concepts. While it is not necessary for an iCR to use the SCA/SRA as its internal model of itself, it certainly must have some model, or it will be incapable of reasoning about its own internal structure and adapting or modifying its radio functionality.

7.5.1 SWR and SDR Architecture Principles

Hardware-defined radios such as the typical AM/FM broadcast receiver convert radio to audio using radio hardware, such as antennas, filters, analog demodulators and the like. SWR is the ideal radio in which the analog-to-digital converter (ADC) and digital-to-analog converter (DAC) convert digital signals to and from radio frequencies (RF) directly, and all RF channel modulation, demodulation, frequency translation and filtering are accomplished digitally. For example, modulation may be accomplished digitally by multiplying sine and cosine components of a digitally sampled audio signal (called the 'baseband' signal, e.g., to be transmitted) by the sampled digital values of a higher frequency sine wave to up-convert it, ultimately to RF.

Figure 7.10 shows how SDR principles apply to a cellular radio base station. In the ideal SWR, there would be essentially no RF conversion, just ADC/DAC blocks accessing the full RF spectrum available to the (wideband) antenna elements. Today's SDR base stations approach this ideal by digital access (DAC and ADC) to a band of spectrum allocations, such as 75 MHz allocated to uplink and downlink frequencies for third-generation services. In this architecture, RF conversion can be a substantial system component, sometimes 6% of the cost of the hardware, and not amenable to cost improvements through Moore's law. The ideal SDR would access more like 2.5 GHz from, say 30 MHz to around 2.5 GHz, supporting all kinds of services in television (TV) bands, police bands, air traffic control bands – you name it. Although considered radical when introduced in 1991 [14] and popularized in 1995 [15], recent regulatory rulings are encouraging the deployment of such 'flexible spectrum' use architectures.

This ideal SWR may not be practical or affordable, so it is important for the radio engineer to understand the trade-offs (again, see [20] for SDR architecture trade-offs). In

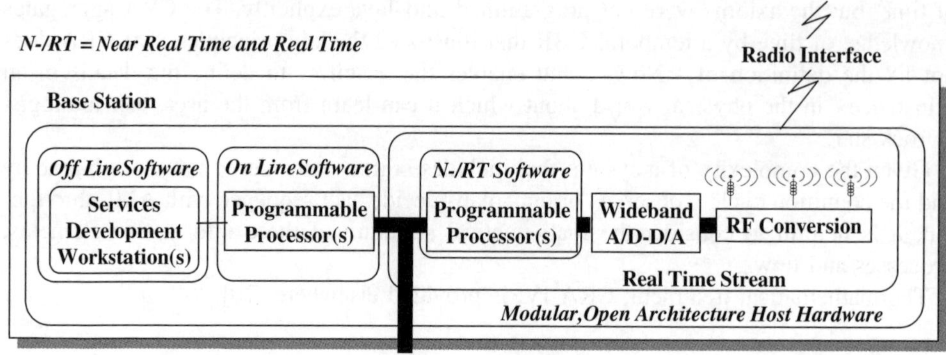

Figure 7.10 SWR principle applied to cellular base station. © 1992 Dr. Joseph Mitola III, used with permission

particular, the physics of RF devices (e.g., antennas, inductors, filters) makes it easier to synthesize narrowband RF and intervening analog RF conversion and intermediate frequency (IF) conversion. Given narrowband RF, the hardware-defined radio might employ baseband (e.g., voice frequency) ADC, DAC and digital signal processing. The programmable digital radios (PDR) of the 1980s and 90s used this approach. Historically, this approach has not been as expensive as wideband RF (antennas, conversion), ADCs and DACs. Handsets are less amenable to SWR principles than the base station (Figure 7.11). Base stations access the power grid. Thus, the fact that wideband ADCs, DACs and DSP consume many watts of power is not a major design driver. Conservation of battery life, however, is a major design driver in the handset.

Thus, insertion of SWR technology into handsets has been relatively slow. Instead, the major handset manufacturers include multiple single-band RF chip sets in a given handset. This has been called the Velcro radio or slice radio.

Since the ideal SWR is not readily approached in many cases, the SDR has comprised a sequence of practical steps from the baseband DSP of the 1990s towards the ideal SWR. As the economics of Moore's law and of increasingly wideband RF and IF devices allow, implementations move upward and to the right in the SDR design space (Figure 7.12).

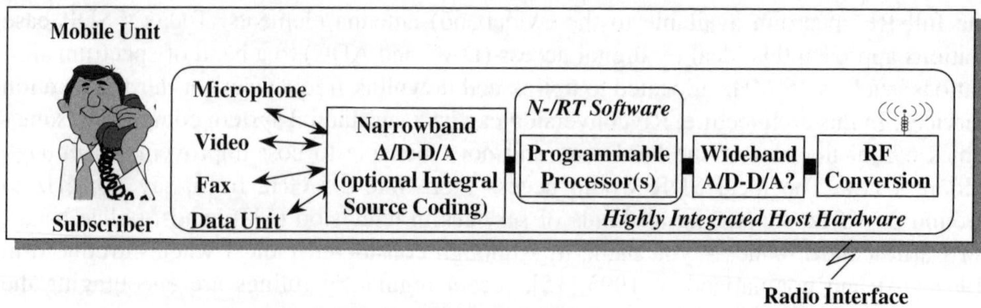

Figure 7.11 Software radio principle – 'ADC and DAC at the antenna' may not apply. © 1992 Dr. Joseph Mitola III, used with permission

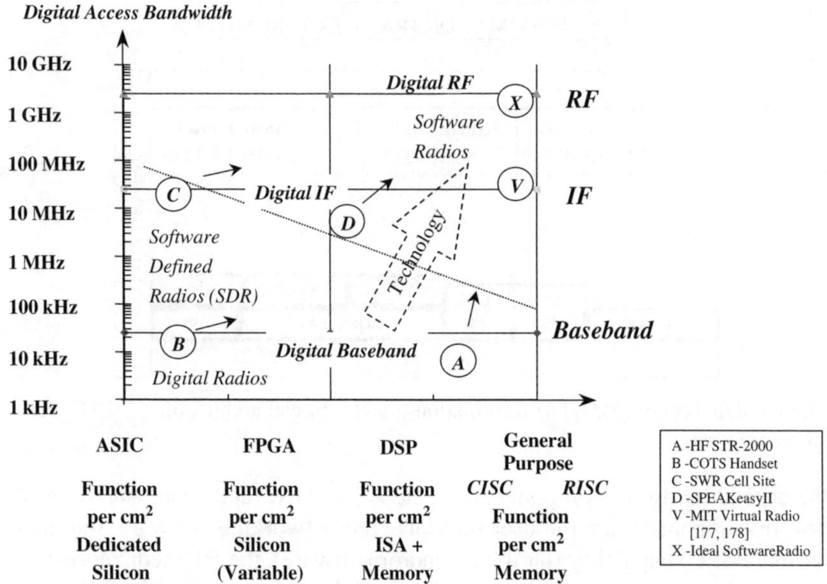

Figure 7.12 SDR design space shows how designs approach the ideal SWR. © 1996–2003 Dr. Joseph Mitola III, used with permission

This space consists of the combination of digital access bandwidth and programmability. Access bandwidth consists of ADC/DAC sampling rates converted by the Nyquist criterion or practice into effective bandwidth. Programmability of the digital subsystems is defined by the ease with which logic and interconnect may be changed after deployment. Application-specific integrated circuits (ASICs) cannot be changed at all, so the functions are 'dedicated' in silicon. Field programmable gate arrays (FPGAs) can be changed in the field, but if the new function exceeds some parameter of the chip, which is not uncommon, then one must upgrade the hardware to change the function, just like ASICs. Digital signal processors (DSPs) are typically easier or less expensive to program and are more efficient in power use than FPGAs. Memory limits and instruction set architecture (ISA) complexity can drive up costs of reprogramming the DSP. Finally, general-purpose processors, particularly with reduced instruction aet architectures (RISC) are most cost-effective to change in the field. To assess a multiprocessor, such as a cell phone with a CDMA-ASIC, DSP speech codec and RISC microcontroller, weight the point by equivalent processing capacity.

Where should one place an SDR design within this space? The quick answer is so that you can understand the migration path of radio technology from the lower left towards the upper right, benefiting from lessons learned in the early migration projects captured in *Software Radio Architecture* [20].

This section contains a very brief synopsis of the key SDR knowledge you will need in order to follow the iCR examples of this text.

7.5.2 Radio Architecture

The discussion of the software radio design space contains the first elements of radio architecture. It tells you what mix of critical components are present in the radio. For

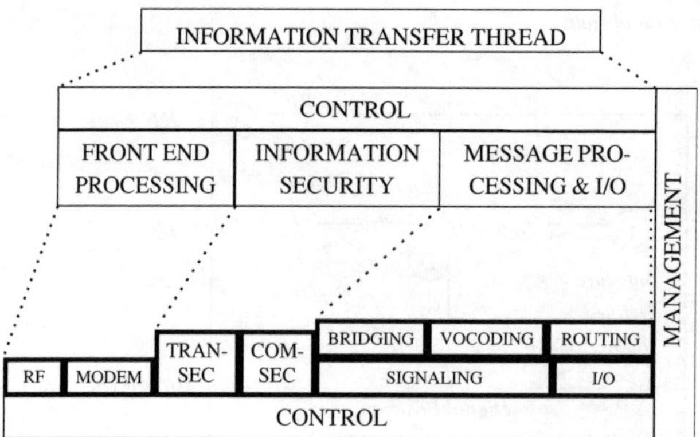

Figure 7.13 SDR Forum (MMITS) information transfer thread architecture. ©1997 SDR Forum, used with permission

SDR, the critical hardware components are the ADC, DAC and processor suite. The critical software components are the user interface, the networking software, the information security (INFOSEC) capability (hardware and/or software), the RF media access software, including the physical layer modulator and demodulator (modem) and media access control (MAC), and any antenna-related software such as antenna selection, beamforming, pointing and the like. INFOSEC consists of transmission security, such as the frequency hopping spreading code selection, plus communications security encryption.

The SDR Forum defined a very simple, helpful model of radio in 1997, shown in Figure 7.13. This model highlights the relationships among radio functions at a tutorial level. The CR has to 'know' about these functions, so every CR must have an internal model of a radio of some type. This one is a good start because it shows both the relationships among the functions and the typical flow of signal transformations from analog RF to analog or with SDR, digital modems and on to other digital processing including system control of which the user interface is a part.

This model and the techniques for implementing a SWR and the various degrees of SDR capability are addressed in depth in the various texts on SDR [9] [23] [31], [34].

7.5.3 The SCA

The US DoD developed the software communications architecture (SCA) for its joint tactical radio system (JTRS) family of radios.

The architecture identifies the components and interfaces shown in Figure 7.14. The APIs define access to the physical layer, to the MAC layer, to the logical link layer (LLC), to security features, and to the input/output of the physical radio device. The physical components consist of antennas and RF conversion hardware that are mostly analog and that therefore typically lack the ability to declare or describe themselves to the system. Most other SCA-compliant components are capable of describing themselves to the system to enable and facilitate plug and play among hardware and software components. In addition, the SCA embraces POSIX and CORBA.

The model evolved through several stages of work in the SDR Forum and Object Management Group (OMG) into a UML-based object-oriented model of SDR (Figure 7.15).

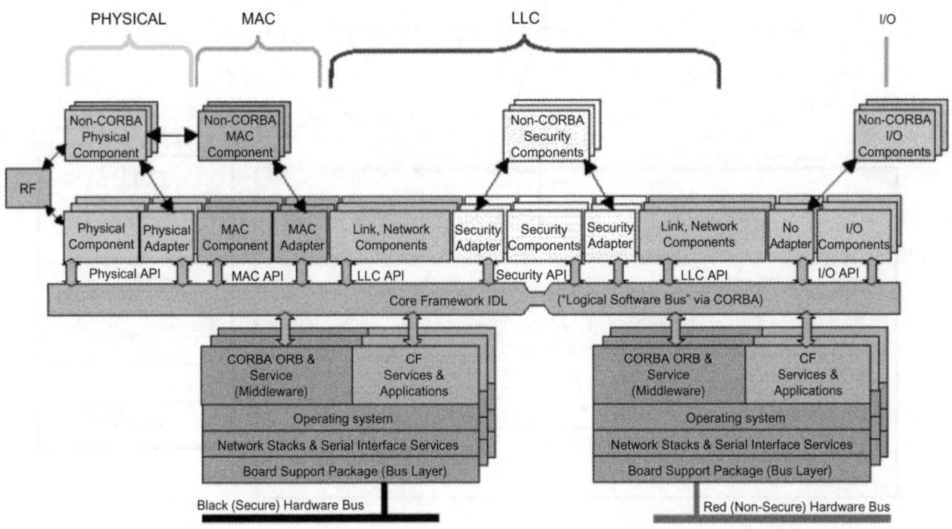

Figure 7.14 JTRS SCA Version 1.0. © 2004 SDR Forum, used with permission

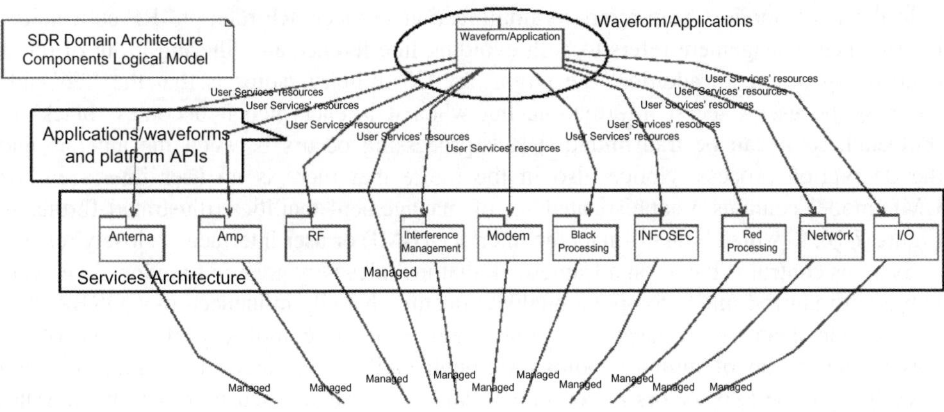

Figure 7.15 SDR Forum UML model of radio services. © 2004 SDR Forum, used with permission

Waveforms are collections of load modules that provide wireless services, so from a radio designer's perspective, the waveform is the key application in a radio. From a user's perspective of a wireless PDA, the radio waveform is just a means to an end, and the user doesn't want to know or have to care about waveforms. Today, the cellular service providers hide this detail to some degree, but consumers sometimes know the difference between CDMA and GSM, for example, because CDMA works in the US, but not in Europe. With the deployment of the third generation of cellular technology (3G), the amount of technical jargon consumers will need to know is increasing. So the CR designer is going to write code (e.g., Java code) that insulates the user from those details, unless the user really wants to know.

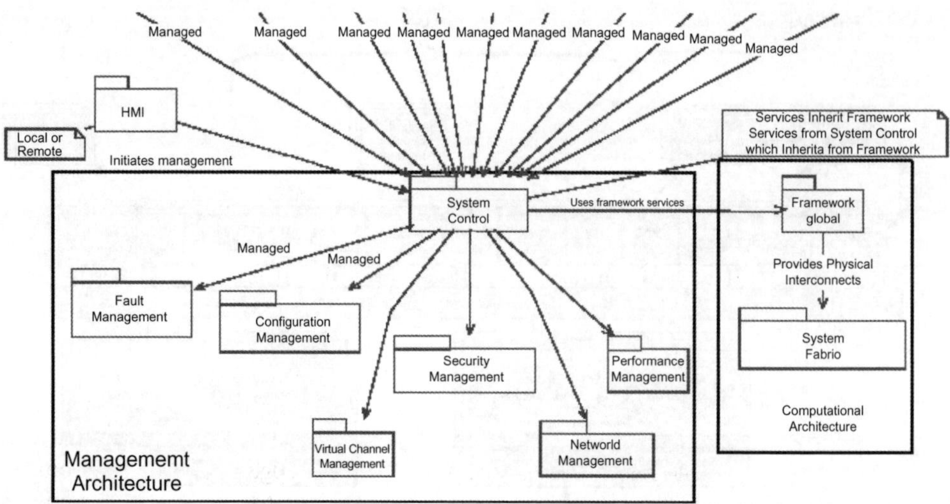

Figure 7.16 SDR Forum UML management and computational architectures © 2004 SDR Forum, used with permission

In the UML model, Amp refers to amplification services, RF refers to RF conversion, interference-management refers to both avoiding interference and filtering it out of one's band of operation. In addition, the jargon for US military radios is that the 'red' side contains the user's secret information, but when it is encrypted it becomes 'black' or protected, so it can be transmitted. Black processing occurs between the antenna and the decryption process. Notice also in the figure that there is no user interface. The UML model contains a sophisticated set of management facilities, illustrated further in Figure 7.16, to which human–machine interface (HMI) or user interface is closely related.

Systems control is based on a framework that includes very generic functions like event logging, organized into a computational architecture, heavily influenced by CORBA. The management features are needed to control radios of the complexity of 3G and of the current generation of military radios. Although civil sector radios for police, fire and aircraft lag these two sectors in complexity and are more cost-sensitive, baseband SDRs are beginning to insert themselves even into these historically less technology-driven markets.

Fault management features are needed to deal with loss of a radios processors, memory or antenna channels. CR therefore interacts with fault management to determine what facilities may be available to the radio given recovery from hardware and/or software faults (e.g., error in a download). Security management is increasingly important in the protection of the user's data by the CR, balancing convenience and security which can be very tedious and time consuming. The CR will direct virtual channel management (VCM) and VCM will learn from the VCM function what radio resources are available, such as what bands the radio can listen to and transmit on and how many it can do at once. Network management does for the digital paths what VCM does for the radio paths. Finally, SDR performance depends on the availability of analog and digital resources, such as linearity in the antenna, millions of instructions per second (MIPS) in a processor and the like.

7.5.4 Functions-Transforms Model of Radio

The self-referential model of a wireless device used by the CRA and used to define the RKRL and to train the CRA is the function-transforms model illustrated in Figure 7.17. In this model, the radio knows about sources, source coding, networks, INFOSEC and the collection of front-end services needed to access RF channels. Its knowledge also extends to the idea of multiple channels and their characteristics (the channel set), and that the radio part may have both many alternative personalities at a given point in time, and that through evolution support, those alternatives may change over time.

Since CR reasons about all of its internal resources, it also must have some kind of computational model of analog and digital performance parameters and how they are related to features it can measure or control. MIPS, for example, may be controlled by setting the clock speed. A high clock speed generally uses more total power than a lower clock speed, and this tends to reduce battery life. The same is true for the brightness of a display. The CR only 'knows' this to the degree that it has a data structure that captures this information and some kind of algorithms, preprogrammed and/or learned, that deal with these relationships to the benefit of the user. Constraint languages may be used to express interdependencies, such as how many channels of a given personality are supported by a given hardware suite, particularly in failure modes. CR algorithms may employ this kind of structured reasoning as a specialized knowledge source (KS) when using case-based learning to extend its ability to cope with internal changes.

The ontological structure of the above may be formalized as follows:

```
<SDR>
    <Sources/><Channels/><Personality>
            <Source-Coding-Decoding/><Networking/><INFOSEC/>
<Channel-Codec><Modem/><IF-Processing/><RF-Access/>< /Channel-Codec>
</Personality>
    <SDR-Platform/><Evolution-Support/>
</SDR>
```

Example 7.4 Defined SDR subsystem components

While this text does not spend a lot of time on the computational ontology of SDR, semantically based dialogs among iCRs about internal issues like downloads may be

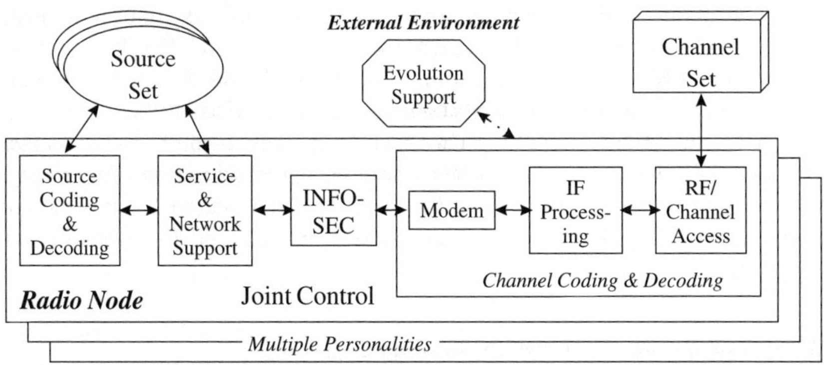

Figure 7.17 Functions-transforms model of a wireless node. © 1996 Dr. Joseph Mitola III, used with permission

mediated by developing the RXML above to more fully develop the necessary ontological structures.

7.5.5 Architecture Migration: From SDR to iCR

Given the CRA and contemporary SDR architecture, one must address the transition of SDR, possibly through a phase of iCRs toward the ideal CR. As the complexity of handheld, wearable and vehicular wireless systems increase, the likelihood that the user will have the skill necessary to do the optimal thing in any given circumstance goes down. Today's cellular networks manage the complexity of individual wireless protocols for the user, but the emergence of multiband multimode iCR moves the burden for complexity management towards the PDA. The optimization of the choice of wireless service between the 'free' home WLAN and the for-sale cellular equivalent moves the burden of radio resource management from the network to the WPDA.

7.5.6 Cognitive Electronics

The increasing complexity of the PDA–user interface also accelerates the trend towards increasing the computational intelligence of personal electronics. iCR is in some sense just an example of a computationally intelligent personal electronics system. For example, using a laptop computer in the bright display mode uses up the battery power faster than when the display is set to minimum brightness. A cognitive laptop could offer to set the brightness to low level when it was turned on in battery-powered mode. It would be even nicer if it would recognize operation aboard a commercial aircraft and therefore automatically turn the brightness down. It should learn that my preference is to set the brightness low on an aircraft to conserve the battery. A cognitive laptop shouldn't make a big deal over that, and it should let me turn up the brightness without complaining. If it had an ambient light sensor or ambient light algorithm for an embedded camera, it could tell that a window shade is open, so I have to deal with the brightness. By sensing the brightness of the *on-board aircraft* scene and associating my control of the brightness of my display with the brightness of the environment, a hypothetical cognitive laptop could learn do the right thing in the right situation.

How does this relate to the CRA? For one thing, the CRA could be used 'as-is' to increase the computational intelligence of the laptop. In this case, the self is the laptop and the PDA knows about itself as a laptop, not as a WPDA. It knows about its sensors suite, which includes at least a light level sensor if not a camera through the data structures that define the <Self/>. It knows about the user by observing keystrokes and mouse action as well as by interpreting the images on the camera, e.g., to verify that the Owner is still the user since that is important to building user-specific models. It might build a space–time behavior model of any user or it might be a one-user laptop. Its actions then must include the setting of the display intensity level. In short, the CRA accommodates the cognitive laptop with suitable knowledge in the knowledge structures and functions implemented in the map sets.

7.5.7 When Should a Radio Transition Towards Cognition?

If a wireless device accesses only a single RF band and mode, then it is not a very good starting point for cognitive radio. It's just too simple. Even as complexity increases, as long as the user's needs are met by wireless devices managed by the network(s), then embedding computational intelligence in the device has limited benefits. In 1999, Mitsubishi and AT&T announced the first 'four-mode handset.' The T250 operated in TDMA mode on 850 or 1900 MHz, in the first-generation analog mobile phone system (AMPS) mode on 850 MHz, and in cellular digital packet data (CDPD) mode on 1900 MHz. This illustrates early development of multiband, multimode, multimedia (M3) wireless. These radios enhanced the service provider's ability to offer national roaming, but the complexity was not apparent to the user since the network managed the radio resources in the handset.

Even as device complexity increases in ways that the network does not manage, there may be no need for cognition. There are several examples of capabilities embedded in electronics that typically are not heavily used. Do you use your laptop's speech recognition system? What about its IRDA port? If you were the typical user circa 2004, you didn't use either capability of your Windows XP laptop all that much. So complexity can increase without putting a burden on the user to manage that complexity if the capability isn't central to the way in which the user employs the system.

For radio, as the number of bands and modes increases, the SDR becomes a better candidate for the insertion of cognition technology. But it is not until the radio or the wireless part of the PDA has the capacity to access multiple RF bands that cognition technology begins to pay off. With the liberalization of RF spectrum use rules, the early evolution of iCR may be driven by RF spectrum use etiquette for ad hoc bands such as the FCC use-case. In the not-too-distant future, SDR PDAs could access satellite mobile services, cordless telephone, WLAN, GSM and 3G bands. An ideal SDR device with these capabilities might affordably access three octave bands from 0.4 to 0.96 GHz, (skip the air navigation and GPS band from. 96 to 1.2 GHz), 1.3 to 2.5 GHz, and from 2.5 to 5.9 GHz (Figure 7.18). Not counting satellite mobile and radio navigation bands, such radios would

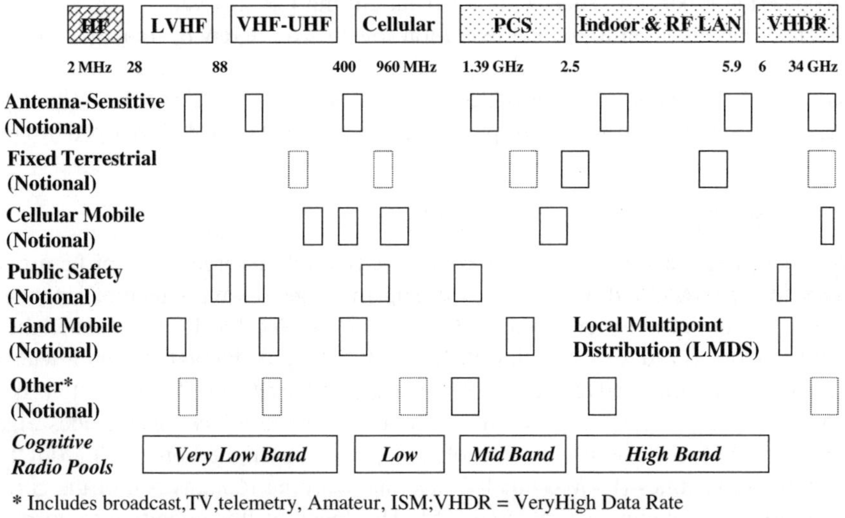

* Includes broadcast,TV,telemetry, Amateur, ISM;VHDR = VeryHigh Data Rate

Figure 7.18 Fixed spectrum allocations versus pooling with cognitive radio. © 1997 Dr. Joseph Mitola III, used with permission

have access to over 30 mobile sub-bands in 1463 MHz of potentially shareable outdoor mobile spectrum. The upper band provides another 1.07 GHz of shareable indoor and RF LAN spectrum. This wideband radio technology will be affordable first for military applications, next for base station infrastructure, then for mobile vehicular radios and later for handsets and PDAs. When a radio device accesses more RF bands than the host network controls, it is time for CR technology to mediate the dynamic sharing of spectrum. It is the well-heeled conformance to the radio etiquettes afforded by cognitive radio that makes such sharing practical [5].

7.5.8 Radio Evolution Towards the CRA

Various protocols have been proposed by which radio devices may share the radio spectrum. The US FCC Part 15 rules permit low power devices to operate in some bands. In 2003, a Rule and Order (R&O) made unused television (TV) spectrum available for low power RF LAN applications, making the manufacturer responsible for ensuring that the radios obey this simple protocol. DARPA's NeXt Generation (XG) program developed a language for expressing spectrum use policy [12]. Other more general protocols based on peek-through to legacy users have also been proposed [35].

Does this mean that a radio must transition instantaneously from the SCA to the CRA? Probably not. The simple six-component iCR architecture may be implemented with minimal sensory perception, minimal learning and no autonomous ability to modify itself. Regulators want to hold manufacturers responsible for the behaviors of such radios. The simpler the architecture, the simpler the problem of explaining it to regulators and of getting concurrence among manufacturers regarding open architecture interfaces that facilitate technology insertion through teaming. Manufacturers who fully understand the level to which a highly autonomous CR might unintentionally reprogram itself to violate regulatory constraints may decide they want to field aware-adaptive (AA) radios, but may not want to take the risks associated with a self-modifying CR just yet.

Thus, one can envision a gradual evolution towards the CRA beginning initially with a minimal set of functions mutually agreeable among the growing community of iCR stakeholders. Subsequently, the introduction of new services will drive the introduction of new capabilities and additional APIs, perhaps informed by the CRA developed in this text.

7.5.9 Cognitive Radio Architecture Research Topics

The cognition cycle and related inference hierarchy imply a large scope of hard research problems for cognitive radio. Parsing incoming messages requires natural language text processing. Scanning the user's voice channels for content that further defines the communications context requires speech processing. Planning technology offers a wide range of alternatives in temporal calculus [28], constraint-based scheduling [10], task planning [42], causality modeling [26], and the like. Resource allocation includes algebraic methods for wait-free scheduling protocols [14], open distributed processing (ODP), and rarallel virtual machines (PVM). Finally, machine learning remains one of the core challenges in artificial intelligence research [15]. The focus of this cognitive radio research, then, is not on the development of any one of these technologies per se. Rather, it is on the organization of cognition tasks and on the development of cognition data structures needed to integrate contributions from these diverse disciplines for the context-sensitive delivery of wireless services by software radio.

Learning the difference between situations in which a reactive response is needed versus those in which deliberate planning is more appropriate is a key challenge in machine learning for CR. The CRA framed the issues. The CRA goes further, providing useful KSs and related ML so that the CR designer can start there in developing good engineering solutions to this problem for a given CR applications domain.

7.5.10 Industrial Strength iCR Design Rules

The CRA allocates functions to components based on design rules. Typically design rules are captured in various interface specifications including APIs and object interfaces, such as Java's JINI/JADE structure of intelligent agents [3]. While the previous section introduced the CRA, this section suggests additional design rules by which user domains, sensory domains and radio knowledge of RF band knowledge may be integrated into industrial-strength iCR products and systems.

The following design rules circumscribe the integration of cognitive functions with the other components of a wireless PDA within the CRA:

1. The cognition function should maintain an explicit (topological) model of space-time
   (a) of the user
   (b) of the physical environment
   (c) of the radio networks, and
   (d) of the internal states of the radio, the <self/>.
2. The CRA requires each CR to predict in advance an upper bound on the amount of computational resources (e.g., time) required for each cognition cycle. The CR must set a trusted (hardware) watchdog (e.g., timer) before entering a cognition cycle. If the watchdog is violated, the system must detect that event, log that event and mark the components invoked in that event as nondeterministic.
3. The CRA should internalize knowledge as procedural skills, e.g., serModels.
   (a) the CRA requires each CR to maintain a trusted index to internal models and related experience, and
   (b) each CR must preclude cycles from its internal models and skills graph because a CRA conformance requires reliable detection of cycles to break cycles (e.g., via timer) to avoid Gödel–Turing unbounded resource use endemic to self-referential Turing-capable computational entities like iCRs.
4. Context that references space, time, RF, the <User/> and the <Self/> for every external and internal event could be represented formally using a topologically comprehensive, extensible, and logically sound model of space-time context.
5. Each CR conforming to the CRA should include an explicit grounding map, M, that maps its internal data structures on to elements sensed in the external world represented in its sensory domains, including itself. If the CR cannot map a sensed entity to a space-time-context entity with specified time allocated to attempt that map, then the entity should be designated 'UNGROUNDABLE,' to identify introspection goals.
6. The model of the world might best follow a formal treatment of time, space, radio frequency, radio propagation and the grounding of entities in the environment.

7. Models could be represented in an open architecture radio knowledge representation language suited to the representation of radio knowledge (e.g., a semantic web derivative of RKRL). That language shall support topological properties and inference (e.g., forward chaining) but must not include unconstrained axiomatic first order predicate calculus which per force violates the Gödel–Turing constraint.
8. The cognition functions really must maintain location awareness, including:
   (a) the sensing of location from global positioning satellites
   (b) sensing position from local wireless sensors and networks
   (c) and sensing precise position visually
   (d) location shall be an element of all contexts
   (e) the cognition functions shall estimate time to the accuracy necessary to support the user and radio functions
   (f) the cognition functions shall maintain an awareness of the identity of the PDA, of its Owner, of its primary user, and of other legitimate users designated by the owner or primary user.
9. The cognition functions shall reliably infer the user's communications context and apply that knowledge to the provisioning of wireless access by the SDR function.
10. The cognition function shall model the propagation of its own radio signals with sufficient fidelity to estimate interference to other spectrum users:
    (a) it shall also assure that interference is within limits specified by the spectrum use protocols in effect in its location (e.g., in spectrum rental protocols)
    (b) it shall defer control of the <Self/> to the wireless network in contexts where a trusted network manages interference.
11. The cognition functions should model the domain of applications running on the host platform, sufficient to infer the parameters needed to support the application. Parameters modeled include QoI, QoS, data rate, probability of link closure (grade of service), and space-time-context domain within which wireless support is needed.
12. The cognition functions should configure and manage all the SDR assets to include hardware resources, software personalities and functional capabilities as a function of network constraints and use context.
13. The cognition functions should administer the computational resources of the platform. The management of software radio resources may be delegated to an appropriate SDR function (e.g., the SDR Forum domain manager). Constraints and parameters of those SDR assets shall be modeled by the cognition functions. The cognition functions shall assure that the computational resources allocated to applications, interfaces, cognition and SDR functions are consistent with the user communications context.
14. The cognition functions must represent the degree of certainty of understanding in external stimuli and in inferences. A certainty calculus shall be employed consistently in reasoning about uncertain information. Uncertainty research can be a rich source of tools for representing certainty and uncertainty.
15. The cognition functions should recognize and react immediately to preemptive actions taken by the network and/or the user. In case of conflict, the cognition functions shall defer the control of applications, interfaces and/or SDR assets to the Owner, to the

network or to the primary user, according to appropriate priority and operations assurance protocol.

7.6 Summary and Future Directions

The progeny of TellMe® seem headed to a purse or belt near you to better sense and perceive your needs for communications services so that you can take fuller advantage of the technology by, well, by doing nothing but letting the technology adapt to you. In 2005, the technology was not capable of sensing, perceiving and adapting to the user, but enabling technologies in machine speech and vision were maturing. Because of the FCC's rulings encouraging CR, many 'cognitive radio' products capable only of sniffing TV channels and employing unoccupied radio spectrum were appearing in the marketplace.

7.6.1 Architecture Frameworks

Often technical architecture frameworks of the kind presented in this chapter accelerate the state of practice by catalyzing work across industry on plug and play, teaming and collaboration. The thought is that to propel wireless technology from limited spectrum awareness towards valuable user awareness, some architecture like the CRA will be needed. In short, the CRA articulates the functions, components and design rules of next-generation cognitive radios. Each of the different aspects of the CRA contributes to the dialog:

1. The functional architecture identifies components and interfaces for cognitive radios with sensory and perception capabilities in the user domain, not just the radio domain.
2. The cognition cycle identifies the processing structures for the integration of sensing and perception into radio: observe (sense and perceive), orient (react if necessary), plan, decide, act and learn.
3. The inference hierarchies suggest levels of abstraction helpful in the integration of radio and user domains into the synthesis of services tailored to the specific user's current state of affairs given the corresponding state of affairs of the radio spectrum in space and time.
4. The introduction to ontology suggests an increasing role for semantic web technologies in making the radios smarter, initially about radio and over time about the user.
5. Although not strictly necessary for CR, SDR provides a very flexible platform for the regular enhancement of both computational intelligence and radio capability, particularly with each additional Moore's law cycle.
6. Finally, this chapter has introduced the CRA to the reader interested in the cutting edge, but has not defined the CRA. The previous section suggested a few of the many aspects of the embryonic CRA that must be addressed by researchers, developers and markets in the continuing evolution of SDR towards ubiquitous and really fun iCRs.

7.6.2 Industrial Strength Architecture

Although the CRA provides a framework for APIs, it doesn't specify the details of the data structures or of the maps. Thus, the CRA research prototype emphasizes ubiquitous learning via serModels and case-based reasoning (see [22]), but it doesn't implement critical features that would be required in consumer-class iCRs. Other critical aspects of such industrial-strength architectures include more capable scene perception and situation interpretation specifically addressing:

1. ***Noise***, in utterances, images, objects, location estimates and the like. Noise sources include thermal noise, conversion error introduced by the process of converting analog signals (audio, video, accelerometers, temperature, etc.) to digital form, error in converting from digital to analog form, preprocessing algorithm biases and random errors, such as the accumulation of error in a digital filter, or the truncation of a low energy signal by threshold logic. Dealing effectively with noise differentiates a tutorial demonstration from an industrially useful product.
2. ***Hypothesis management***, keeping track of more than one possible binding of stimuli to response, dialog sense, scene, etc. Hypotheses may be managed by keeping the N-best hypotheses (with an associated degree of belief), by estimating the prior probability or other degree of belief in a hypothesis, and keeping a sufficient number of hypotheses to exceed a threshold (e.g., 90 or 99% of all the possibilities), or keeping multiple hypotheses until the probability for the next most likely (second) hypothesis is less than some threshold. The estimation of probability requires a measurable space, a sigma algebra that defines how to accumulate probability on that space, proof that the space obeys the axioms of probability and a certainty calculus that defines how to combine degrees of belief in events as a function of the measures assigned to the probability of the event.
3. ***Training corpora and training interfaces***, the reverse flow of knowledge from the inference hierarchy back to the perception subsystems. The recognition of the user by a combination of face and voice could be more reliable than single-domain recognition either by voice or by vision. In addition, the location, temperature and other aspects of the scene may influence object identification. Visual recognition of the Owner outdoors in a snow storm, for example, is more difficult than indoors in an office. While the CR might learn to recognize the user based on weaker cues outdoors, access to private data might be constrained until the quality of the recognition exceeds some learned threshold.
4. ***Nonlinear flows***, although the cognition cycle emphasizes the forward flow of perception enabling action, it is crucial to realize that actions may be internal, such as advising the vision subsystem that its recognition of the user is in error because the voice does not match and the location is wrong. Because of the way the cognition cycle operates on the self, these reverse flows from perception to training are implemented as forward flows from the perception system to the self, directed towards a specific subsystem such as vision or audition. There may also be direct interfaces from the CWN to the CR to upload data structures representing *a priori* knowledge integrated into the UCBR learning framework.

7.6.3 Conclusion

In conclusion, iCR seems likely to leverage the semantic web, but the markets for services layered on practical radio networks will shape that evolution away from the verbose and general purpose and towards the efficient and focused characterization of content rich with context and increasingly independent of form. Although many information-processing technologies from eBusiness solutions to the semantic web are relevant to iCR, the integration of audio and visual sensory perception into SDR with suitable cognition architectures remains both a research challenge and a series of increasingly interesting radio systems design opportunities. A CRA that is broadly supported by industry could facilitate such an evolution.

References

[1] Carbonell, J. (2004) 'Phraselator' presentation at Fordham University, December.

[2] Cognitive Vision (2004) June, Palo Alto, CA.

[3] Das, S.K. *et al.* (1997) Decision making and plan management by intelligent agents: theory, implementation, and applications. *Proceedings of Autonomous Agents*, www.acm.org.

[4] Eriksson, H.-E. and Penker, M. (1998) *UML Toolkit*, John Wiley & Sons, Inc., New York.

[5] Esmahi, L. *et al.* (1999) Mediating conflicts in a virtual market place for telecommunications network services. *Proceedings of the 5th Baiona Workshop on Emerging Technologies in Telecommunications*, Universidade de Vigo, Vigo, Spain.

[6] Hennie, R. (1997) *Introduction to Computability*, Addison-Wesley, Reading, MA.

[7] http://en.wikipedia.org/wiki/Multiple-input_multiple-output_communications.

[8] http://www.cyc.com/cycdoc/upperont-diagram.html.

[9] Jondral, F. (1999) *Software Radio*, Universität Karlsruhe, Karlsruhe, Germany.

[10] Koser *et al.* (1999) 'read.me', www.cs.kun.nl, March, University of Nijmegen, The Netherlands.

[11] Mahonen, P. (2004) *Cognitive Wireless Networks*, RWTH Aachen, Aachen, Germany.

[12] Marshall, P. (2003) *Remarks to the SDR Forum*, September, SDR Forum, New York.

[13] Michalski, R., Bratko, I. and Kubat, M. (1998) *Machine Learning and Data Mining*, John Wiley & Sons, Inc., New York.

[14] Mitola III, J. (1992) Software radio: survey, critical evaluation and future directions, *Proceedings of the National Telesystems Conference*, IEEE Press, New York.

[15] Mitola III, J. (1995) Software radio architecture. *IEEE Communications Magazine*, May.

[16] Mitola III, J. (1998) Email to Professor Gerald Q. Maguire, recommending the title of the Licentiate as Cognitive Radio, January.

[17] Mitola III, J. (1998) Software radio architecture: a mathematical perspective. *IEEE JSAC*, April.

[18] Mitola III, J. (1999) Cognitive radio for flexible mobile multimedia communications. *Mobile Multimedia Communications (MoMUC 99)*, November, IEEE Press, New York.

[19] Mitola III, J. (1999) *Cognitive radio: model-based competence for software radio*, Licentiate Thesis TRITA-IT AUH 99:04, KTH, The Royal Institute of Technology, Stockholm, Sweden, August.

[20] Mitola III, J. (2000) *Cognitive radio: an integrated agent architecture for software defined radio*, KTH, The Royal Institute of Technology, Stockholm, Sweden, June.

[21] Mitola III, J. (2000) *Software Radio Architecture*, John Wiley & Sons, Inc., New York.

[22] Mitola III, J. (2006) *Cognitive Radio Architecture*, John Wiley & Sons, Inc., New York.

[23] Mitola III, J. and Zvonar, Z. (1999) *Software Radio Technologies*, IEEE Press, New York.

[24] Mowbray, T. and Malveau, R. (1997) *CORBA Design Patterns*, John Wiley & Sons, Inc., New York.

[25] Order FCC 03-322 (2003) Notice of proposed rule making, December, United States Federal Communications Commission, Washington, DC.

[26] PC-KIMMO Version 1.0.8 for IBM PC (1992) 18 February.

[27] Pearl, J. (2000) *Causality: Models, Reasoning, and Inference*, Morgan-Kaufmann, San Francisco.

[28] Phillips, C. (1997) Optimal time-critical scheduling, STOC 97, www.acm.org.

[29] Pisano, A. (1999) *MEMS 2003 and beyond*, www.darpa.mil/mto/mems.

[30] *Proceedings of the Dagtsuhl Workshop on Cognitive Radios and Networks* (2004) October, RWTH Aachen, Germany.

[31] Reed, J. (2002) *Software Radio: A Modern Approach to Radio Engineering*, Prentice Hall, Englewood Cliffs, NJ.

[32] SnePS (1998) ftp.cs.buffalo.edu:/pub/sneps/.

[33] TellMe Networks (2005) www.tellme.com, Mountain View, CA.

[34] Tuttlebee, W.H. (2002) *Software Defined Radio: Enabling Technologies*, John Wiley & Sons, Inc., New York.
[35] Watzman, A. (2002) *Robotic Achievements: GRACE Successfully Completes Mobile Robot Challenge at Artificial Intelligence Conference*, Carnegie Melon Views, Pittsburg, PA.
[36] Wireless World Research Forum (www.wwrf.com).
[37] www.ieee-dyspan.org.
[38] www.jtrs.mil.
[39] www.omg.org.
[40] www.omg.org/UML.
[41] www.sdrforum.org.
[42] XTAG Research Group (1999) *A Lexicalized Tree Adjoining Grammar for English*, Institute for Research in Cognitive Science, University of Pennsylvania, Philadelphia, PA.
[43] Zue, V. (2005) *Speech Understanding System*, MIT Press, Boston.
[44] Stuart. Kaufman, *At Home in the Universe* (Santa Fe, NM: The Santa Fe Institute) 1992

8

The Wisdom of Crowds: Cognitive Ad Hoc Networks

Linda Doyle and Tim Forde
CTVR at the University of Dublin, Trinity College, Dublin, Ireland

8.1 Introduction

Cognitive ad hoc networks are a natural endpoint for the evolution of the contemporary mobile, ad hoc network (MANET). In order for ad hoc networks to fulfill the promise that their title implies, they need to move beyond being merely mobile, peer-to-peer, distributed systems. For an ad hoc network to deliver a truly ad hoc networking service, it needs to evolve to a stage where the network can discover networking solutions beyond those tested in the laboratory, and which allow it to provide a networking service which is *fit for purpose*.

In this chapter, we suggest that the cognition cycle, as envisioned by Mitola [34], be adopted by the network-wide system. To do this we differentiate between node and network-level cognitive processing and show that there is need to investigate how a cognitive ad hoc network would go about orienting, deciding, planning and learning. While network-level cognition may be more readily incorporated into infrastructure-based networks, the transient and distributed nature of mobile ad hoc networks is a challenging environment for any kind of collective activity, be it agreement or learning. In order to meet the challenge, it is necessary to look beyond the traditional methods used in distributed systems. Therefore, we present a discussion of some observed social conventions which enable loosely associated groups of people to pool knowledge so as to collectively learn and to make decisions which would otherwise require the intervention of a centralized expert. The title of the chapter is in fact borrowed from one of the titles referenced in the discussion, namely the book by James Surowiecki entitled *The Wisdom*

Cognitive Networks: Towards Self-Aware Networks Edited by Qusay H. Mahmoud

of Crowds [39]. The characteristics of the social conventions, discussed by Surowiecki and others appear suited to the cognitive ad hoc networking domain.

The last part of the chapter focuses on the potential role of cognitive ad hoc networks in future communication systems. The application area chosen to explore and highlight the roles for cognitive ad hoc networks is dynamic spectrum access (DSA) and three different DSA regimes are discussed in that context.

The design of robust cognitive ad hoc networks is demanding. This chapter does not attempt to present solutions to what is a complex and large challenge. It does, however, endeavor to delineate a path that can begin to address some of the challenges involved.

8.2 Towards Ad Hoc Networks

Many networking systems, algorithms and techniques lay claim to the notion of having ad hoc attributes or employing ad hoc techniques, and through such labeling these systems impute that they are somehow more capable, more flexible and more prepared for dynamic communications challenges than competing systems and standards. But a networking system that can truly call itself ad hoc should be more than a network which allows plug-and-play master/slave connectivity, such as that provided by Bluetooth [19], or which allows peer-to-peer network formation such as that enabled by the IEEE 802.11 MAC [19]. Within these standards the scope for quasi-ad hoc behavior is tightly constrained. So, before we even begin to discuss cognitive ad hoc networks, we take a fresh look at what it means to be *simply* ad hoc.

Returning to basics, the Latin phrase *ad hoc*, translates literally to the English phrase *to this*. In contemporary usage, it is often used as an adjective, meaning *for this purpose; to this end; for the particular purpose in hand or in view*. One of the more familiar usages of the phrase is in the context of institutional ad hoc committees. Such committees are generally established by parliaments, unions, corporations and other bodies on a temporary basis to address unforeseen issues and events, to serve as forums for debate and decision making and to allow their institutions to respond to new or changed circumstances. Such ad hoc committees are required when no permanent committee structure exists from day to day, as the cost is prohibitive or the need for one is not perceived, or when existing permanent committees are not constituted or staffed in a manner suited to deal with the issue at hand.

It is important to note that when such ad hoc committees are formed, they are assembled to deal with the issue at hand and are *fit for that purpose*, and no other purpose. Once the task has been completed, the committee is disbanded and the resources (i.e., people) are released for other uses.

Generally speaking then, the need for such ad hoc entities stems from the advent of problems, the identification of tasks and the demand for services which are beyond the scope of existing systems. In many cases it is more cost-effective to address a problem when it arises, marshalling available resources there and then, rather than having pre-existing structures ready for an event or need that may never arise.

Bearing that general definition of what it means to be *ad hoc* in mind, we find that there are two definitions of an ad hoc network: a definition emanating from the ideal and a definition informed by reality. In the first case, that of a definition informed by the ideal of ad hoc behavior, ad hoc networks could be described as networks that form for a specific purpose, for as long as that purpose exists. The notion of being able to come into being for a specific purpose, and indeed to be able to cease being when the specific purpose no longer exists, is a very powerful and attractive notion. It embodies a sense of dynamic responsiveness as well as a sense of fitness for purpose. However, that definition is obviously too general to be translated by an engineer into an algorithm, a protocol or a networking system. When informed by reality, the term *ad hoc network* is often used as shorthand for mobile, wireless, peer-to-peer, multihop networks. It is not that the MANET (mobile ad hoc network) community, the people working at the coal-face of ad hoc network research and development, set out to define ad hoc networks as such, but the target hardware and specific set of problems that brought about a renewed interest in the area dictated that the definition should encompass the attributes that distinguish this type of network from the existing planned, infrastructure-based networking systems. Each part of that definition corresponds to a challenging attribute that must be considered in the development of any ad hoc algorithm or protocol. However, that list is not a sufficient description of an ad hoc network.

The concept of an *ad hoc network* is not entirely novel. Ad hoc packet radio networks that did not rely on pre-existing infrastructure were developed for military purposes in the 1970s by the United States Department of Defense Applied Research Projects Agency (DARPA) in such projects as PRNET [21, 20], and SURAN [2]. However, the late 1990s saw the advent of more powerful and more portable computing mobility devices and the development of a range of wireless transmission media, such as IEEE 802.11 and HiperLAN. These advances spurred a renewed interest in the development of ad hoc networks within the wider research community, leading to a dramatic growth in the MANET community.

The MANET research has yielded a plethora of protocols and algorithms for the various layers of the communications stack. In the main, the focus of most parties was on tackling the problems which are identified by the oft-cited attributes used to describe the ad hoc network, i.e., mobility, peer-based, distributed. In many cases, the approach taken to design protocols for ad hoc networks was based on tweaking and optimizing the protocols that existed for the infrastructure networks. So, at the routing layer, the three protocols that graduated to the IETF standardization process are mobility-optimized versions of classical routing techniques – AODV [36] (distance-vector), DSR [18] (source-routed) and OLSR [17] (link-state).

The drive to optimize the performance of these protocols in networking *windtunnels* (i.e., simulators) has not served the research community well. Ad hoc networks are not equivalent to infrastructure networks, whether wired or wireless. While it is possible to model a relatively well-behaved, predictable and static network in a simulator so that the results of simulation will bear some relation to reality when the new algorithm is deployed in a real network, the same does not hold true for ad hoc networks. Simulations in which the network population is bounded, and in which the mobility modeling of the nodes is tightly regulated do not represent ad hoc networks. A mobile, peer-to-peer, distributed network is an inherently *free* network – free from control, wire, infrastructure and hierarchy. The transient behavior of an ad hoc network is not captured in a simulator, partly because the means to address that transience has not yet been addressed.

Ad hoc networks are dynamic organisms, pursuing dynamic goals in the face of dynamic constraints. That dynamism is the most pertinent characteristic of mobile, wireless, peer-to-peer, distributed networks. However, link and topological changes brought about by mobility are only one of the sources of transience experienced in an ad hoc network. The purpose of the network is also subject to change at the whim of the users partaking in the network. Both the literature, and our experience, has demonstrated that different protocols are more suited to particular types of traffic, whether it is multicast-like traffic, sensor-sink traffic, peer-to-peer streaming, etc. Furthermore, the interaction of transport, MAC and routing schemes in ad hoc networks is heavily dependent on the physical distribution of the nodes, i.e. local node densities, and the distribution and type of traffic. It requires an ad hoc response to make the network fit for purpose. We identify three distinct phases that a system must pass through in order to provide an ad hoc response to any event.

- Phase one: the system must be able to identify and understand what is being demanded of it, what its current purpose is. That is easier said than done in an ad hoc network. Remember that an ad hoc network will most likely take the form of a distributed system. As such there is no network-wide, omniscient entity to ascertain what demands are being placed on the network. Nodes may only have local or regional knowledge about the state of the network. Consequently, it is a challenge for each node to understand the network-wide demands on the network.
- Phase two: the system needs to identify the resources that can be used to provide an ad hoc response to those demands. Again, the lack of network-wide information impedes the ready reckoning of this information.
- Phase three: the network must marshal its resources in response to the demands so that it is fit for the task that has been identified.

In short, an ad hoc network requires an ability to marshal resources on the fly, without a reliance on *a priori* knowledge.

8.3 A Cognitive Ad Hoc Network

A cognitive network can be described as a network that is capable of knowing and understanding the context within which it finds itself and is capable of processing and acting on that knowledge as it sees fit. It could be said that cognitive networks also embody a sense of dynamic responsiveness as actions are typically taken in response to changing circumstances and changing (radio/node/network) resource availability. It is true also that cognitive networks embody a sense of fitness for purpose, as the the response of a cognitive network is typically focused on making best use of whatever limited (or abundant) resources are available for the (communication) purpose in hand. Looked at in this way, it is clear that similar philosophies underlie both ad hoc and cognitive networks.

The responsiveness in a cognitive network is, however, much more involved than in the case of ad hoc networks, as cognitive networks must be capable of learning and planning and to do so need a greater sense of self-awareness. It could be argued that a fully functioning cognitive network is the natural evolution endpoint of an ad hoc network.

8.3.1 The Cognitive Cycle

Since being introduced by Mitola [34], the operation of cognitive radios has been frequently envisioned by the cognition cycle. The cognition cycle is a state machine that shows the stages in the cognitive process for a cognitive radio and a simplified version

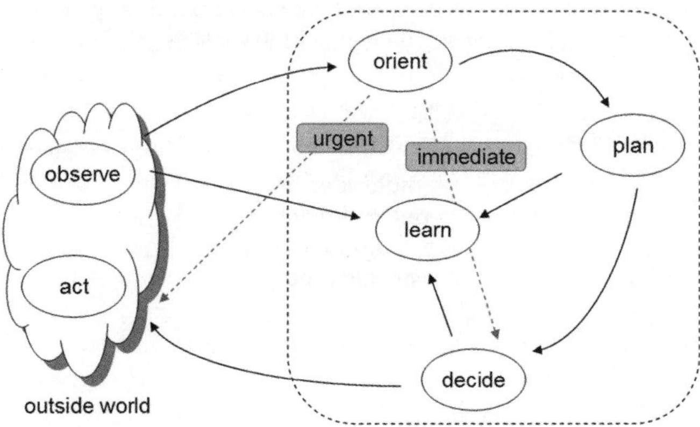

Figure 8.1 The cognition cycle

is shown in Figure 8.1. In simple terms, a radio receives information about its operating environment – the outside world. This corresponds to the **Observe** state. This information is then evaluated to determine its importance during the **Orient** state. Based on this evaluation, the radio can either react immediately and enter the **Act** state, or it can determine its various options in a more considered manner during the **Decide** state, or it can **Plan** for the longer term before deciding and acting. Throughout the process, the radio is using these observations and decisions to improve the operation of the radio and to **Learn**. The cognition cycle, though only approximating the process has proved a very useful framework within which to discuss the concept of the cognitive radio and is used as a basis for a large amount of work in the area [37, 33, 14].

In network terms, the cognition cycle tends to model the process that occurs at a node. It does not fully capture the network elements of the process. To frame the operation for a cognitive network, therefore, the cognition cycle needs to be expanded.

8.3.2 Cognition in Networks

Figure 8.2 is an attempt to redraw the cognition cycle and in doing so, to include the network element of the process. To this end, two individual nodes in the network, node x and node y, are depicted. The key point to be taken from Figure 8.2 is the fact that two distinct levels of processing exist. Let us denote the first level as *node-level cognitive processing* and the second level as *network-level cognitive processing*. In other words any node in a cognitive network can, in a standalone fashion, make decisions about how to react to a given situation or a collective decision can be taken by multiple nodes.

In Figure 8.2, the node-level Orient stage is concerned with the determination of whether the decision to be taken is *unilateral* or *multilateral*. Unilateral decisions can be taken by a node alone and involve evaluating actions that only affect the individual node making and executing the decision (node-level cognitive processing). An example of such a unilateral decision is one regarding local parameter values for the size of the route cache. Multilateral actions are ones that have a network-wide effect, such as a decision regarding the choice of routing protocol (network-level cognitive processing). The

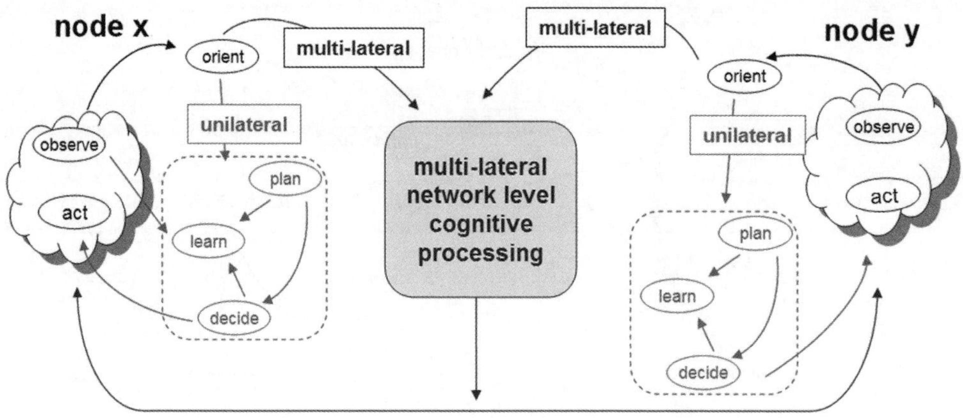

Figure 8.2 The cognition cycle in cognitive networks

processing depicted in the center of Figure 8.2 is intended to capture this notion of the multilateral, the collective and the network-wide jurisdiction.

8.3.2.1 Centralized, Infrastructure-based Networks

Figure 8.3 expands on Figure 8.2 and depicts a specific cognitive network scenario involving an infrastructure-based network. In this scenario some god-like entity exerts ownership and control over the nodes within its range. A base station or an access point is an example of such an entity. All nodes in the diagram are within radio range of the base station.In this case, the network-level cognitive processing is located within the base station or access point. This is emphasized in Figure 8.3 by associating a network-level cognition cycle specifically with the base station. Information and preferences can be collected and

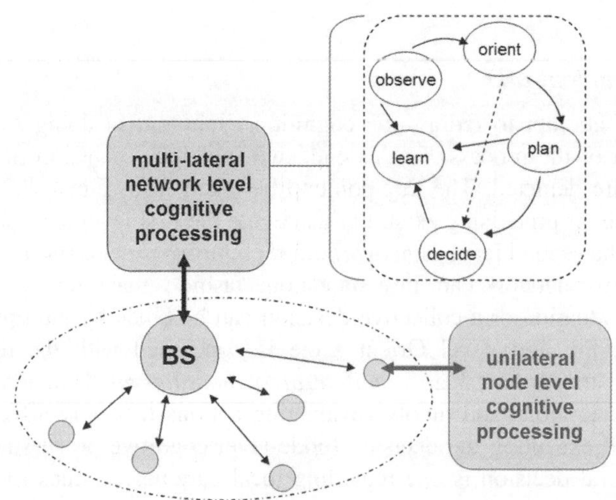

Figure 8.3 The structure of a cognitive infrastructure-based network

aggregated from participating nodes, who perform their own node-level cognitive processing, but whose observations are combined at the network-level Observe stage. The base station or controlling node can then move through the cycle. In the network-level Orient stage the base station determines whether immediate or long-term action is needed to address the *collective* needs of the participating nodes. The network-level Decide and Plan stages can involve input from all participating nodes and once a course of action has been selected the base station can also control and synchronize the execution of the selected action by all of the nodes in the cell. Of course, the strategy that the controlling base station uses can vary and the collective needs of the nodes in the network can be served on the basis of discussion with all nodes that leads to some kind of a consensus or on the basis of a more dictatorial approach.

The IEEE 802.22 [6] wireless regional area network is a real example of a system that could be described by Figure 8.3. The transition to digital television has, and will further result, in empty and unused bands of spectrum which have been allocated for broadcast TV channels – such empty spectrum is known as *white space*. Making the best use of this available spectrum without at the same time infringing on the incumbents (e.g., TV service providers) is a challenge. The IEEE 802.22 working group has been working towards developing an air interface based on cognitive radios for unlicensed operation in the TV broadcast bands. In order to protect the incumbent services the IEEE 802.22 system follows a strict master/slave relationship with a base station (BS) as the master and the customer premises equipment (CPE) nodes as the slaves within the cell of the base station. The BS manages the sensing that each CPE node might perform to identify white space. Distributed sensing is carried out in the sense that CPEs distributed throughout a given cell take sensing measurements, i.e., perform node-level cognitive processing. However, the sensing is coordinated, managed and interpreted by the BS in a centralized fashion. The observations and analysis performed by each node feeds to the central BS so that decisions can be made by the BS on how to avoid interfering with incumbents. Hence, all nodes can influence and play a role in the collective decision through performing their own local cognitive analysis, but the decision is taken in a centralized fashion.

8.3.2.2 Distributed, Ad Hoc Networks

Figure 8.4, on the other hand, depicts a standard ad hoc network: an on-hierarchical, peer-to-peer, distributed system. The range of each node is indicated by a dotted circle. Here the multilateral cognitive processing is completely distributed. What this means in effect is that the process of making multilateral decisions takes place in a distributed fashion. In this case both the node-level processing and the network-level processing are distributed throughout the network.

The idea of network-level cognitive processing is clear for the case of the network in Figure 8.3. In that scenario, the BS can gather all observations, all preferences, all viewpoints from the participating nodes and a global view can be ascertained and acted upon. In the ad hoc case, network-level cognitive processing is more challenging.

In an ad hoc network, it is generally not feasible, or indeed possible in many cases, to obtain a global view of the network. A node in a distributed system may, at any given time, only know of the existence of its immediate neighbors – its knowledge horizon can be quite limited. For the purposes of this discussion, we define the neighbors of an ad hoc

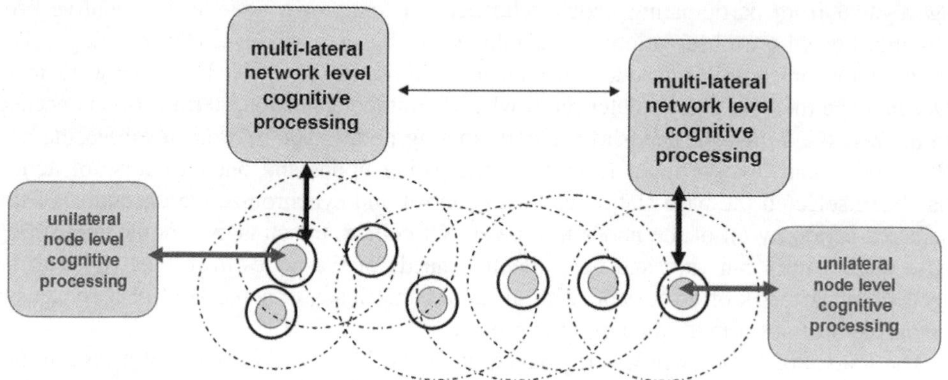

Figure 8.4 The structure of a cognitive ad hoc network

node to be all nodes from which it can successfully receive packets. This is illustrated in Figure 8.5. In the figure, node 1 and node 9 are of interest. Node 1 has neighbors 2, 3 and 4. You will note that while node 5 is within radio range of node 1, node 1 is not within range of node 5. This is an example of an asymmetric link and hence node 5 is not a neighbor of node 1, given our stringent neighborhood definition. In Figure 8.5, node 9 has two neighbors, nodes 7 and 8.

As the neighbors of a given node in an ad hoc network are not static, Figure 8.6 shows the same set of nodes and the changed neighborhood relationships at some later time t_2. In this figure node 1 is now left with no neighbors at all, while node 9 has neighbors 2, 3, 4, 6, 7, 8, 10 and 11.

What this means is there will be, in essence, varying pockets of local understanding and local knowledge that apply to some definition of a local neighborhood around a node and the stages of the network-level cognitive processing illustrated in Figure 8.4 must be understood in this context.

However, the global (or network-wide) aspect must not be forgotten. Figure 8.7 attempts to conceptually capture the issue. In the case of multilateral decisions in an ad hoc network, one of the two situations, illustrated in Figure 8.7, would be the typically desired endpoint. Image (a) shows a multi-textured surface while image (b) shows a smooth surface. The smooth surface corresponds to a global consensus being reached by the ad hoc network

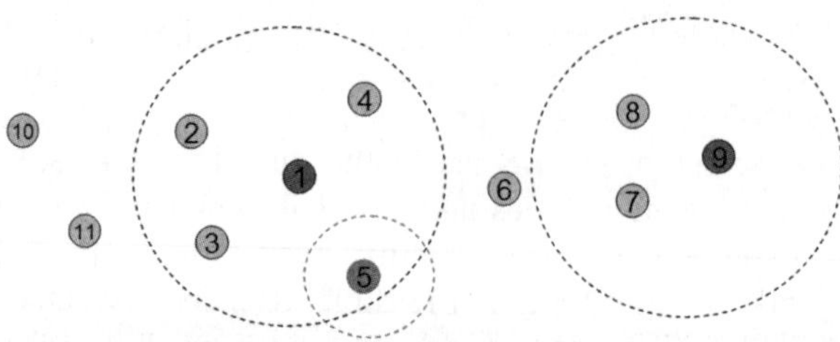

Figure 8.5 A selection of ad hoc nodes at time t_1. The nodes of interest are nodes 1 and 9. Their radio ranges are indicated by a dotted line

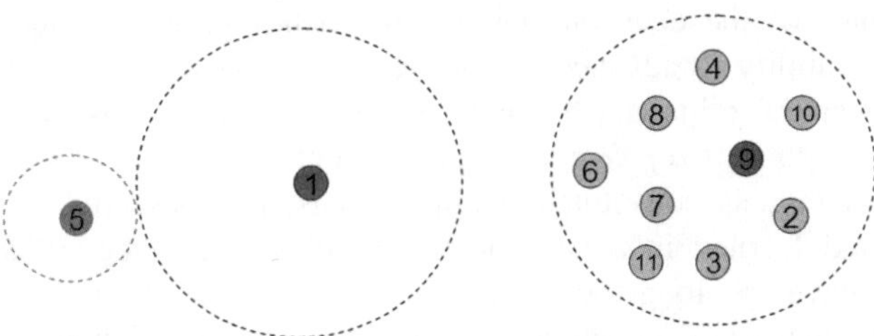

Figure 8.6 A selection of ad hoc nodes at time t_2. The nodes of interest are nodes 1 and 9. Their radio ranges are indicated by a dotted line

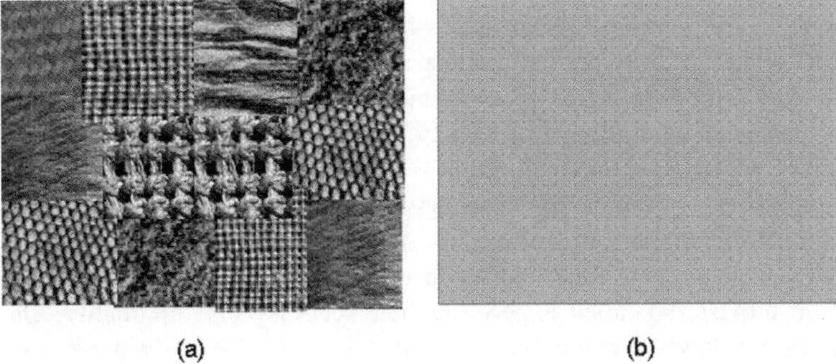

Figure 8.7 Image (a) shows a varied textured surface and image (b) a smooth surface. These images conceptually summarize the cognitive ad hoc network's end goals

while the textured surface represents the notion of a solution that permits a fragmentary approach based around different local requirements and objectives.

Not unlike many other processes targeted to ad hoc networks, distributed consensus, or decision making, is challenging in this environment. As discussed earlier, a node in an ad hoc network may only be aware of its own neighbors and may have no sense of the size and extent of the network. The links in the network can be asymmetrical and the quality of the links between neighbors can vary. There is no central clock and any decision-making process must therefore be entirely carried out in an asynchronous manner. The lack of a central clock also means that the process must have some mechanism for determining whether the process of reasoning, making judgments and resolving conflicts is still ongoing or whether conflict has been resolved. The well-known Fischer–Lynch–Patterson impossibility result clearly delineates the limits of deterministic decision making in systems of the kind characterized by open, mobile ad hoc networks [10]. The challenges associated with cognitive ad hoc networks are therefore even larger than those associated with cognitive infrastructure-based networks.

8.4 The Wisdom of Crowds

If we are to fully embrace the concept of network-level cognition for ad hoc networks, then it is necessary to expand and deepen the pool of decision-making experience from which solutions are drawn. Looking beyond the traditional sources, we have turned our attention to studies of the decision-making approaches used in different groupings in society. Society's ability to advance, to innovate, to address the unforeseen events that have occurred over the millennia has not been based on the existence of some handy guidebook or on a pool of *a priori* knowledge gathered in a laboratory before the Big Bang. Rather, society, as constituted by individuals, has observed, oriented, planned, decided, acted and learned in a very organic fashion over time. If cognitive ad hoc networks are to develop along the lines of the cognition-cycle framework outlined by Mitola [34], then the actual techniques by which cognitive ad hoc networks observe, orient, plan, decide, act and learn must become as flexible and organic as those techniques used by society in its cognition cycle.

Among the most promising sources of inspiration are the works of Everett Rogers, *The Diffusion of Innovations* [38], and both James Surowiecki's [39] and Charles Mackay's [29] discussions and analyses of the means by which the human masses have both successfully and unsuccessfully pooled the disparate and distributed *wisdom of crowds*.

Both Mackay and Surowiecki refer to the collection or grouping of individuals or massing of people in certain parts of society as the *crowd*. The crowd manifests itself in many different ways and in many different societal situations. The crowd can literally be a crowd on the street. It may be the investors in a stockmarket. It may be the workers at a corporation or the academics in a university. James Surowiecki, in his book, *The Wisdom of Crowds* [39], documents time and again the ability of the crowd to make decisions that rival, or surpass, so-called experts in their accuracy and optimality. On the other hand, Charles Mackay's *Extraordinary Popular Delusions and the Madness of Crowds* has documented the errant behavior of the crowd, most notably the irrational behavior that concluded in the South Sea Bubble of 1720.

Both Surowiecki and Mackay suggest that the difference between wisdom and madness revolves around the ability of individuals to balance the influence they place on the public information, i.e., that gleaned from the masses, over their own private information, i.e., information based on their own first-hand observations. The *wisdom of crowds* refers to fact that the knowledge of a group of *average* people can rival or surpass the wisdom of an expert. Of course, it is only possible to leverage this wisdom from the crowd if the crowd exhibits certain traits (independence and diversity) and behaviors (speculation). Independence and diversity are seen as important for successful decision making, i.e., distilling the wisdom from the madness in the crowd. Independence is important as the mistakes of individuals will not correlate if independent means and methods are used to make a judgment or to gather information. So a group's collective judgment won't be in error unless there is systematic individual error making.

On the other hand, diversity is important as there is a need to encourage speculative ideas or solutions. If there is diversity among entrepreneurs, i.e., those who go about introducing new ideas to the group, then there will be meaningful differences between the choices that they introduce and the group has a better chance of finding a successful solution from among *many different* proposals. Independence and diversity can manifest themselves in a cognitive ad hoc network so long as it is an open, heterogeneous network.

The means by which innovative ideas are evaluated, and possibly adopted, through communities has been extensively studied by Everett Rogers [38]. In an earlier work, we developed a decision-making system for ad hoc networks based on the concept of diffusing an innovation (or change) throughout the network. In our system, the ad hoc network layer had the choice of using three different routing protocols, each of which was optimal for a given set of network conditions [11]. An innovation in that system was a proposal to run a different protocol when the network characteristics warranted such a change.

In that work, we addressed some of problems posed by the Fischer–Lynch–Patterson impossibility result. In fact, we found that the use of *soft* distributed decision-making techniques was an elegant and light-weight way to circumvent that impossibility result. Our solution involved a continuous phase of soft persuasion in which the classically defined actors of an innovation diffusion process, e.g. early adopters and the late majority, exchanged private information (i.e. first-hand local observation) and public information (i.e. second-hand, network-disseminated observations) through the crowd. The diffusion of innovations generally operates as a sophisticated information cascade in which certain actors in the process guard against herding tendencies which could lead to rash and ill-informed group decisions.

However, a major drawback of our system, when contemplated from the point of view of cognitive ad hoc networks, was that the system had to know, *a priori*, which routing protocols suited the prevailing conditions being experienced by the nodes. By conducting experiments under a range of different conditions we ascertained which protocols yielded the optimal results, e.g., highest throughput or lowest overhead, for a given set of node mobility, node density and traffic conditions. Our system enabled both the network and the nodes to *observe* the network conditions. However, a weakness of our system, when viewed from the perspective of facilitating cognitive ad hoc networking, was that it was unable to cope outside of the networking conditions that we had the foresight to plan for.

A cognitive evolution of that system could see nodes experimenting with new solutions. If the ad hoc nodes had access to a component-based routing layer [27] and a component-based software-radio-based physical layer [30], then the possibility of speculating and innovating new solutions is present. In that context, the planning, deciding and learning stages of the network-level cognition cycle could be designed to innovate using these basic building blocks which must be intelligently assembled to provide both network-layer and physical-layer functionality. As Surowiecki notes, a successful system will generate many solutions, but more importantly, it will recognize the losers and kill them off fast.

Surowiecki cites Kaivan Munshi's study [35], which, like that of Rogers, concerned the decision-making processes used by heterogeneous communities to trial, and possibly adopt, new technologies. In this study, the focus was on the adoption of new strains of wheat and rice by Indian farmers during the so-called Green Revolution of the 1940s through to the 1960s. The wheat farmers and rice farmers used differing techniques to evaluate the possible adoption of new crops.

In the wheat-growing regions, the land conditions were fairly uniform and crop yields did not generally differ much from farm to neighboring farm. Referring back to Figure 8.7 in Section 8.3.2.2, the notion of texture and smoothness becomes relevant again. Rather than those textures representing the uniformity or fragmented nature of a decision-making process as they did in that section, for now Figure 8.7(b) should be seen to represent the case in which the neighboring land conditions are fairly uniform, as depicted by the smooth texture. Figure 8.7(a) depicts the opposite case of a swathe of land in which the conditions change from field to field. When a wheat farmer adopted a new seed, the success or failure of the crop was watched closely by his neighboring farmers. If the seeds prospered, then it was a good indicator that the new seed would do well on neighboring farms which had similar land conditions. So, observing that his neighbor did well and knowing that they shared the same land conditions would allow the farmer to intelligently imitate his neighbor.

However, in the rice-growing areas, there was considerable variation in the condition of the land from farm to farm, Figure 8.7(a), and there were substantial differences in the amount of rice that neighboring farmers yielded. Consequently, the fact that one neighbor did well with a new type of rice could not be taken as an indicator that the same would hold on a neighboring farm. So, in this situation, the option of imitating one's neighbor was not possible. Rather, the farmers engaged in crop trials and experiments with the new rice variety on a small patch of their own land, so that they could prove to themselves that the new variety was a better option, i.e., they did not rely on any public information. In terms of the cognitive ad hoc network, it should be just as important that a node (farmer) could understand the nature of its radio/network environment (farm land) with respect to that of its neighbors. As shown in that example, it was the farmers' understanding of the relationship between each other's land that informed them as to what technique they would use to decide.

8.5 Dynamic Spectrum: Scenarios for Cognitive Ad Hoc Networks

The technical challenges of designing cognitive ad hoc networks, which have the type of characteristics that are described in the previous section of this chapter, are vast. The natural question that arises, is whether a need for such complex communications systems will exist. This question is the focus of the final part of the chapter. To answer it we look to dynamic spectrum management and access and explore the proposed roles for cognitive ad hoc networks in this field.

The current spectrum regulatory regime, in the main, takes what is known as a *command and control* approach to the management of spectrum. Regulators specify which band of frequencies can be used for which services and which technologies should be used to deliver these services. The cycle involved in allocating spectrum for certain usages, specifying the characteristics of the usage and awarding or granting licenses for use can be slow and hence the ability of the command and control approach to keep apace with technological innovation can be quite limited. Added to this there is growing evidence that while on the one hand there is a perceived scarcity of spectrum due to the increasing demand for mobile RF services, on the other hand measurements have shown spectrum usage levels of no more than 10% being common [32]. Hence there is a need for spectrum management regimes that can keep apace with technology innovation and that facilitate more efficient use of the spectrum.

The field of *dynamic spectrum access* focuses on new and very dynamic methods for managing spectrum that move away from this traditional command and control means of regulation. A dynamic spectrum access based regulatory regime would embrace liberalization of spectrum usage, be technology neutral, encourage techniques for spectral efficiency, support innovation and open the spectrum up to a greater number of potential players. To facilitate this kind of vision, cognitive radios and cognitive networks will be needed, especially for the purposes of interference management. To examine the role of cognitive ad hoc networks, in a world of dynamic spectrum access, we divide dynamic spectrum access approaches into a number of broad categories, namely approaches that coexist with existing regimes (Section 8.5.1) and market-based regimes that involve the trading of exclusive usage rights and commons regimes (Section 8.5.2 and Section 8.5.3, respectively).

8.5.1 Cognitive Ad Hoc Networks and Current Regimes

Cognitive radio is typically associated with the opportunistic use, by what are termed *secondary* users, of unused parts of *primary* users' spectrum. The secondary users are typically unlicensed users who have the ability to sense *white spaces* in the licensed primary users' spectrum and provided they do not cause harmful interference to primary users, use the spectrum for their own needs. The primary users have priority access to the spectrum. Hence, secondary users making use of white spaces, must vacate the bands when a primary user requires the spectrum. This type of opportunistic use is also termed *overlay* use as secondary users overlay their services opportunistically on top of the primary users. This kind of usage has already been mentioned in Section 8.3.2.1. The IEEE 802.22 system for making use of the TV white spaces, which result from the transition to digital television, is an example of a system that overlays new services on top of an existing regime and that makes use of cognitive-sensing techniques to ensure no infringements of the rights of the incumbents (e.g. TV service) occur. A spectrum *underlay* is also possible. This is a secondary easement for low-power devices that allows users to operate in the noise floor of the primary, licensed spectrum user. Ultrawide-band (UWB) devices are often cited as the type of devices that could be used to provide underlay services [3]. The vision for this kind of opportunistic approach to spectrum access is one in which overlay and underlay techniques will allow secondary users to fit in around existing primary user regimes, using whatever technologies are suitable, and in doing so make optimal use of the spectrum.

Cognitive ad hoc networks have a role to play in this type of vision. To examine this, the following diagram is reproduced from the paper 'A NeXt Generation/Dynamic Spectrum Access/Cognitive RadioWireless Networks: A Survey' [1]. Figure 8.8 depicts the authors' view of next-generation network architectures and is a useful summary diagram for the purposes of the current discussion.

In Figure 8.8, the secondary users form what the authors' term *xG networks*. In the diagram the authors have depicted a set of xG users functioning in infrastructure mode, in which an xG base station controls access to the white spaces and a set of xG users functioning in ad hoc mode. While the paper does not focus on the details of the ad hoc mode operation, this is an example of where a cognitive ad hoc network is needed. Without the centralized system in which a base station, in this case through a spectrum broker, gets access to spectrum and subsequently manages the sharing of the spectrum among the participating xG users associated with the base station, a distributed cognitive assessment (distributed sensing) of the available spectrum is needed and distributed cognitive decisions on how to make use of the white-space opportunities are necessary. All of the challenges for cognitive ad hoc networks mentioned in this chapter come to bear in this complex example.

The UWB underlay example is also a prime example of where cognitive ad hoc networks are needed. The idea for underlay use, as stated above, is that low-power devices are used in order to keep within a certain noise floor. Low-powered devices by their nature will have lower ranges than similar devices using higher powers. The natural way to extend the range of any such devices will be in the form of multihop ad hoc networks.

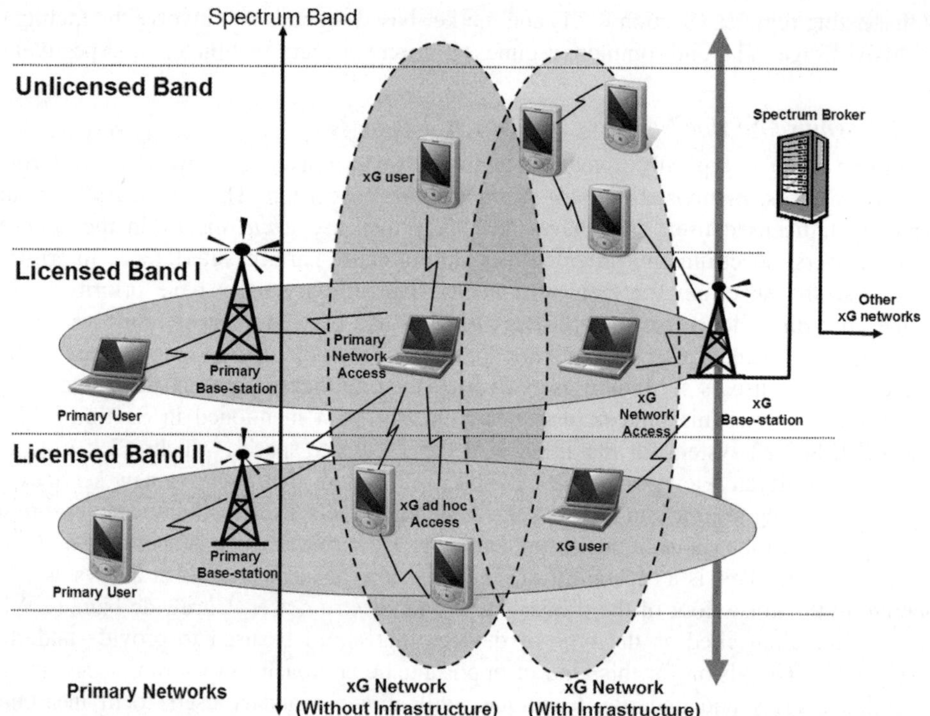

Figure 8.8 Representation of a next-generation architecture, by Ian F. Akyildiz, Won-Yeol Lee, Mehmet C. Vuran and Shantidev Mohanty

The cognitive properties of the network will be needed for interference management, i.e., to ensure that the operation of the network does not exceed the noise floor threshold. In a distributed system there may be pockets of the ad hoc network that contribute to exceeding the noise floor threshold and corrective action would be needed to address this situation. It is possible to envisage this happening in multiple ways, either through for example the rerouting of traffic around the problem zone or the selective deactivation of nodes.

8.5.2 Cognitive Ad Hoc Networks and Market-based Regimes

Rather than use the underlay and overlay opportunistic approaches outlined in the previous section, there is also the opportunity to embrace what are termed market-based approaches. There is an emerging international consensus that market regimes will play much larger roles in the future of spectrum management and that a significant move away from the current command and control approach is inevitable. The arguments for this move are well established and regulators such as the FCC,[1] Ofcom,[2] and the EU,[3] have already indicated a move towards the use of more flexible trading regimes.

[1] Notice of proposed rule-making (FCC 00–402):promoting efficient use of spectrum through elimination of barriers to the development to secondary markets, Federal Communications Commission, November 2000.

[2] Spectrum framework review statement, The UK Office for Communications, June 2005.

[3] The RSPG opinion on secondary trading of rights to use radio spectrum (rspg04-54 final version), EU Commission Radio Spectrum Policy Group, November 2004. frequencies.

In this instance we are concerned with the trading of exclusive usage rights. In simple terms an exclusive usage rights approach involves the commoditizing or quantizing of spectrum into blocks of usage rights that provide exclusive access to spectrum. Ideally, such a market for trading this commodity would allow spectrum consumers to assemble packages of spectrum that suit their intended network technologies and projected services. The term *spectrum consumer* has been chosen to emphasize the fact that the nature of future spectrum users or businesses built around fluid spectrum are not preempted. Rather, we consider that some agent will want to acquire spectrum for some, possibly still unforeseen, use. We currently would understand such a market actor in terms of entities such as cellular network operators, TV companies and wireless broadband providers, who could then provide services of their choice, using technologies of their choice to their *customers*.

The definition of the rights associated with exclusive usage of spectrum is a challenging problem. At the most basic level blocks of usage rights are considered as having frequency, space and time dimensions. In reality these dimensions are not complex enough to capture usage rights as electromagnetic radiation cannot be contained in neat parcels. Hatfield [13] does an excellent job of summarizing many of the main contributions to this debate and traces the development of the definition of spectrum usage rights through from Coase [4, 5], to De Vany [8, 9, 7], to Kwerel [25, 24, 23, 22] and to Matheson [31], before going on to further the discussion himself. While these discussions are key to progress in this area, they are not the focus of this chapter. Instead, we focus on the characteristics of networks that would support an exclusive usage rights market.

On first glance, an exclusive usage rights approach tends to lend itself to centralized infrastructures as the spectrum consumers of the rights would manage their spectrum for the geographical space in which they own the rights, over the time period for which the rights are valid. However, this needs to be examined in the context of evolving technology.

In the early stages of an exclusive rights market, it is probable that networks will exist which are much like today's networks and operate over a limited range of frequencies. As technology advances, networks will become more flexible making greater use of software-defined radio, wideband frequency agile RF frontends, wideband antennas, self-organizing and self-planning algorithms and general cognitive capabilities. The consequences of these developments will mean that networks will be capable of operation in multiple modes, at a wider range of frequencies, availing of a wider range of communication techniques than before. Hence, it is envisaged that spectrum will become more fungible, that is one band of spectrum can *take the place of* another band of spectrum for a given *spectrum consumer*. With this in mind, Figure 8.9 is of interest.

In Figure 8.9(a) the access points/base stations of a cellular-like network, that is operating at a certain range of frequencies, are depicted. In Figure 8.9(b) the access points/base stations of another cellular network operating at a higher frequency are shown. The number of access points/base stations needed in case (a) is much smaller than the number needed in case (b) to cover the same geographical area. This is because of the decrease in radio range that is associated with the increase in frequency of operation, as indicated by the dotted lines surrounding each access point/base station.

Hence, in order for a spectrum consumer that uses a cellular-like approach to switch from frequency range (a) to frequency range (b), as depicted in Figure 8.9, more access points/base stations will be needed. And a reduction in the number of access points/base

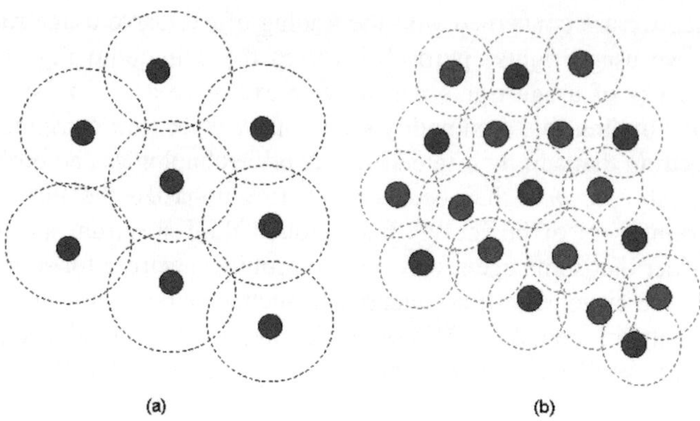

Figure 8.9 Network uses frequency range f_1. (b) Network uses frequency range $f2 > f1$

stations would be needed to facilitate a decrease in frequency range. (The increase in access points/base stations could be achieved by multiple infrastructure owners acquiring the access rights together for example.)

The increasing and decreasing of the network of access points/base stations can been viewed as an ad hoc network that is increasing and decreasing in size in response to the changing dynamics of the frequency allocation. While this is a more contained challenge than the opportunistic usage described in Section 8.5.1, the idea of using a cognitive ad hoc network, operating albeit at a less dynamic level, to self-organize and self-plan an ad hoc response for the new frequency range is central to this idea.

8.5.3 Cognitive Ad Hoc Networks and Commons Regimes

At the other end of the spectrum from the exclusive usage rights approach are commons regimes. A commons regime supports unlicensed access to the spectrum, much like in the case of the current ISM bands. A commons offers nonexclusive shared use, i.e., in the case of the commons there is no concept of primary and secondary users. All users are equal and access to the spectrum is open to all, subject to obeying the etiquette of that commons.

The main criticism leveled at a commons approach is the potential for, what is termed, the *tragedy of the commons* that results from overuse of what is essentially a free resource. Currently the goal for researchers who advocate a commons approach is to be able to specify some kind of etiquette or behaviors that can be used to govern the commons and so avoid the tragedy, while avoiding using heavy-handed means to achieve that goal. In the ISM bands, simplistic mechanisms for control of the commons exist, namely there are strict power restrictions on devices that operate in these bands. In the commons regime of the dynamic spectrum access world, a much more flexible approach is envisaged.

The commons approach by its very nature has an ad hoc network philosophy at its core [28]. A commons does not have a centralized controlling entity. Multihop, peer-to-peer communication may be the central mode of operation. There is lack of *a priori* knowledge of who is participating in the network. Distributed collaboration of some sort will be needed if etiquettes are to be used and some level of interference management so as to avoid the tragedy of the commons. Hence the need for cognitive ad hoc networks is once again underlined and in fact a commons presents significant challenges for such networks. While it is challenging, *the wisdom of crowds* should offer up some solutions.

8.5.4 The Future

The scenarios laid out in the Sections 8.5.1, 8.5.2 and 8.5.3 are scenarios which will form part of the short- and longer term future of communication systems. There is a global consensus emerging for the need for significant change from the command and control approach to spectrum management and there is a significant role for cognitive ad hoc networks in many of the alternative options. While the potential roles have only been briefly illustrated at the end of this chapter, it is clear that dynamic spectrum management and access networks will be a key driver for cognitive ad hoc networks.

8.6 Summary and Conclusions

There are two definitions of ad hoc: the real and the ideal. Unfortunately, engineering pragmatism has somewhat stifled the understanding of what it means to be an ad hoc network. The discussion in this chapter has sought to refocus the definition of ad hoc around its true meaning in order to explore the real potential of ad hoc networking and, in doing so, to illustrate that the natural endpoint for the evolution of ad hoc networks is a fully cognitive ad hoc network. It is important that the wider definition of ad hoc network be embraced for the field to grow.

The use of the cognition cycle for defining the stages in the cognitive processes of a cognitive radio has been expanded to take account of the stages of a cognitive network. In doing this, a clear distinction has been made between node-level cognitive processing and network-level cognitive processing. The objective of splitting the processing into two components has been to allow a clear focus on the challenges associated with network level processing for ad hoc networks. The major challenges are in the design of a framework for network-level processing that is not limited for use in predefined scenarios but that can evolve to cope with any situation that arises. Inspiration from other knowledge domains, such as the social science domains mentioned in this chapter, can help develop a better understanding of the framework we seek to design.

Ad hoc networks are not currently in widespread use. While impromptu business meetings and disaster area communication provisioning are commonly cited as example usage scenarios for such networks, there is still limited evidence that ad hoc networks are used in any substantial capacity worldwide. It could be argued that the true usefulness of ad hoc networks has yet to be recognized as well as realized. The communication systems of the future will be more flexible, they will be adaptive and they will be highly dynamic (such as in the case of dynamic spectrum access networks described in this chapter). The cognitive ad hoc network captures the sense of dynamic responsiveness, the sense of fitness for purpose, the sense of evolution and sustainability that will be needed in communication systems that can meet the challenges of dynamic spectrum access among other challenges in the future.

The challenge going forward is to design these networks to leverage the advantages gained from heterogeneous, distributed, disaggregated nodes rather than seeing the differing functionality, capabilities, influences, connectedness of the network participants as being restrictions. The challenge going forward is to tap into the wisdom of crowds.

References

[1] Akyildiz, I.F., Lee, W.Y., Vuran, M.C. and Mohanty, S. (2006) NeXt generation/dynamic spectrum access/ cognitive radio wireless networks: a survey. *Computer Networks: The International Journal of Computer and Telecommunications Networking*, **50**(13), 2127–59.

[2] Beyer, D. (1990) *Accomplishments of the DARPA SURAN program*. Conference Record of the Military Communications Conference (MILCOM), pp. 855–62, September 30–October 3.

[3] Cabric, D., Chen, M.S.W., Sobel, D., Wang, S. *et al.* (2006), Novel radio architectures for UWB, 60 GHz, and cognitive wireless systems. *EURASIP Journal on Wireless Communications and Networking*, April.

[4] Coase, R.H. (1959) The Federal Communications Commission. *Journal of Law & Economics*, **2**, 1–40.

[5] Coase, R.H. (1962) The Interdepartment Radio Advisory Committee. *Journal of Law & Economics*, **5**, 17–47.

[6] Cordeiro, C.M., Challapali, K., Birru, D. and Shankar, S. (2005) IEEE 802.22: an introduction to the first wireless standard based on cognitive radios. *Proceedings of the 1st IEEE International Symposium on New Frontiers in Dynamic Spectrum Access Networks (DySPAN),* pp. 328–37, November 8–11.

[7] De Vany, A.S., Eckert, R.D., Meyers, C.J., O'Hara, D.J. *et al.* (1969) A property system for market allocation of the electromagnetic spectrum: a legal–economic–engineering study. *Stanford Law Review*, **21**, 1499–561.

[8] De Vany, A. (1998) Implementing a market-based spectrum policy. *Journal of Law & Economics*, **41**, 627–46.

[9] De Vany, A. (1998) Property rights in the electromagnetic spectrum. In Peter Newman (ed.), *The New Palgrave Dictionary of Economics and the Law*, 3rd edn, pp. 167–71, Macmillan, London.

[10] Fischer, M.J., Lynch, N.A. and Paterson, M.S. (1985) Impossibility of distributed consensus with one faulty process. *Journal of the Association for Computing Machinery*, **32**(2), 374–82.

[11] Forde, T.K., Doyle, L.E. and O'Mahony, D. (2006) Ad hoc innovation: distributed decision-making in ad hoc networks. *IEEE Communications Magazine*, **44**(4), 131–7.

[12] Hamacher, C. (2002) Guard bands for coexistence of UMTS and DVB in a hybrid radio system. *Proceedings of the IST Mobile Communications Summit*, pp. 261–5.

[13] Hatfield, D.N. and Weiser, P.J. (2005) Property rights in spectrum: taking the next step. *Proceedings of the* 1st *IEEE International Symposium on New Frontiers in Dynamic Spectrum Access Networks (DySPAN)*, pp. 43–56, November 8–11.

[14] Haykin, S. (2005) Cognitive radio: brain-empowered wireless communications. *IEEE Journal on Selected Areas in Communications*, **23**(2), 201–20.

[15] Hazlett, T. (2006) The spectrum-allocation debate: an analysis. *IEEE Internet Computing*, September/ October, pp. 52–8.

[16] Jackson, C. (2005) Limits to decentralization: the example of AM radio broadcasting or was a common law solution to chaos in the radio waves reasonable in 1927? *Proceedings of the Telecommunications Policy Research Conference (TPRC)*.

[17] Jacquet, P., Mhlethaler, P., Clausen, T., Laouiti, A. *et al.* (2001) Optimized link state routing protocol for ad hoc networks. *Proceedings of the IEEE International Multi Topic Conference (INMIC)*, pp. 62–8, December 28–30.

[18] Johnson, D.B. and Maltz, D.A. (1996) Protocols for adaptive wireless and mobile networking. *IEEE Personal Communications*, **3**(1), 34–42.

[19] Jordan, R. and Abdallah, C.T. (2002) Wireless communications and networking: an overview. *IEEE Antennas and Propagation Magazine*, **44**(1), 185–93.

[20] Jubin, J. and Tornow, J.D. (1987) The DARPA packet radio network protocols. *Proceedings of the IEEE*, **75**(1), 21–32.

[21] Kahn, R.E., Gronemeyer, S.A., Burchfiel, J. and Kunzelman, R.C. (1978) Advances in packet radio technology. *Proceedings of the IEEE*, **66**(11).

[22] Kwerel, E.R. and Felker, A.D. (1985) *Using auctions to select FCC licensees*. FCC Office of Plans and Policy, Working Paper No. 16, Washington, DC.

[23] Kwerel, E.R. and Williams, J.R. (1992) *Changing channels: voluntary reallocation of UHF television spectrum*. FCC Office of Plans and Policy, Working Paper No. 27, Washington, DC.

[24] Kwerel, E.R. and Williams, J.R. (1998) Free the spectrum: market-based spectrum management. In T.W. Bell and S. Singleton (eds.), *Regulators' Revenge: The Future of Telecommunications Deregulation*, pp. 101–11, Cato Institute, Washington, DC.

[25] Kwerel, E.R. and Rosston, G.L. (2000) An insider's view of FCC spectrum auctions. *Journal of Regulatory Economics*, **17**, 253–89.

[26] Kolodzy, P. (2006) Interference temperature: a metric for dynamic spectrum utilization. *International Journal of Network Management*, **16**(2), 103–13.

[27] Lee, M.J., Zheng, J., Hu, X. and Juan, H. *et al.* (2006) A new taxonomy of routing algorithms for wireless mobile ad hoc networks: the component approach. *IEEE Communications Magazine*, **44**(11), 116–23.

[28] Lehr, B. and Crowcroft, J. (2005) Managing shared access to a spectrum commons. *Proceedings of the 1st IEEE International Symposium on New Frontiers in Dynamic Spectrum Access Networks (DySPAN)*, pp. 420–44, November 8–11.

[29] Mackay, C. (2000) *Extraordinary Popular Delusions and the Madness of Crowds*, Templeton Foundation Press.

[30] Mackenzie, P. (2004) *Software and reconfigurability for software radio systems* (Ph.D dissertation) Trinity College, Dublin, Ireland.

[31] Matheson, R.J. (2005) Flexible spectrum use rights tutorial. *Proceedings of the NTIA International Symposium of Advanced Radio Technology*, March.

[32] McHenry, M. and McCloskey, D. (2004) New York City spectrum occupancy measurements, September, http://www.sharedspectrum.com. Accessed January 16, 2007.

[33] Mishra, S.M., Cabric, D., Chang, C., Willkomm, D. *et al.* (2005) A real-time cognitive radio testbed for physical and link layer experiments. *Proceeedings of the 1st IEEE International Symposium on New Frontiers in Dynamic Spectrum Access Networks (DySPAN)*, pp. 562–7, November 8–11.

[34] Mitola, J., III and Maguire, G.Q., Jr. (1999) Cognitive radio: making software radios more personal. *IEEE Personal Communications* (see also *IEEE Wireless Communications*), **6**(4), 13–18.

[35] Munshi, K. (2004) Social learning in a heterogeneous population: technology diffusion in the Indian green revolution. *Journal of Development Economics*, **73**(1), 185–215.

[36] Perkins, C.E. and Royer, E.M. (1999) Ad-hoc on-demand distance vector routing. *Proceeedings of the 2nd IEEE Workshop on Mobile Computing Systems and Applications (WMCSA)*, pp. 90–100, February 25–26.

[37] Rieser, C.J., Rondeau, T.W., Bostian, C.W. and Gallagher, T.M. (2004) Cognitive radio testbed: further details and testing of a distributed genetic algorithm based cognitive engine for programmable radios. *Proceedings of the IEEE Military Communications Conference (MILCOM)*, vol. 3, pp. 1437–43, October 31–November 3.

[38] Rogers, E.M. (1996) *Diffusion of Innovations*, 4th edn, Free Press, London.

[39] Surowiecki, J. (2004) *The Wisdom of Crowds: Why the many are smarter than the few*, Doubleday.

[40] White, L.J. (2000) *Propertyzing the electromagnetic spectrum: why it's important, and how to begin.* The Progress & Freedom Foundation Telecommunications Reform Project, October.

9

Distributed Learning and Reasoning in Cognitive Networks: Methods and Design Decisions

Daniel H. Friend, Ryan W. Thomas, Allen B. MacKenzie and Luiz A. DaSilva*

Virginia Polytechnic Institute and State University, USA

9.1 Introduction

The emerging research area of cognitive networks offers a potential solution for dealing with the increasing complexity of communications networks by empowering networks with decision-making capabilities. A key feature of a cognitive network as described in [50] is the cognitive process, which is responsible for the learning and reasoning that occurs in the cognitive network. There has been little progress in establishing the underlying mechanisms for the cognitive process and the trade-offs involved in selecting and implementing these mechanisms. This chapter attempts to reveal the learning and reasoning mechanisms within the cognitive process as well as some of the design decisions involved. With respect to reasoning and learning methods, we restrict our focus to methods that appear to be applicable to cognitive networks and do not attempt to evaluate all possible methods for reasoning and learning.

Since the same term may have different meanings depending upon who is using it, we first establish our definition of a cognitive network. We adopt the following definition from [49]: 'a cognitive network is a network with a cognitive process that can

* This material is based upon work supported by the National Science Foundation under Grant Nos. 0448131 and 0519959 and by a Bradley Graduate Fellowship from the Bradley Department of Electrical and Computer Engineering at Virginia Tech. Any opinions, findings and conclusions or recommendations expressed in this material are those of the authors and do not necessarily reflect the views of the sponsors.

Cognitive Networks: Towards Self-Aware Networks Edited by Qusay H. Mahmoud

perceive current network conditions, and then plan, decide, and act on those conditions. The network can learn from these adaptations and use them to make future decisions, while taking into account end-to-end goals.' The perception, decision, action and learning aspects of this definition are all elements of learning and reasoning. These capabilities are essential to ensure that the decisions made by a cognitive network will improve the network performance as measured against the end-to-end goals.

The terms learning and reasoning are difficult to pinpoint. Following the line of thought in [15], we consider reasoning to be the immediate decision process that must be made using available historical knowledge as well as knowledge about the current state of the world. The primary responsibility of reasoning is to choose a set of actions. Learning is a long-term process consisting of the accumulation of knowledge based on the perceived results of past actions. Cognitive nodes learn by enriching the knowledge base to improve the efficacy of future reasoning. It is not always easy to separate reasoning and learning along these lines since the two may be tightly coupled. We will endeavor to point out cases in which they cannot be easily separated as they are encountered.

Our emphasis in this chapter is specifically on distributed learning and reasoning versus centralized methods. The standard motivations for distributed algorithms, such as scalability, robustness and parallel processing gains, are applicable to cognitive networks, but there are additional motivations for choosing distributed solutions. The primary purpose of a cognitive network is to serve the data transfer needs of users and applications. The volume of traffic that would be generated by nodes sending observations of the network state to a central repository for cognitive processing would significantly detract from the ability of the network to service users and applications. Cognitive processing at each node allows information exchanges to occur at higher levels of abstraction, which implies lower communication overhead. Also, there are scenarios where the network already contains distributed cognitive processing nodes, such as when there are cognitive radios present. In cases such as these, centralized solutions waste this resource and make integration of the cognitive radio and cognitive network concepts more difficult.

Having established our position on the nature of learning and reasoning in a cognitive network and justified our focus on distributed processing, we will proceed to discuss frameworks for learning and reasoning in Section 9.2, where we also give our reasons for choosing multiagent systems (MAS) as our framework for learning and reasoning in cognitive networks. Section 9.3 goes into greater detail on the methods that might be used for distributed learning and reasoning in a cognitive network. We then turn our attention in Section 9.4 to the sensor and actuator functions which close the cognitive loop and give our perspective on how existing research in sensor networks can be used to benefit cognitive networks. Section 9.5 is devoted to high-level design decisions and their interrelation with distributed learning and reasoning in a cognitive network. Section 9.6 summarizes our conclusions and closes with some comments on key open issues related to the subjects addressed.

9.2 Frameworks for Learning and Reasoning

As a starting point for investigating systems that are capable of learning and reasoning, it is helpful to consider general cognitive architectures. More specifically, we are interested in cognitive architectures that are intended for implementation in a computing device.

A variety of such architectures exist with varying degrees of complexity. At the higher complexity end of the scale are the so-called universal theories of cognition such as ACT-R [3], Soar [30] and ICARUS [26]. Each of these architectures seeks to capture all of the components of the human mind and the interactions necessary for cognition. Toward the other end of the complexity scale are the OODA and CECA loops [7]. These two architectures do not attempt to incorporate all of the elements of human thinking but are intended to model the decision-making process to provide a basis for improved command and control in military decision making. A cognitive architecture that is closely related to the OODA loop is the cognition cycle proposed in [34], which is the basis for many of the architectures used in cognitive radio research.

9.2.1 Linking Cognitive Architectures to Cognitive Network Architectures

While it is tempting to convert the most human-like cognitive architectures to cognitive network architectures, we must not forget that the purpose of the cognitive network is to exchange data between users and applications within the network. Therefore, it may be appropriate to simplify a human-like cognitive architecture to remove the elements that are superfluous to the achievement of that purpose. On the other hand, some cognitive architectures may be oversimplified and lack key elements for a successful cognitive network architecture. An example of this may be the OODA loop, which in its simplest incarnation [7] does not include learning. It is conceivable for a cognitive network to be designed without the capacity to learn; however, cognitive networks are most likely to be applied to complex problems that are difficult or infeasible to completely characterize at design time. Therefore, we expect learning to play a central role in a cognitive network.

Recently, there have been proposals for cognitive network architectures in [32], [46] and [50]. The architectures presented in [32] and [46] are based on the cognition cycle of Mitola [34] and focus on the inner workings of a cognitive node. In [46], Mahonen *et al.* propose that a toolbox of methods from machine learning, mathematics and signal processing be available for matching to the needs of the decision-making process. The architecture in [50], reproduced in Figure 9.1, is presented from a network level and shows the interaction between end-to-end goals, a cognitive process that is composed of one or more cognitive elements, and a software adaptable network (SAN). The architecture is general enough to allow both centralized and distributed implementations of the cognitive process.

9.2.2 The Cognitive Network as a Multiagent System

Discussion of cognition, learning and reasoning in a computational environment inevitably leads to a discussion of artificial intelligence. Our interest in distributed methods points us to distributed artificial intelligence (DAI). The DAI field originally consisted of two research agendas: cooperative distributed problem solving (CDPS) and multiagent systems (MAS) [16]. However, the scope of MAS has increased to the point that DAI is now considered by some to be a subset of MAS. Although MAS has no universally accepted definition, we will adopt the definition in [23], which relies on three concepts: *situated, autonomous* and *flexible*. 'Situated' means that agents are capable of sensing and acting upon their environment. The agent is generally assumed to have incomplete knowledge, partial control of the environment, or both limitations [47]. 'Autonomous' means that

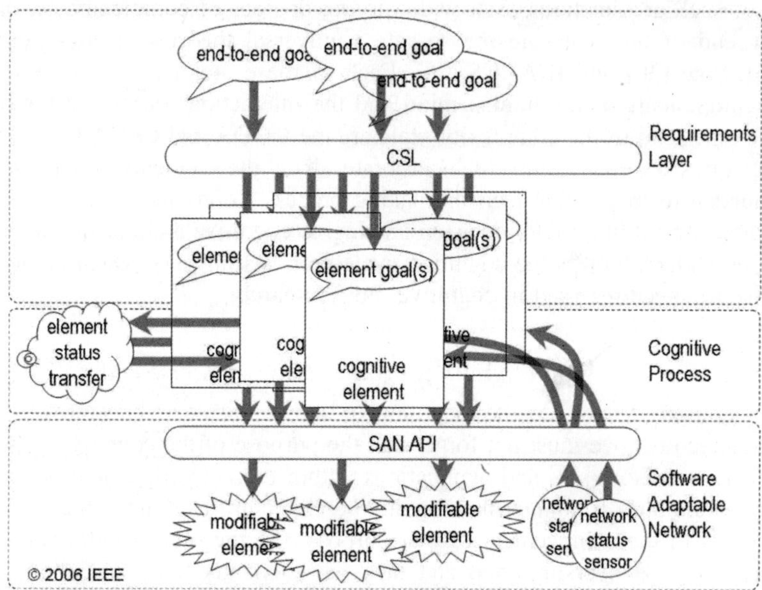

Figure 9.1 Representative cognitive network architecture [50] Reproduced by permission of © 2006 IEEE

agents have the freedom to act independently of humans or other agents though there may be some constraints on the degree of autonomy each agent has. 'Flexible' means that agents' responses to environmental changes are timely and proactive and that agents interact with each other and possibly humans as well in order to solve problems and assist other agents.

There is an additional concept that separates CDPS from the rest of MAS research. In addition to the three concepts describing an MAS, a CDPS also possesses inherent group *coherence* [55]. In other words, motivation for the agents to work together is inherent in the system design. This simplifies the problem but leaves the challenging issue of how to ensure that this benevolence is put to good use.

In [15], Dietterich describes a standard agent model consisting of four primary components: observations, actions, an inference engine and a knowledge base. In this agent model, reasoning and learning are a result of the combined operation of the inference engine and the knowledge base. By our definition, reasoning is the immediate process by which the inference engine gathers relevant information from the knowledge base and sensory inputs (observations) and decides on a set of actions. Learning is the longer term process by which the inference engine evaluates relationships, such as between past actions and current observations or between different concurrent observations, and converts this to knowledge to be stored in the knowledge base. This model fits well within most of the cognitive architectures previously mentioned, and we shall use it as our standard reference for our discussion of distributed learning and reasoning.

Returning to the cognitive network architecture in Figure 9.1, we will resolve this architecture with the agent model just described. Sensory inputs (network status sensors) in the cognitive network architecture are received from nodes in the network through the SAN application programmer's interface (API). Actions are taken on the modifiable elements using the SAN API as well. Inference engines and knowledge bases are contained within the cognitive elements. In distributed implementations, each cognitive element consists of multiple agents (i.e., network nodes), which learn and reason cooperatively to achieve the common element goals. Interactions between groups of agents belonging to different cognitive elements can be opportunistic, allowing competition between cognitive elements.

At this point it is clear that a cognitive network can be framed as an MAS. This is perhaps no surprise to the reader due to the generality of the MAS definition, but we will continue a step further and explain why a cognitive network is more aligned to a CDPS framework. A key component of the definition of a cognitive network is the existence of end-to-end goals that drive its behavior. These goals provide a common purpose for each agent and motivate them to work together. Referring to Figure 9.1, the greatest alignment occurs among agents within the same cognitive element since they have the same element goals. Across cognitive elements, the element goals are derived from the network's end-to-end goals. This implies that there is some degree of coherence across cognitive elements.

9.3 Distributed Learning and Reasoning within an MAS Framework

Having established the framework within which we will treat learning and reasoning for a cognitive network, we turn our attention to methods that allow agents to learn and reason and the factors involved in selecting a method. The number of methods that we could discuss is much larger than this chapter can accommodate, so we have selected methods for which we can provide some motivation for use in cognitive network development.

9.3.1 Elements of Distributed Reasoning

The primary objective of reasoning within a cognitive network is to select an appropriate set of actions in response to perceived network conditions. This selection process ideally incorporates historical knowledge available in the knowledge base (often referred to as short-term and long-term memories) as well as current observations of the network's state.

Often reasoning is categorized as either *inductive* or *deductive*. Inductive reasoning forms hypotheses that seem likely based on detected patterns whereas deductive reasoning forgoes hypotheses and only draws conclusions based on logical connections. Due to the size of the cognitive network state space, which grows combinatorially with the number of network nodes, the cognitive process must be capable of working with partial state information. Since the cognitive process always sees a limited view of the network state, it is difficult to draw certain conclusions as required by deductive reasoning. The approach in inductive reasoning of generating a best hypothesis, based on what is known, is more conducive to the limited observations available to the cognitive process.

Reasoning (or decision making) can also be categorized as *one-shot* or *sequential* [15]. A one-shot decision is akin to single-shot detection in a digital communications receiver. A final action is selected based on immediately available information. Conversely, sequential reasoning chooses intermediate actions and observes the response of the system following each action. Each intermediate action narrows the solution space until a final action can be chosen. This is a natural approach to problem diagnosis and seems particularly useful for failure diagnosis in a cognitive network. Sequential reasoning may also be especially useful for proactive reasoning where time constraints are more relaxed and there are only indications of an impending problem. However, when immediate action is needed, such as a response to congestion at a particular network node, a one-shot decision is more expedient.

The formulation of our approach to distributed reasoning in a cognitive network is partly driven by the depth of our understanding of relationships between the parameters we can control and the observations that we have available to us. Treating the network as a system, our actions are the inputs to the system and the observations are the outputs. If we cannot mathematically represent relationships between the inputs and outputs, then we may select a reasoning method that is capable of dealing with uncertainty such as Bayesian networks. This approach relies heavily on the learning method to uncover the relationships between inputs and outputs. If we can link inputs and outputs mathematically, then we can use methods based on the optimization of objective functions such as distributed constraint reasoning or metaheuristics. In this case, learning may be used to reduce the time required to find optimum or suboptimum solutions.

9.3.2 Methods for Distributed Reasoning

9.3.2.1 Distributed Constraint Reasoning

Distributed constraint reasoning is usually broken into distributed constraint satisfaction problems (DisCSP) and distributed constraint optimization problems (DCOP). DisCSP differs from DCOP in that DisCSP attempts to find any of a set of solutions that meets a set of constraints while DCOP attempts to find an optimal solution to a set of cost functions. The cost functions in DCOP essentially replace the constraints in DisCSP, which allows for a more general problem-solving method. Since DCOP always admits a solution whereas DisCSP may not be satisfiable, we focus our attention on the application of DCOP to cognitive networks.

Following the definition given in [35], each agent in an MAS DCOP has control over a variable x_n. Each variable has a finite and discrete domain D_n. A set of cost functions $f_{i,j}$ are defined over pairs of variables x_i and x_j. The sum of the cost functions provides an objective function that the DCOP seeks to minimize.

A well-known algorithm for performing DCOP is Adopt [35]. The goals of Adopt are to require only local communication, allow flexibility in the accuracy of the solution to speed execution time, provide theoretical bounds on the worst-case performance, and allow completely asynchronous operation. Adopt first forms a tree structure based on the cost functions. This tree structure is used to determine which agents communicate and what messages will be transmitted between agents. Agents then iteratively choose solutions for their variables based on local knowledge and exchange these partial solutions in the search for a global solution. Upper and lower bounds on the cost functions are computed and continually refined by the agents as new information is received. Tracking the lower and upper bounds allows early termination of the algorithm when some amount of error in the solution is tolerable.

The desirable characteristics for using Adopt in a cognitive network are asynchronous message passing, autonomous control of each agent's variable, and the flexibility of choosing solution accuracy. Additional advantages of Adopt are its potential for robustness to message loss and its ability to reduce interagent communications through a timeout mechanism[36]. These are especially important for wireless cognitive networks. Two limitations of DCOP algorithms are that cost functions are functions of only two variables and that each agent is assigned only one variable. However, procedures for extension to n-ary cost functions and multiple variables per agent are presented in [35] (without evaluation) with further work on n-ary cost functions in [6].

9.3.2.2 Distributed Bayesian Networks

Bayesian networks (BNs) are a method of reasoning under uncertainty. The uncertainty may be a result of limited observations, noisy observations, unobservable states, or uncertain relationships between inputs, states and outputs within a system [56]. All of these causes for uncertainty are common in communications networks. In particular, the ability of cognitive networks to potentially control parameters at different layers in the protocol stack gives rise to concern over interactions between different protocol layers, interactions that are currently not well understood [4]. BNs provide a means for dealing with this interaction uncertainty probabilistically.

BNs decompose a joint probability distribution (JPD) over a set of variables (i.e., events) into a set of conditional probability distributions defined on a directed acyclic graph (DAG). Each node represents a variable in the JPD. The directionality of the edges in the DAG represents parent–child relationships where, conditioned on the knowledge of its parents, a child is independent of all other nondescendents in the network [56]. Each node contains a conditional probability distribution which is the probability distribution of the variable at that node conditioned on the variables at all parents of the node. The JPD can then be reconstructed as the product of all of the conditional probability distributions.

BNs are an example of a method that incorporates both reasoning and learning. Learning in a BN is accomplished through belief updating. Belief updating is the process of modifying the conditional probability distributions based on observations of variables in the JPD. Knowledge is contained within the conditional probability distributions as well as the DAG structure itself. Reasoning in a BN consists of producing probability estimates of sets of variables based on the current JPD and possibly observed variables as well. BNs only satisfy part of our definition of reasoning since they do not specify a method for selecting an action. To completely satisfy our definition of reasoning, BNs may be paired with a decision-making method such as multi-criteria decision making (MCDM) (see [53] for an example).

A well-known method for distributed BNs is the multiply sectioned Bayesian network (MSBN) [56]. The MSBN is constructed by sectioning the BN into subdomains that are then assigned to agents in an MAS. Each agent's subdomain is organized as a junction tree in order to simplify belief updating. The overall network structure is known as a linked junction forest. Organizing the original DAG as a linked junction forest provides guarantees that cannot be made with arbitrary decompositions of the DAG.

There are two guarantees provided by the MSBN framework that are of particular interest to its application in a cognitive network. First, an agent is only allowed to directly communicate with other agents that share one or more of its internal variables. This ensures that communications are localized with respect to the variable dependencies. Second, the global JPD is consistent with the agents' conditional JPDs, which means that the MSBN is equivalent to a centralized BN.

Perhaps the first question that arises when considering an MSBN for application to a cognitive network is what the variables should be. There are two scenarios for the types of variables in the network. The variables may inherently be tied to each agent (in which case dynamic cognitive networks such as mobile wireless networks are more easily dealt with), or a subset of the variables may be independent of the number of agents in the MAS (in which case these variables must be protected from loss as agents enter and leave the cognitive network). It is likely that the dependence structure is unknown at design time and therefore must be learned (see Section 9.3.4.2 on learning the DAG structure of a BN). Also, a high average node degree in the DAG implies greater complexity in the computation of conditional distributions as well as increased communication between subdomains in the MSBN. A sparse DAG is ideal from both a computational as well as a communications standpoint.

9.3.2.3 Parallel Metaheuristics

Some of the most complex reasoning methods fall under the category of metaheuristics. A metaheuristic is an optimization method that teams simpler search mechanisms with a higher level strategy which guides the search. Metaheuristics are approximate algorithms as opposed to exact algorithms and commonly employ randomized algorithms as part of the search process [2]. This means that they are not guaranteed to find a globally optimum solution and that, given the same search space, the metaheuristic may arrive at a different solution each time it is run. Metaheuristics are commonly applied to problems for which finding an exact solution is infeasible, such as NP-hard problems. For these types of problems, the time required to find a globally optimum solution grows exponentially in the dimension of the search space. Metaheuristics trade off solution optimality for a feasible computation time. Since metaheuristics are adept at handling complex problems, they are of particular interest for cognitive networks in which the size of the search space over which reasoning is performed may grow exponentially in the number of agents. We identify some promising parallel metaheuristics and the particular versions of these parallel metaheuristics that are most suitable to distributed reasoning.

9.3.2.3.1 Parallel Genetic Algorithms

Parallel genetic algorithms (PGAs) are members of a class of biologically inspired metaheuristics called evolutionary algorithms (EAs). This class also includes evolutionary programming and evolution strategies. Because all three types of algorithms share a similar approach, we will focus only on PGAs. The PGA evolves a population of candidate solutions through crossover and mutation operations. Parameters of the search space are encoded in chromosomes which are represented by binary vectors. The objective function is used to evaluate candidate solutions (i.e., chromosomes) and produce a fitness for each candidate solution. The fitness values are used to decide which members of the population survive and/or are crossbred. Randomness in the crossover and mutation operations seeks to explore the search space and provide diversity within the population.

Parallel implementations generally fall into three categories: master–slave, island (also distributed or coarse-grain) and cellular (or fine-grain) [2]. The PGA is centrally controlled in the master–slave model while the island and cellular models give agents more autonomy. We dismiss master–slave PGAs from consideration for distributed reasoning since these algorithms require centralized control. Furthermore, we believe that island

PGAs hold more promise for cognitive network applications than cellular PGAs because the migration policy can be controlled whereas cellular PGAs require local communication at all agents for each generation of the PGA. Migration in island models occurs infrequently, which implies reduced communications between agents when compared to cellular PGAs. Also, the migration in island PGAs can be asynchronous [31], which is a better fit for MAS than the synchronous requirements of cellular PGAs. Additional flexibility is available in the injection island genetic algorithm (iiGA), which allows island populations to have different resolutions in the binary encoding of parameters in a chromosome [31]. This may be useful in a cognitive network with heterogeneous nodes that have different computational abilities or different amounts of energy that can be expended on computation.

9.3.2.3.2 Parallel Scatter Search

Another member of the EA family is scatter search. Scatter search differs from a genetic algorithm in that new generations are constructed (deterministically or semi-deterministically) rather than randomly generated via mutation and crossover. The basic procedure for a scatter search is to select an initial population (usually randomly), choose a reference set of good and diverse solutions, and then enter an iterative loop of combining and improving solutions to create a new population from which a new reference set is selected [18]. If the reference set can no longer be improved, the algorithm may add diversity to the reference set and continue iterating.

Parallel scatter search (PSS) is a recent development. Three initial approaches to parallelization given in [18] are synchronous parallel scatter search (SPSS), replicated combination scatter search (RCSS), and replicated parallel scatter search (RPSS). SPSS and RCSS are centrally controlled algorithms that require synchronous operation and the distribution of the reference set at each iteration. Therefore, they are undesirable for use in distributed reasoning in a cognitive network. RPSS is similar to island PGAs in that multiple populations exist simultaneously. Because there is no mechanism specified in [18] for exchanging subsets of the population (i.e., migration), we conclude that the algorithm is asynchronous, which is desirable for cognitive network applications.

A potential drawback of PSS is that the combination and improvement operations are problem dependent [21], which restricts the flexibility of a PSS. Particularly, it may be difficult to adapt a PSS to changes in the objective function. Since allowing changes in the end-to-end goals of a cognitive network implies changing the objective function, a PSS-based cognitive network may be restricted to having fixed end-to-end goals.

9.3.2.3.3 Parallel Tabu Search

One of the most prominent metaheuristics is tabu search. The key elements of tabu search are a short-term memory called a tabu list, an intermediate-term memory, a long-term memory, and a local transformation that defines the neighborhood of the current solution [19]. The use of memory in the search process is a major factor in the success of tabu search and distinguishes it from the metaheuristics we have discussed. The tabu list prevents the search from backtracking into territory that has been previously explored. The intermediate-term memory directs the search toward promising areas of the search space (intensification), and the long-term memory directs the search toward unexplored regions of the search space (diversification). By using these memories, tabu search learns from its exploration of the search space. When this learning is tied to the agent's knowledge base for use in future searches, tabu search is an example of a method that incorporates both learning and reasoning.

A taxonomy for parallel tabu search (PTS) is given in [12]. This taxonomy defines two classes of algorithms that have distributed control and asynchronous communications: *collegial* and *knowledge collegial*. 'Collegial' identifies those PTSs in which a process (or agent in our context) simply communicates promising solutions to other processes that make up the PTS. This class is similar to island PGAs. 'Knowledge collegial' is more complex in that agents communicate promising solutions but also try to infer global properties from the communicated solutions such as common characteristics of good solutions.

9.3.2.4 Distributed Multiobjective Reasoning

The discussion so far has implicitly assumed that the cognitive network has a single goal (or objective function). It is also likely that the cognitive network will have multiple, potentially competing goals. Reasoning then becomes a multiobjective problem. Multiobjective optimization attempts to find members of the Pareto-optimal set, which is a set of solutions in which no single parameter can be improved without a corresponding decrease in the optimality of another parameter. Multiobjective DCOP is a recent development. Results in this area can be found in [5], [6] and [27]. Multiobjective optimization is a well-studied topic for evolutionary computation with some results for distributed multiobjective genetic algorithms in [9] and [11]. Bayesian networks can be teamed with multi-criteria decision making (MCDM) to handle multiple objectives [53].

9.3.3 Elements of Distributed Learning

Thus far we have discussed some powerful methods for reasoning based on current knowledge. However, these methods are severely impaired when there is no knowledge base available to guide reasoning. Teaming learning with reasoning makes the core of the cognitive process complete.

Learning methods may be classified as *supervised, unsupervised* or *rewards-based* [42]. Supervised learning requires that the learning method be trained using known inputs and outputs provided by an expert. The expert is usually human, and the process of generating training data sets is generally laborious, if not infeasible. Rewards-based learning uses a feedback loop in which the feedback consists of measurements of the utility (i.e., rewards) of prior actions. Rewards-based methods must then infer the correctness of prior actions based on the utility measurements. Unsupervised learning operates in an open loop without any assistance from an expert or utility measurements. Due to the difficulty of applying supervised and unsupervised learning methods to MASs, the majority of research on learning in an MAS environment has focused on rewards-based learning [42]. It is reasonable to use rewards-based learning for distributed learning in cognitive networks as well since the possible variations in the network environment are too large to make supervised learning feasible, whereas performance measurements (i.e., observations) are readily available at all layers of the protocol stack.

One of the major issues encountered in developing a rewards-based learning system for an MAS is the credit assignment problem (CAP). The CAP is the problem of determining how to assign credit or blame to agents for the results of prior actions. It can be broken into two parts: the correlation of past actions with observations and the assignment of credit. The first stage is particularly difficult in a cognitive network because of the variation in response time to changes made at different layers of the protocol stack. For example, a change in transmit power may be observed within milliseconds as degraded signal-to-interference-plus-noise ratio (SINR) at neighboring nodes. The same change in transmit power may also result in congestion apparent at the transport layer; however, the response time of the congestion observation is likely to be on the order of seconds.

For MASs in general, there are two levels of credit assignment that must be made: *inter-agent* CAP and *intra-agent* CAP [55]. Inter-agent CAP deals with assigning credit to agents based on the level of responsibility that each agent has for taking the actions. Since an MAS may not be cooperative, one agent may be more responsible for a particular action than any other and should therefore be assigned more credit or more blame. An example of this is the ACE algorithm [54] in which agents compete for the right to take actions based on estimates of their action's usefulness. Since only winning agents in ACE are allowed to implement their actions, credit should only be assigned to winning agents. The inter-agent CAP in a cognitive network may be simpler if agents take actions based on global agreement. In this case, a uniform credit assignment may be appropriate. This leaves intra-agent CAP, which is the process of determining the knowledge or inference that led to the action under consideration. To perform intra-agent CAP, a short-term history of past actions and their bases must be kept until actions can be correlated with their results. Agents must determine whether the action was due to inference from observations, knowledge in the knowledge base, or some combination of the two and then respond appropriately.

9.3.4 Methods for Distributed Learning

Methods for learning cannot be considered independently from the reasoning method. As we have seen with Bayesian networks, the coupling is sometimes so tight as to make the distinction between learning and reasoning difficult. While not all feasible pairings of learning and reasoning methods are this closely coupled, there is an inescapable dependency that necessitates their joint selection. The dependency comes from the need to apply what the learning algorithm has stored in the knowledge base. If this knowledge is not used, then learning is superfluous. This dependency motivates us to consider three methods of distributed learning that we pair with the three methods of distributed reasoning discussed previously. These pairs are listed in Table 9.1.

In line with our discussion of rewards-based learning, we will discuss Q-learning and case-based reasoning. Reinforcement learning, of which Q-learning is an example, has been a popular subject for learning in MASs. For our discussion, we have paired Q-learning with DCOP. Case-based reasoning may seem misplaced for a section on

Table 9.1 Summary of reasoning and learning pairs

Reasoning method	Learning method
DCOP	Q-learning
Bayesian network	Metaheuristic for learning the DAG
Metaheuristic combinatorial optimization	Case-based reasoning

learning, but, in fact, it is a framework that includes both reasoning and learning. We will discuss learning within the case-based reasoning framework and pair it with a meta-heuristic for reasoning.

Bayesian networks do not fit cleanly within the supervised, unsupervised and rewards-based classification of learning methods. This stems from being a probabilistic network. Learning in Bayesian networks can be thought of as occurring in two stages: first, learning of the network structure and initial conditional distribution estimates; and second, continual belief updating of the conditional distributions. In this section we focus on the first stage since the second stage is part of the previously discussed MSBN algorithm.

9.3.4.1 Distributed Q-learning

Q-learning (and reinforcement learning in general) models the world as a discrete-time, finite state, Markov decision process where actions result in an observable scalar reward [48]. Q-learning seeks to determine a policy that will maximize the expected reward. The optimal policy is learned by an iterative process of first selecting an action based on the current state, a set of available actions and a probability distribution, and then updating the current policy based on the reward for the selected action (see [52] for a tutorial on Q-learning).

In distributed Q-learning, agents can be classified as performing joint action learning (JAL) or independent learning (IL) [10]. In JAL the agents are fully aware of the actions of all agents, whereas in IL the agents are only aware of their own actions. The JAL case implies communication between agents to share their actions. The IL approach is clearly beneficial when trying to minimize communication overhead in a cognitive network. However, convergence to the optimum learned policy is much more difficult to guarantee. An approach to this problem for stochastic environments is given in [28], where implicit coordination is used to overcome the lack of information about other agents' actions and converge to the optimal joint policy.

9.3.4.2 Distributed Learning of Bayesian Networks

As we mentioned in the discussion of BNs, the structure of the dependency graph may not be known *a priori*. An obvious consequence of this is that the conditional distributions are unknown as well. This issue is normally addressed by learning the structure of the DAG and then estimating the conditional distributions. The process requires a set of data consisting of past combinations of inputs and outputs of the system. This data set is not training data in the sense of a supervised learning method because there is no concept of a correct output for any particular input. It consists of samples of the system's behavior.

The ability to learn the structure and conditional distributions of a BN allows a cognitive network to construct a set of beliefs about the network by observing the behavior of the network. This could be beneficial for networks that may be deployed in a variety of scenarios but cannot be fully characterized beforehand. One example of this is wireless networks for emergency response, where the geographic location for deployment of the network is not known until a disaster occurs. An additional benefit of learning a BN is that prior knowledge (such as that generated by an expert) can be incorporated into the BN to improve the learning process [8].

Learning a BN imposes a heavy computational burden, and this has led researchers to find ways to parallelize the process. For structures of moderate dimensionality, an asynchronous complete search such as in [25] may be feasible. For larger networks, researchers have turned to metaheuristics such as ant colony optimization [14], variable neighborhood search [13], and evolutionary algorithms [37] to provide approximate solutions to the search for BN structure. However, work on distributed metaheuristics for learning BNs seems to be in the early stages with one of the first results being reported in [41]. Due to exponential growth in the size of the DAG structure's search space, further research is needed in learning BNs with distributed metaheuristics to apply BNs to cognitive network applications of moderate size.

9.3.4.3 Distributed Case-Based Reasoning

Case-based reasoning (CBR) is a combination of reasoning and learning. In CBR, the knowledge base (or case base) contains cases that are representations of past experiences and their outcomes. Dynamic networks will inevitably lead to disparity in the contents of agents' case bases. However, the structured contents of the case base can easily be shared among agents. This allows agents that have recently joined the network to benefit from the experience of other agents. This sharing of case bases also makes the cognitive network more robust to loss of agents, since the case base is essentially a distributed, overlapping database.

A four-stage cycle for CBR is presented in [1], with the stages being *retrieve, reuse, revise* and *retain*. When a new case is encountered, the retrieve stage selects the most similar case from the knowledge base. The reuse stage combines the new case with this similar case to form a proposed solution. The efficacy of the proposed solution is evaluated in the revise stage, and the case is repaired if it failed to achieve the desired outcome. The final stage, retain, extracts any meaningful results and either enters the learned case into the knowledge base or updates existing cases. Based on this decomposition of CBR, reasoning consists of the retrieve and reuse stages while learning consists of the revise and retain stages.

Distributed case-based reasoning is discussed in [43] and [44] where agents have individual responsibility for maintaining their case bases as well as sharing cases with other agents to improve global solutions. We conceptually pair case-based reasoning with a distributed metaheuristic (this combination has been used in [29] for a single agent scenario using the genetic algorithm as the metaheuristic). The distributed metaheuristic revises retrieved cases using information from past searches. This allows the metaheuristic to partition the search space so that it can either focus on finding a better solution in a region that has already produced good solutions or avoid trying to optimize dimensions of the search space that have little effect on solution quality.

The learning stages (revise and retain) are generally problem-specific when using case-based reasoning. For the cognitive network problem, learning may consist of evaluating how well a solution achieved the end-to-end goals. A similar problem to credit assignment occurs when evaluating the success of a solution. The response time of the network to the actions of the cognitive process may have different delays just as with Q-learning. This makes learning more difficult when there are multiple end-to-end goals (i.e., multiobjective optimization), especially when the goals apply to different protocol layers. Learning may also extend to the parameters used by the metaheuristic, in which case the cognitive network is learning how to search more efficiently.

9.4 Sensory and Actuator Functions

The MAS characteristic of 'situated' encompasses the ability of agents to sense their environment as well as take actions on their environment. The ability to sense the environment provides the reasoning method with up-to-date information on the state of the network. These sensory inputs also provide the feedback necessary for learning. Actuator functions provide the ability to enact the solution or decision reached after reasoning. Therefore, sensory and actuator functions are vital to learning and reasoning.

Sensory functions lead to parallels between cognitive networks and sensor networks. Sensor networks are deployed to observe changes in an environment. In a cognitive network, the environment consists of the network itself. The cognitive network can be conceptually thought of as having a sensor network embedded within it that interacts with the cognitive elements through the SAN API (see Figure 9.1) to provide the status of the network. Sensor networks are usually considered to be limited in terms of available energy and processing power at each sensor node. The energy constraint drives sensor networks to limit communication in the network. In the cognitive network, it is also desirable to limit communication in the network, though the reason is to allow the communication channels in the network to serve users and applications as much as possible.

The processing limitations are less applicable to the sensors within cognitive networks. However, the cost of computation versus communication is leading some researchers to consider performing in-network computation [20] to reduce the amount of communication necessary to obtain the information of interest. It was pointed out in [20] that applications in sensor networks are often interested in aggregate queries such as averages or maximums. This is also relevant to cognitive networks where the observation of interest may be the average delay for flows in the network or the maximum transmit power over all the nodes. Thus, it is equally valuable to perform in-network computation in cognitive networks in order to reduce the required communication needed to obtain aggregates.

One point of difference between typical sensor networks and a cognitive network is that there is usually a single point in a sensor network that requires the aggregate observation, whereas in a cognitive network multiple nodes may want to know this result. One question that may be asked is whether this problem is even scalable in the number of nodes in the network. It has been shown in [45] that (using joint data compression) it is scalable for local observations to be globally distributed. In-network computation then eliminates redundant data and further reduces the load on the network.

Two methods that may be considered as types of in-network computation and data compression, respectively, are *filtering* and *abstraction* [50]. Filtering gives agents some autonomy over what local observations should be communicated to one or more agents. The learning process will ideally provide a means for agents to learn which observations have strictly local impacts and which observations need to be shared. When the nature of an observation is such that geographical proximity leads to correlation in agents' observations, agents may learn that only a subset of the cognitive network needs to report this observation. Abstraction is a common method of representing data in a more compact form. This is analogous to data compression in that redundant information has been removed.

The actuator functions are the means for the cognitive network to change its state as a result of reasoning and learning. Changes are enacted in the modifiable elements using the SAN API and must be coordinated by agents so that actions are coherent. Some changes, such as modulation type, addressing scheme and MAC protocol, will require a degree of synchronization to prevent interruption to application sessions. In the worst case, poorly implemented actuator functions may lead to network instability. In this regard, research in distributed control may be beneficial.

9.5 Design Decisions Impacting Learning and Reasoning

Regardless of the learning and reasoning methods employed by the cognitive network, there are several design decisions that impact the network's performance. These decisions provide a set of 'axes' upon which different designs can be classified and categorized independent of the cognitive process implementation. Understanding how these decisions affect various network objectives can lead to sets of design space trade-offs, providing the network engineer with guidelines for selecting and incorporating specific cognitive features [51].

Referring to Figure 9.1, cognitive elements are the distributed decision-making and implementation entities in the network and act as part of a larger cognitive process that attempts to attain a set of end-to-end goals that are defined by the operators, users, applications or resources of the network. Cognitive elements in many instances meet the definition for an agent when they are situated, autonomous and flexible. For instance, in the context of a wireless network, cognitive networks may consist of multiple cognitive radios with the cognitive radios acting as the cognitive elements. These radios could be described as agent-like. However, cognitive elements may not always have so direct a hardware mapping. For instance, several network devices may distribute the operation of a single cognitive element, such as the case where a cognitive element utilizes a distributed reasoning algorithm. In these cases, flexibility and situatedness may require that the cognitive element consist of many agents, distributed across the network hardware. A useful analogy here may be one of the ant-eater and the ant-hill. Both could be considered cognitive elements, with the ant-hill sentience emerging from the individual actions of the ants in the hill and the ant-eater sentience coming directly from the mammal. In a similar manner, some cognitive elements may be like the ant-eater, distinct agent-elements, while others may be like the ant-hill, consisting of multiple agents.

There are two levels of objectives in a cognitive network: cognitive element-level goals and end-to-end network objectives. Examples of element-level goals include: minimizing node energy consumption, selecting radio spectra for maximum channel capacity, reducing MAC collisions, selecting network routes that maximize connectivity lifetime, and selecting optimal data codecs that reduce processing load. End-to-end objectives, on the other hand, are system-wide in nature and include such goals as identifying network faults and repairing them, maximizing network connectivity and lifetime, reducing the cost of operation, maintaining QoS considerations and providing network security.

9.5.1 Game Theory for Cognitive Networks

Game theory is a particularly useful framework for investigating cognitive networks, since it models the decision-making processes of groups of rational agents. Game theory can give some insight into the convergence properties, fixed points and system behaviors of the network. It also provides a convenient notational framework for describing the goals, behaviors and action spaces of the multiple autonomous cognitive elements of the network. For similar reasons, game theory has been used to describe and model cognitive radio behavior [22] [39]. For the remainder of this section, we will use a repeated game theoretic model to formalize the discussion of design decisions for cognitive elements.

We begin by representing a cognitive network by the tuple $C = <M, A, U>$. This tuple consists of the set of $M = \{1, 2, \ldots, n\}$ cognitive elements. Each element, through cognition, chooses some *action* a_i from the set of all possible actions A_i available to that element. Thus the action space for the network is $A = \times_{i \in M} A_i$. All action choices at a particular instant are held in an action vector $\boldsymbol{a}$. Since it is a repeated game, we will denote the ordering of the decisions by the superscript. Thus node i chooses to play an action a_i^t at stage t. The element-level goals are denoted by the set of functions U, where the function $u_i : A \rightarrow R$ represents the *utility* for element i (and higher values represent improved attainment of the goal). An objective function called *cost* quantifies the end-to-end objectives of the network. It is denoted by the function $C : A \rightarrow R$ and describes the performance of a network in achieving these goals for a specific action vector. This function is determined from all the objectives of the network and, unlike the utility function, is improved as it is decreased. A series of action choices by an element is called a *strategy*.

9.5.2 Critical Design Decisions

The three-tier model of cognition motivates our selection of design decisions. This model was first suggested by David Marr [33] in his work on computer vision and then generalized for other cognitive tasks in [24]. It describes the general functionality required of a cognitive entity. The model consists of the *behavioral* layer, the layer of cognition that dictates what motivates the decisions of the elements; the *computational* layer, which decides on the course of action for an element; and the *neuro-physical* layer, which enacts these decisions into the environment.

From this model, we identify critical design decisions at each layer. At the behavioral layer, the designer must decide whether the cognitive elements of the network act in selfish or altruistic manners. Should action choices only be in their own best interest or should actions reflect the end-to-end objectives? Computationally, the designer must decide how much information the cognitive elements need. Can cognitive elements operate with some degree of ignorance of the network state or do they require complete network state information? The neuro-physical layer can control some subset of the network functionality. Does functionality in every instance need cognitive control, or will limited cognitive control over network functionality suffice?

9.5.2.1 Behavior

Examining the first design decision, the selfishness of a cognitive element can be defined by the degree that the elements pursue their individual goals. Selfish actions arise from the element-level goals. Using our game theory notation, we define a selfish action as one in which

$$u_i(a_i^{t+1}, \tilde{a}_{-i}{}^{t+1}) \geq u_i(a_i{}^t, \tilde{a}_{-i}{}^{t+1})$$

where the above notation means 'the utility for element i's next action, given that every other element plays the actions that element i believes them to (represented by $\tilde{a}_{-i}^{t+1}$), is no less than the utility for element i's current action.' The belief part of the notation for the other elements' actions is important, since ignorance may make this action vector different than the actual actions the other elements are employing. The network exhibits the feature of *selfishness* when every element plays a selfish strategy, meaning that all cognitive elements only make selections that continuously improve their individual goals.

Selfish behavior can achieve end-to-end goals in two cases: when the selfish behavior is aligned with the end-to-end goal or when incentive structures are used to force an alignment between the two. Alignment between the element goals and end-to-end goals occurs when the ordering of the cognitive element utility space follows the same ordering as the end-to-end objectives. A cost function with that property is called *monotonic*. A monotonic cost function preserves the partial ordering of the utility space and occurs when for any $\boldsymbol{a}, \boldsymbol{b} \in A$ if $\boldsymbol{u}(\boldsymbol{a}) \prec \boldsymbol{u}(\boldsymbol{b})$ then $C(\boldsymbol{b}) < C(\boldsymbol{a})$ and if $C(\boldsymbol{b}) \leq C(\boldsymbol{a})$ then $\boldsymbol{u}(\boldsymbol{a}) \preceq \boldsymbol{u}(\boldsymbol{b})$. Here the relational operator $\prec$ means that the right side strictly dominates the left, in our case if $\boldsymbol{u} \prec \boldsymbol{u}'$, then $u_i < u_i'$ for every i. The relational operator $\preceq$ means that the left side is dominated or unordered in comparison to the right side. Unordered means that for some $\boldsymbol{u}$ and $\boldsymbol{u}'$, there exists at least one i such that $u_i < u_i'$ and at least one j such that $u_j > u_j'$. Example cost functions that are monotonic include the following, as well as variations with linear transformations of the utility functions:

- $C(\boldsymbol{a}) = -\sum_{i \in M} u_i(\boldsymbol{a})$
- $C(\boldsymbol{a}) = -\min_{i \in M} u_i(\boldsymbol{a})$
- $C(\boldsymbol{a}) = -\max_{i \in M} u_i(\boldsymbol{a})$
- $C(\boldsymbol{a}) = -\text{average}_{i \in M} u_i(\boldsymbol{a})$

Element goals can be 'forced' into alignment with the end-to-end goals when an incentive is used. Using punishments and/or rewards will modify the utility of the cognitive elements. Transforming the cognitive element utilities can create a monotonic cost function. The area of game theory that examines goal alignment is often called mechanism design [40]. Mechanism design determines incentives or creates rules that give cognitive elements no desire to diverge from the network objectives. In particular, the VGC mechanism can be used to determine the reward and punishment structure for specific classes of games.

In some networks, however, the underlying assumptions of game theory may not make sense. The typical game theoretic assumption of myopic rational behavior assumes that autonomous players will act in a manner that selects only strategies that are better or best response in nature. However, cognitive networks have a higher level end-to-end objective than just the cognitive elements' objectives. Because of this, we define an altruistic action as occurring when both of the following hold:

$$u_i(a_i^{t+1}, \tilde{a}_{-i}^{t+1}) < u_i(a_i^t, \tilde{a}_{-i}^{t+1})$$
$$C(a_i^{t+1}, \tilde{a}_{-i}^{t+1}) \leq C(a_i^t, \tilde{a}_{-i}^{t+1})$$

This occurs when a cognitive element selects an action that is not in its best interest but is in the interest of the end-to-end objectives of the network. A network behavior exhibits the feature of altruism if at least one element selects an altruistic action at some point in the sequence of decisions. Altruistic actions effectively break the myopic rationality of the game theoretic model, since they assume that an entity will sacrifice their own goals. However, this is not an unreasonable assumption, as many cognitive element processes may chase one particular local goal, so long as this achieves the end-to-end objectives. If the local goals are not moving the network closer to the end-to-end objectives, the cognitive elements will act altruistically.

Some altruistic actions may be accomplished autonomously. However, it may be necessary to collaborate in order to achieve the end-to-end objectives. Collaboration, by definition, requires some sort of signaling between cognitive elements. This signaling provides a mechanism for bargaining, negotiation or submission, which in turn allows the network to reach its end-to-end objectives.

There are many mechanisms for collaborating. A selfish form of collaboration occurs in negotiation. Negotiation involves compromising in such a way that all elements benefit (preserving to some degree the assumption of myopic rational behavior), but not as much as they might. While the exact negotiation process can take on many different algorithms, John Nash provides a set of axioms to identify one possible outcome of a negotiation process [38]. His Nash bargaining solution depends on U being a convex, closed and upper bounded subset of $\Re^n$. In this case, $\boldsymbol{u}^*$ is a Nash bargaining solution if the following axioms hold:

- Individual rationality: for every element, u_i^* must be greater than the utility outcome if bargaining fails.
- Pareto optimality: $\boldsymbol{u}^*$ is a Pareto optimal utility vector.
- Invariance to affine transformations: $\boldsymbol{u}^*$ must be invariant if affinely scaled.
- Independent of irrelevant alternatives: if the domain is reduced to a subset of the domain that contains $\boldsymbol{u}^*$, then the solution is still $\boldsymbol{u}^*$.
- Symmetry: $\boldsymbol{u}^*$ does not depend on the labels given to the cognitive elements. If two elements are identical, they will achieve the same u_i^*.

The first two axioms are rationality constraints – the achieved utility should be worth the negotiation process and should be efficient. The last three axioms are fairness constraints, meaning that the negotiation solution should provide a fair benefit to all who participate. If these axioms are satisfied, then the outcome of any negotiation process will be the same unique utility vector.

The concept of a correlated equilibrium [17] is attractive from a collaboration point of view. In order to achieve a correlated equilibrium, cognitive elements utilize an impartial random signal to determine their action. A classic example of a correlated equilibrium is a traffic intersection with a traffic light. Without the traffic light, the intersection would

Table 9.2 Utility matrix for transmission window example

	Element 2 window size		
Element 1 window size		**Short**	**Long**
	Short	(6,6)	(2,7)
	Long	(7,2)	(0,0)

be a chaotic mess, but it would find some (inefficient) Nash equilibrium. With the traffic light, the system is still at equilibrium, since no car has any reason to disregard the instructions of the light. However, this is a more efficient equilibrium from both a driver and traffic flow perspective. From a cognitive network perspective, a correlated equilibrium could be a simple way of forcing a collaboration that meets the end-to-end objectives.

As an example, imagine a network that desires to maximize the average amount of information transferred between wireless nodes. Assume there are two cognitive elements in the network, attempting to choose a transmission window size for a mutual receiver. The wireless environment makes it so that each transmitter is hidden from the other. If both choose short transmission windows, the probability of the windows colliding is low, but each window is inefficient, containing little information and much overhead. If both choose long transmission windows, the probability of collisions is high and even less information makes it through. If one chooses long and one chooses short, the probability of collisions is higher, but the transmitter with the long window still expects to get more information through than with the short window. Some possible utilities for this scenario are quantified in Table 9.2, with the notation (u_1, u_2).

Without a random signal, the game has two Nash equilibria at (7,2) and (2,7) (an average utility of 4.5). If there is a random signal with three equally likely action pairs: (short, short), (short, long) and (long, short), an action pair is chosen at random, and each element is only informed of its particular action. Each element will play the action as given since the expected utility of the action is greater than the alternative. For instance, if element 1 is told to play (short), then it knows that there is a 50% chance that element 2 was told to play (short) and a 50% chance that element 2 was told to play (long). If element 1 was to play (long) then its expected utility would be 3.5 versus 4 if it plays the requested action. We can now see that the average utility of this correlated equilibrium is 5 which is an improvement over the Nash equilibrium.

9.5.2.2 Computational State

Other than the signaling information needed for collaboration, the cognitive process requires some level of information about the system state and other cognitive element actions to perform its computations and make its decisions. Ignorance in a cognitive network can come from one of two places: incomplete information and imperfect information. Incomplete information means that a cognitive element does not know the goals of the other elements or how much the other elements value their goals. Collaboration is a mechanism for nodes to overcome incomplete information, since goals often have to be shared to come to a solution.

Imperfect information, on the other hand, means that the information about the other users' actions is unknown in some respect. We define Y_i as the set of signals that cognitive element i observes to learn about the actions of all elements in the network (including itself). This signal could be gathered from self-observation, collected from information shared by other cognitive elements, or collated from both. If element i observes $y_i \in Y_i$, then the probability that action vector $\boldsymbol{a}$ caused this signal is denoted by $P[\boldsymbol{a}|y_i]$. In this work, we will assume all elements know the goals of the other elements. Thus the feature of ignorance occurs when at any stage in the decision sequence:

$$\forall i \in M, \exists y_i \in Y_i \text{s.t.} P[\boldsymbol{a}_{-i}|y_i] < 1 \forall \boldsymbol{a}_{-i} \in A$$

Some degree of ignorance occurs when at least one cognitive element has a signal with some uncertainty as to its causing actions. Quantifying this degree of ignorance remains an open problem. Ignorance can occur because of any of the following conditions:

- Uncertain information: the signal, as measured, has some random uncertainty in it.
- Missing information: the signal is missing the action of at least one other element.
- Indistinguishable information: the signal may indicate one of several actions for the network, and it is impossible to distinguish between them.

If none of these conditions is present, then the network operates under fully observable actions. This feature is called knowledge and occurs when for every stage of the decision sequence:

$$\exists \boldsymbol{a}_{-i} \in A \text{s.t.} P[\boldsymbol{a}_{-i}|y_i] = 1 \forall y_i \in Y_i; i \in M$$

Ignorance can be caused from a variety of sources in a cognitive network. Measurements of other elements actions may contain errors, there may be unreliable communication between the elements, or it may not be possible to measure or communicate the action choices of other elements. Previous research into multiagent systems has assumed a reliable and structured communications architecture, as well as a homogeneous set of agents with fixed interfaces. Cognitive elements must attempt to provide multiagent-like behaviors in a much less reliable environment.

9.5.2.3 Cognitive Control

Up to this point, we have implicitly considered that there is a symmetric amount of cognitive control in the network. This means that if a cognitive element has control over the actions of one modifiable aspect of the network, that modifiable aspect is under cognitive control across the entire network. For instance, if the modifiable element is the transmission power of the radio, we have assumed every radio's transmission power is controlled by a cognitive element. Reducing cognitive control changes this symmetry. If k is the number of instances of the modifiable aspect in the network, and x is the total number of instances of the modifiable aspect under cognitive control, the feature of partial control occurs when, for some modifiable aspect:

$$x/k < 1$$

Otherwise, the system has the feature of full control when, for all modifiable aspects:

$$x/k = 1$$

As an example, if the transmission power of a wireless network is the modifiable aspect of the network, it may not be necessary to control this feature cognitively at every radio. Instead, having just a few cognitive elements with intelligent power control may be enough to meet the end-to-end objectives. This kind of limited control can be thought of as *coherent* control – if enough cognitive elements place their collective weight behind a problem through coherent action selection, they can influence the end-to-end performance of the network.

Another kind of partial control could be called *induced* control. If non-cognitive nodes control their functionality according to some static algorithm, a cognitive network may be able to induce desired functionality in these radios through the actions of the cognitive elements. A practical example of this might be in TCP. Because of the wireless environment, the congestion control algorithms of TCP (such as slow start or congestion avoidance) may not provide adequate performance. Rather than place the congestion control parameters at every node directly under cognitive control, a limited number may have cognitive control. Congestion adaptation is the result of interaction between nodes, meaning that the cognitive elements may be able to induce desired behaviors at non-cognitive nodes through their choice of action. For TCP, this could be accomplished by changing the handshakes and acknowledgements (for instance, through the use of proxies) in such a manner that 'tricks' the non-cognitively controlled TCP into behaving in certain manners.

These critical design features provide guidelines for determining the implementation of the cognitive network. By examining these trade-offs in light of the end-to-end objectives, engineers can choose appropriate cognitive element goals as well as reasoning and learning mechanisms.

9.6 Conclusion

This chapter has discussed distributed learning and reasoning for cognitive networks. We have shown how a cognitive network fits into the multiagent system and cooperative distributed problem-solving contexts. We have described the application of three classes of distributed reasoning methods to a cognitive network and provided justification for their use, as well as potential drawbacks. Each class of distributed reasoning methods has been paired with a distributed learning method, and the advantages and disadvantages of each learning method have been described. Sensor and actuator functions have been presented as an integral part of learning and reasoning in a cognitive network. Finally, we have investigated behavior, computational state and cognitive control as areas encompassing important design decisions.

Cognitive networking is rich with opportunities for novel research. The areas of learning and reasoning are particularly ripe and central to the success of cognitive network implementations. Most distributed learning and reasoning methods described in this chapter assume that communication is reliable and error-free and that information obtained is accurate. In real-world networks, particularly wireless ones, ensuring reliable communication may be difficult. Also, information may be noisy, inaccurate or purposely falsified by malicious nodes. Thus, a major opportunity exists for developing learning and reasoning methods that are robust and that incorporate measures of the reliability of data.

Other contributors to robustness needing further research are coordination among nodes for reasoning and learning and synchronization of configuration changes that result from reasoning. Coordination is especially important when the topology of the network is dynamic, perhaps due to mobility. New nodes must be incorporated into the structure of the distributed reasoning and learning algorithms. Knowledge transfer ensures that nodes entering the network can quickly learn from others' experiences and that nodes leaving the network do not create a void in the knowledge of those remaining. Even without a dynamic network, failure to synchronously enact changes may interrupt communications and artificially change the network topology.

Just as the term 'cognition' has a spectrum of definitions in different research circles, 'learning' and 'reasoning' have a spectrum of possible implementations. There is a tendency to expect the learning and reasoning processes to be complex; however, simplicity is a virtue both in implementation and analysis. We have purposely included learning and reasoning pairs (see Table 9.1) that differ greatly in their complexity and flexibility. Still needed is an understanding of what kinds of problems are appropriate for learning and reasoning pairs as well as the trade-offs for each (e.g., implementation difficulty, processing and communications overhead, robustness).

The aforementioned 'critical design decisions' lead to many interesting research questions. For instance, the price of ignorance currently considers only knowledge and the lack of it. A more continuous range for the price of ignorance is needed to allow greater flexibility in making design trades. Along these lines, a network-centric measure of information content is needed, not in the sense of network information theory, but from the standpoint of the applicability of information given the amount of control that the cognitive network has over the network state. For these measures to be applied, analytical models will most likely be needed. Research available in distributed control may be helpful, but the nonlinear (and perhaps stochastic) nature of learning and reasoning are sure to complicate the task of analytical modeling.

References

[1] Aamodt, A. and Plaza, E. (1994) Case-based reasoning: foundational issues, methodological variations, and system approaches. *AI Communications*, **7**(1), 39–59.

[2] Alba, E. (ed.) (2005) *Parallel Metaheuristics: A New Class of Algorithms*, John Wiley & Sons, Inc., Hoboken, NJ, USA.

[3] Anderson, J.R., Bothell, D., Byrne, M.D., *et al.* (2004) An integrated theory of the mind. *Psychological Review*, **111**(4), 1036–60.

[4] Barrett, C., Drozda, M., Marathe, A. and Marathe, M. (2002) Characterizing the interaction between routing and MAC protocols in ad-hoc networks. *Proceedings of the 3rd ACM International Symposium on Mobile Ad-hoc Networking and Computing*, June 9–11, Lausanne, Switzerland.

[5] Bowring, E., Tambe, M. and Yokoo. M. (2005) Distributed multi-criteria coordination in multi-agent systems. *Proceedings of Workshop on Declarative Agent Languages and Technologies at the 4th International Joint Conference on Agents and Multiagent Systems*, July 25–29, Utrecht, Netherlands.

[6] Bowring, E., Tambe, M. and Yokoo. M. (2006) *Multiply-constrained DCOP for distributed planning and scheduling*. American Association of Artificial Intelligence Spring Symposium on Distributed Plan and Schedule Management, March 27–29, 2006, Stanford, CA, USA.

[7] Bryant, D.J. (2004) *Modernizing our cognitive model*. The 9th International Command and Control Research and Technology Symposium, September 14-16, Copenhagen, Denmark.

[8] Buntine, W. (1991) Theory refinement on Bayesian networks. *Proceedings of the 7th Conference on Uncertainty in Artificial Intelligence*, July 13–15, Los Angeles, CA, USA.

[9] Cardon, A. and Galinho, T. and Vacher, J.-P. (2000) Genetic algorithms using multi-objectives in a multi-agent system. *Robotics and Autonomous Systems*, **33** (3), 179–190.

[10] Claus, C. and Boutilier, C. (1998) The dynamics of reinforcement learning in cooperative multiagent systems. *Proceedings of the 15th National Conference on Artificial Intelligence*, Madison, WI, USA.

[11] Cochran, J.K., Horng, S.-M. and Fowler, J.W. (2003) A multi-population genetic algorithm to solve multi-objective scheduling problems for parallel machines. *Computers & Operations Research*, **30**(7), 1087–102.

[12] Crainic, T.G., Toulouse, M. and Gendreau, M. (1997) Towards a taxonomy of parallel tabu search heuristics. *INFORMS Journal on Computing*, **9**(1), 61–72.

[13] de Campos, L.M. and Puerta, J.M. (2001) Stochastic local algorithms for learning belief networks: searching in the space of the orderings. *Proceedings of the 6th European Conference on Symbolic and Quantitative Approaches to Reasoning with Uncertainty*, September 19–21, Toulouse, France.

[14] de Campos, L.M., Fernandez-Luna, J.M., Gamez, J.A. and Puerta, J.M. (2002) Ant colony optimization for learning Bayesian networks. *International Journal of Approximate Reasoning*, **31**(3), 291–311.

[15] Dietterich, T.G. (2004) *Learning and reasoning*. Technical Report, School of Electrical Engineering and Computer Science, Oregon State University.

[16] Durfee, E.H. and Rosenschein, J.S. (1994) Distributed problem solving and multiagent systems: comparisons and examples. *Proceedings of the 13th International Workshop on DAI*, Lake Quinalt, Washington, USA.

[17] Fudenberg D. and Tirole, J. (1991) *Game Theory*, MIT Press, Cambridge, MA.

[18] Garcia-Lopez, F., Melian-Batista, B., Moreno-Perez, J.A. and Moreno-Vega, J.M. (2003) Parallelization of the scatter search for the p-median problem. *Parallel Computing*, **29**(5), 575–89.

[19] Gendreau, M. (2003) An introduction to tabu search. In F. Glover and G.A. Kochenberer (eds.), *Handbook of Metaheuristics*, Kluwer.

[20] Giridhar, A. and Kumar, P.R. (2006) Toward a theory of in-network computation in wireless sensor networks. *IEEE Communications Magazine*, **44**(4), 98–107.

[21] Glover, F., Laguna, M. and Marti, R. (2003) Scatter search and path relinking: advances and applications. In F. Glover and G.A. Kochenberer (eds.), *Handbook of Metaheuristics*, pp. 1–36, Kluwer.

[22] Haykin, S. (2005) Cognitive radio: brain-empowered wireless communication. *IEEE Journal on Selected Areas in Communication*, **23**(2), 201–20.

[23] Jennings, N.R., Sycara, K. and Woolridge, M. (1998) A roadmap of agent research and development. *Journal of Autonomous Agents and Multi-Agent Systems*, **1**(1), 7–38.

[24] Johnson-Laird, P.N. (1988) *The Computer and the Mind*, Harvard University Press, Cambridge.

[25] Lam, W. and Segre, A.M. (2002) A parallel learning algorithm for Bayesian inference networks. *IEEE Transactions on Knowledge and Data Engineering*, **14**(1), 93–105.

[26] Langley, P. and Choi, D. (2006) A unified cognitive architecture for physical agents. Proceedings of the 21st National Conference on Artificial Intelligence, July 16–20, 2006, Boston, MA, USA.

[27] Lau, H.C. and Wang, H. (2005) *A multi-agent approach for solving optimization problems involving expensive resources*. ACM Symposium on Applied Computing, March 13–17, Santa Fe, NM, USA.

[28] Lauer, M. and Riedmiller, M. (2004) *Reinforcement learning for stochastic cooperative multi-agent systems. Proceedings of the 3rd International Joint Conference on Autonomous Agents and Multiagent Systems*, July 19–23, New York, USA.

[29] Le, B., Rondeau, T.W. and Bostian, C.W. (2007) Cognitive radio realities. *Wireless Communications and Mobile Computing* (in press).

[30] Lehman, J.F., Laird, J. and Rosenbloom, P. (2006) *A gentle introduction to SOAR, an architecture for human cognition: 2006 update*. Technical Report, University of Michigan, Ann Arbor.

[31] Lin, S.-C., Punch, W.F. and Goodman, E.D. (1994) Coarse-grain parallel genetic algorithms: categorization and new approach. *Proceedings of 6th IEEE Symposium on Parallel and Distributed Processing*, October 26–29, Dallas, TX, USA.

[32] Mahonen, P., Petrova, M., Riihijarvi, J. and Wellens, M. (2006) Cognitive wireless networks: your network just became a teenager. *Proceedings of the 25th Conference on Computer Communications*, April 23–29, Barcelona, Spain.

[33] Marr, D. (1982) *Vision: A Computational Investigation into the Human Representation and Processing of Visual Information*. W.H. Freeman.

[34] Mitola, J. (2000) *Cognitive radio: An integrated agent architecture for software defined radio*. Dissertation. Royal Institute of Technology, Sweden.

[35] Modi, P.J. (2003a) *Distributed constraint optimization for multiagent systems*. University of Southern California. Dissertation.

[36] Modi, P.J. and Ali, S.M. (2003b) Distributed constraint reasoning under unreliable communication. *Proceeding of Distributed Constraint Workshop at 2nd International Joint Conference on Autonomous Agents and Multiagent Systems*, July 14–18, Melbourne, Australia.

[37] Myers, J.W., Laskey, K.B. and DeJong, K.A. (1999) Learning Bayesian networks from incomplete data using evolutionary algorithms. *Proceedings of the Genetic and Evolutionary Computation Conference*, July 13–17, Orlando, FL, USA.

[38] Nash, J. (1953) Two-person cooperative games. *Econometrica*, **21**(1), 128–40.

[39] Neel, J.O., Buehrer, R.M., Reed, J.H. and Gilles, R.P. (2002) Game-theoretic analysis of a network of cognitive radios. *Proceedings of the Midwest Symposium on Circuits and Systems.*

[40] Nisan, N. and Ronen, A. (1999) Algorithmic mechanism design (extended abstract). *Proceedings of STOC.*

[41] Ocenasek, J. and Schwarz, J. (2001) *The distributed Bayesian optimization algorithm for combinatorial optimization.* EUROGEN 2001 – Evolutionary Methods for Design, Optimisation and Control with Applications to Industrial Problems, September 19–21, Athens, Greece.

[42] Panait, L. and Luke, S. (2005) Cooperative multi-agent learning: the state of the art. *Autonomous Agents and Multi-Agent Systems*, **11**(3), 387–434.

[43] Plaza, E. and Ontanon, S. (2003) Cooperative multiagent learning In *Adaptive Agents and Multi-Agent Systems: Adaptation and Multi-Agent Learning*, pp. 1–17. Springer, Berlin.

[44] Prasad, M.V.N. (2000) Distributed case-based learning. *Proceedings of the 4th International Conference on MultiAgent Systems*, July 10–12, Boston, MA, USA.

[45] Scaglione, A. and Servetto, S. (2005) On the interdependence of routing and data compression in multi-hop sensor networks. *Wireless Networks*, **11**(1–2), 149–60.

[46] Sutton, P., Doyle, L.E. and Nolan, KE. (2006) A reconfigurable platform for cognitive networks. *Proceedings of the 1st International Conference on Cognitive Radio Oriented Wireless Networks and Communications*, June 8–10, Mykonos Island, Greece.

[47] Sycara, K.P. (1998) Multiagent systems. *AI Magazine*, **19**(2), 79–92.

[48] Tan, M. (1997) Multi-agent reinforcement learning: independent vs. cooperative agents. In M.N. Huhns and M.P. Singh (eds.), *Readings in Agents*, pp. 487–94, Morgan Kaufmann, San Francisco.

[49] Thomas, R.W., DaSilva, L.A. and Mackenzie, A.B. (2005) Cognitive networks. *Proceedings of IEEE DySPAN 2005*, November, Baltimore, MD, USA.

[50] Thomas, R.W., Friend, D.H., DaSilva, L.A. and Mackenzie, A.B. (2006) Cognitive networks: adaptation and learning to achieve end-to-end performance objectives. *IEEE Communications Magazine*, **44**(12), 51–7.

[51] Thomas, R.W., DaSilva, L.A., Marathe, M.V. and Wood, K.N. (2007) Critical design decisions for cognitive networks. *Proceedings of ICC 2007*, June, Glasgow, Scotland.

[52] Watkins, C. and Dayan, P. (1992) Technical note: Q-learning. *Machine Learning*, **8**(3–4), 279–92.

[53] Watthayu, W. and Peng, Y. (2004) A Bayesian network based framework for multi-criteria decision making. *Proceedings of the 17th International Conference on Multiple Criteria Decision Analysis*, August 6–11, Whistler, BC, Canada.

[54] Weiss, G. (1995) Distributed reinforcement learning. *Robotics and Autonomous Systems*, **15**(1–2), 135–42.

[55] Weiss, G. (ed.) (1999) *Multiagent Systems: A Modern Approach to Distributed Artificial Intelligence*, MIT Press, Cambridge.

[56] Xiang, Y. (2002) *Probabilistic Reasoning in Multiagent Systems*, Cambridge University Press, Cambridge.

10

The Semantic Side of Cognitive Radio

Allen Ginsberg, William D. Horne and Jeffrey D. Poston
The MITRE Corporation, USA

10.1 Introduction

The term *cognitive radio* refers to a number of technological and research trends, some of which appear to bear little relationship to one another, other than being plausibly associated with 'making wireless devices more intelligent.' Before embarking, therefore, on the main theme of this chapter, i.e., the role of formal semantics-based technologies in cognitive radio, it is important to offer a reasoned point of view concerning the intended meaning and scope of this term.

Radio as a form of communication has always involved more than physical phenomena: without explicitly or implicitly agreed upon conventions or protocols, wireless communication is either inefficient or, in the worst case, impossible. From this point of view radio consists of three essential elements: devices, a 'medium' (electromagnetic energy transmitted and/or received through the spectrum), and coordinated systems of conventions. The term 'cognitive,' when used to modify 'radio,' applies to *both* devices and the systems of conventions governing their use in relation to the medium. Cognitive radio is a type of technology that exists within wireless devices and makes them behave in ways that can be said to be intelligent. It is also a way of architecting the systems and conventions surrounding human use of the radio spectrum so as to make it possible for such radio devices to coexist with themselves and other devices. We use the expression *cognitive radio device architecture* to refer to the former, and the term *cognitive radio community architecture* to refer to the latter, and to the connection between the two.

It is important to understand that the connection between these two aspects of radio would continue to exist even if human use of the spectrum evolved into a mix of spectrum access and allocation mechanisms such as 'command and control,' 'commons,' or

Cognitive Networks: Towards Self-Aware Networks Edited by Qusay H. Mahmoud

'exclusive use' (market-based) systems [10] [27]. A world in which smart radios were completely responsible for regulating their behavior, whether based on game theory [25], economic transactions such as auction theory, or some other decision-making process, would, if anything, be a world in which the role of conventional meaning (coordinated systems of conventions) would be of increased importance: each and every radio would have to *internalize, in a consistent way, what it means*, for example, to experience interference, be the possible cause of interference, to take action to avoid interfering, to compute the actions dictated by a strategy or, perhaps more importantly, a policy. Without a shared or common understanding of the 'rules of the game,' so to speak, a spectrum commons or market system could not exist. Additionally, it would almost certainly be necessary for radios to share a common language in which they could communicate special circumstances that would cause the rules of the game to be altered or suspended, e.g., emergency situations.

For the purposes of this chapter, whatever additional connotations the term may evoke, *cognitive radio* means the theories and technologies that make it possible to precisely express and mechanically utilize knowledge concerning both physical radio phenomena and systems of conventions governing radio design and use, e.g., policies and technical specifications. To fully understand the requirements of *precision* in expression of knowledge and the ability of that knowledge to be utilized in a *mechanical* fashion is to embark on an exploration of the *semantic* side of cognitive radio.

To express a precise theory about a domain of knowledge requires having a well-defined system of concepts for the phenomena in question. A domain includes the knowledge and concepts within an area of human endeavor including the terminology used in general discourse or in detailed exchanges by subject matter experts. Domains of knowledge related to cognitive radio may include fundamental physics, radio engineering and policy. The term 'ontology' is now widely used to talk about such a conceptual scheme. Note that the word can be used ambiguously to refer either to a conceptual scheme itself, or to a specification or representation of one in a formal language. (This is similar to the distinction between an *algorithm* and its representation as a program in some programming language.) The *Web Ontology Language* or OWL [33] is a formal language for specifying ontologies, and the discussion of semantics will include certain relevant aspects of OWL. However, this chapter is not intended to be a tutorial or overview of OWL or any other formal language. Our goal is to provide the reader with a foundation for understanding the role of formal semantics, as exemplified in ontology languages such as OWL (and associated rule languages such as Semantic Web Rule Language (SWRL) [16] [31]), in the field of cognitive radio. It should be noted that 'non-OWL' (i.e., non-formal language-based) approaches using XML syntax in developing machine-readable device and community cognitive radio architecture specifications exist and certainly have their uses [26]. To the extent that these approaches also have a precise formal semantic foundation, this chapter should aid in appreciating their role as well.

10.2 Semantics, Formal Semantics and Semantic Web Technologies

Semantics, in its most basic understanding, relates to the meaning of a sign such as the meaning of a word within a language. To get an intuitive feeling for what semantics is, consider the fact that it is possible to design and use artificial languages in which there are very precise rules for how 'symbols' can be concatenated and manipulated without saying anything about what the symbols are or what their configurations represent. Games such as tic-tac-toe or John Conway's 'game of life,' provide an example. In the latter, starting with an initial pattern of dots or other symbols on a two-dimensional grid, the rules of the game completely determine how to generate a sequence of successive patterns. This 'language' can be very entertaining and educational, but, without some conventions as to how to interpret the symbols and patterns, it cannot be used to literally communicate anything at all. Semantics is the additional 'whatever it is' that makes communication via language possible. It is also the word used to talk about the discipline or study of such phenomena. Consequently, as cognitive networks are developed that utilize the increasing processing power of devices such as computers, semantics can be used to enable and enhance the interaction of devices within such a cognitive network.

With this basic understanding, a *semantics for a language* L can be considered to be the basis of a theory of *meaning*: given any syntactically well-formed expression E of L, a semantics for L can be used to describe, characterize and reason about the meaning of E in a *principled* fashion. This characterization may seem to imply that supplying a semantics for a language L is pointless: one would surely *already* have to understand the meanings of expressions in L in order to engage in such an activity, so why bother?

There are a number of answers to this question. First of all, being able to talk about meanings in a *principled* fashion implies that the quest for a semantics of a language is a scientific activity in which intuitions are replaced by rigorous methods. Secondly, a semantics for a language L is of vital importance if the intended users of L are machines, i.e., computers or any device that includes a computer-like processor as a control element. This requirement already exists in the realm of traditional programming. Programs written in platform-independent languages such as C++ or Java, need to be translated into platform-specific languages or sequences of native operations in order to run on a particular target architecture. Without a semantics for these higher level languages each translation software developer would be free to come up with idiosyncratic interpretations leading to highly variable results. Finally, having semantics is a necessary condition for investigating certain logical and computational properties of a language which are crucial in determining its utility for various applications.

As these considerations and example indicate, this chapter focuses on *formal semantics*: theories that provide precise and rigorous characterizations of meaning for a language, i.e., characterizations that can be expressed mathematically. Even with this proviso, however, the range of semantic frameworks and theories is enormous and their exploration is beyond the scope of this chapter. We proceed, therefore, by focusing on the kinds of technologies that have captured the attention of researchers due to interest in the Semantic Web initiative of the World Wide Web Consortium [30], and the general trend toward what might be called 'semantics-based interoperability.' This is also appropriate because many of the semantics-based efforts in cognitive radio use such technologies (e.g., [4], [8], and [24]).

10.2.1 The Word 'Semantics'

One often hears expressions like 'what is the semantics of' applied to particular concepts or constructs rather than to entire languages, e.g., *what is the semantics of a* `read()` *system call*? This is an important use of the term 'semantics' that is related to the way this chapter uses the term. Let L be a language that has a well-defined semantics and let

C be some application domain concept or construct that is not a primitive notion in L. For example, C might be the notion of something being 'part of' something else [1]. To give a semantic account of C relative to L is to construct a set of formulas or 'axioms' in L which capture the meaning of C as required in the context of the given application. Section 10.3.3 provides an example of this concept.

10.2.2 Syntax

A formal semantics for a language L requires that the notion of a *well-formed formula (wff)* of L be formally specified as well, i.e., L must have a formal *syntax*. Furthermore, for a language to be of practical use, it must be possible to construct a parser based upon the syntax that will efficiently determine whether or not an expression belongs to the language and, if it does, will provide an analysis of its syntactic structure (a parse).

With respect to languages of the sort we are concerned with, there is a great diversity of syntactic variations on common themes. For example there are many ways of writing expressions using logical connective symbols corresponding to the English terms 'if...then,' 'and,' 'or,' 'not,' 'for all,' 'some,' etc. Table 10.1 shows an example of this.

Note that the semantics associated with these syntactic variants can be different, although that is not always the case.

Formal syntax can also vary with respect to the way in which various *application vocabulary* components are written. Typically such components include at least individual constants and variables, predicates and relations, and functions. For example, in the area of wireless communication, the predicate 'x is an access point' could be written as *AccessPoint(x)* where *x* is a variable that ranges over individuals in the domain. A relation, such as, 'message type of x is y,' could be written as *MsgTypeIs(x,y)*. This is the style we will use in this chapter. Whenever possible this chapter uses English-based formulations of logical connectives.

10.2.3 Syntax of Description Logics

One of the key components of the semantic web vision is the notion of an *ontology*, the roles of which are discussed in subsequent sections (Sections 10.3–10.6). The *Web Ontology Language* or OWL is a formal language for specifying ontologies including the

Table 10.1 Schematic example of diversity of syntactic styles for 'if...then...'

Style	Community
If A then B	English speaking
A → B	Mathematical logic
B : = A	Logic programming
A => B	Rule-based programming
<ruleml:Imp> <ruleml:body>A</ruleml:body> <ruleml:head>B</ruleml:head> </ruleml:Imp>	Semantic web

associated syntax. The OWL specification is heavily influenced by work in the area of *description logic* (DL) [2]. In fact, OWL comes in three flavors, one of which, OWL-DL, is essentially such a logic, specialized for use in web-based environments. One of the interesting syntactic features of OWL that it shares with DL is the *lack* of individual variables. In DL instead of writing *AccessPoint(x)* one simply writes *AccessPoint*. The latter, referred to as an 'atomic concept name,' is interpreted as representing a set of individuals. Using predicates and individual variables the statement that *all access points are stations* could be written as

$$\text{IF } \mathit{AccessPoint}(x) \text{ THEN } \mathit{Station}(x) \tag{10.1}$$

where it is understood that the variable x ranges over all individuals in some domain of interest. To represent this statement in DL, however, one would write

$$\mathit{AccessPoint} \subseteq \mathit{Station} \tag{10.2}$$

indicating that the set of all individuals in the class *AccessPoint* is a subset of the set of all individuals in the class *Station*. In OWL this could be written as follows:

```
<owl:Class rdf:ID='AccessPoint'>
  <rdfs:subClassOf>
   <owl:Class rdf:ID='Station'/>
  </rdfs:subClassOf>
</owl:Class>
```
(10.3)

This chapter's exposition uses the style of Formula (10.1) in the examples, which can be thought of as a 'pseudo-code' variant of classical or standard *first-order logic (FOL)*, rather than that of Formula (10.2) or Formula (10.3). FOL, both as a formal language and a field of study, is one of the fundamental results of twentieth-century mathematical logic. It goes beyond simple propositional logic (sometimes called 'Boolean algebra') by allowing the internal structure of statements to be represented through use of predicate symbols, variables and quantifiers. For example, in FOL, Formula (10.1) would be written as:

$$(\forall x)(\mathit{AccessPoint}(x) \rightarrow \mathit{Station}(x))$$

where the universal quantifier $(\forall x)$ reads 'for all x.' There are other differences, much more significant between DL and FOL that are explained in Section 10.3.

10.3 Community Architecture for Cognitive Radio

The existing world of spectrum-dependent devices and systems and the associated set of policies and regulations essentially create a community of spectrum users. Inherent with this community is an architecture which includes the conventions, such as protocols and operational procedures, and the spectrum policies and regulations that make it possible for such radio devices to operate and coexist with themselves and other devices. The baseline architecture uses static conventions and policies that are not typically presented in machine-readable formats, thus limiting applications for dynamic spectrum access envisioned by cognitive radios. Knowledge representation and semantics can provide such a *cognitive radio community architecture* with a shared base of constructs that can enable a more flexible and dynamic environment.

Declarative semantics deal with languages for making *statements* about what is or is not the case and typically provide a formal theory of what it means for statements to *follow from* or *entail* other statements. The standard semantics of FOL is a prime example of declarative semantics. OWL and DL are also in this category. All of these languages can be given what is called a *model-theoretic* semantics. The following subsections (Sections 10.3.1–10.3.4) show how such languages provide a mechanism for encoding knowledge of use in cognitive radio applications. To do that, the following focuses on a simple example taken from the IEEE 802.11 specifications [18–21].

10.3.1 Two Varieties of Knowledge Representation and Cognitive Radio

There is a close connection between formal languages, their semantics and the type of knowledge that needs to be represented in particular applications or domains. For example, consider an idealized automobile traffic flow application wherein assumptions such as all traffic signals are obeyed and all equipment functions perfectly are warranted for modeling purposes. In such an application one would want to represent 'laws of nature,' so to speak, such as the fact that a red traffic light will cause automobiles in the vicinity to slow down and halt. In designing such an application one would take a third-person point of view in which events and actions take place, their consequences are known and follow patterns that can be taken into account at design time.

Contrast that type of application with a robot vehicle application that is trying to 'program' a vehicle to be able to drive in traffic. In this case, one would not view the behavior of stopping for traffic lights as a 'law of nature' to be described, but as a prescription or imperative that the robot agent must obey. In designing such an application, or at least in attempting to state the kind of knowledge that could conceivably be used to control the vehicle, one would take a first- (or second-) person point of view in which one views the agent as having control over certain actions and a key factor is the ability to invoke those actions when desired. Instead of encoding a statement such as 'If a red light occurs a vehicle will come to a halt,' one would encode a command such as 'If you see a red light, stop!'

From a semantic point of view *statements* are linguistic entities that purport to describe or refer to some truth or factual aspect of some 'world' of interest. They are said to be *declarative* in nature because they declare something to be the case. Commands, directives and similar linguistic entities (often referred to as 'speech acts'), do not purport to describe some factual aspect of a world, but rather are ways of getting some agent to *do* something; they are said to be *imperative* in nature.

It is clear that the semantic side of cognitive radio includes both declarative and imperative forms of knowledge representation. Everything that comes under the rubric of the *cognitive radio community architecture* concept is very likely to involve mainly declarative forms of knowledge, while *cognitive radio device architecture*, which deals with the behavior of devices from 'their point of view,' so to speak, will certainly involve encoding imperative knowledge.

10.3.2 Model-Theoretic Semantics

A model-theoretic semantics utilizes a *domain* or *universe* of individuals, denoted by Δ. The individual variables in a declarative language are said to 'range over' the individuals of Δ. These individuals can be named by using individual constants as provided by the syntax of the language. From a formal point of view that is really all that matters about Δ. From an application point of view, however, one often views Δ as being populated by individuals of the sorts that are of interest in the application. In the case of the IEEE 802.11 example that includes mainly *stations, access points*, and *messages* (and message parameters). As discussed above, these alternative ways of viewing Δ are reflected by the way in which FOL represents predicates versus the OWL and DL use of concept names. If one considers an ontology as a 'specification of a conceptualization' [15] then thinking of Δ as consisting of certain types of objects that are instances of certain named concepts, is more in tune with ontological methodology. In practice, however, as shown below, what the elements of Δ 'really are' is immaterial, in the sense that it is only the abstract structural relationships among the entities in Δ that come into play.[1]

The second major component of a model-theoretic semantics for a language is called an *interpretation*, denoted ***I***. As the name implies, ***I*** provides a mapping of the vocabulary elements of the language into elements and subsets of Δ. Thus to every individual constant ***I*** assigns a particular element in Δ. To every n-place predicate, ***I*** assigns a set of n-tuples of members of Δ. The pair consisting of $<\Delta, \boldsymbol{I}>$ is known as a *semantic structure.*

Let σ represent a wff of some language. A semantic structure $<\Delta, \boldsymbol{I}>$ is said to *satisfy* σ if, intuitively, the mappings given by ***I*** reflect 'what σ says.' For example, consider a simple wff like *AccessPoint(a)*, where a is an individual constant. Intuitively this wff states that the object named by a is an access point. Let ***I****(a)* be the element of Δ that ***I*** maps a to, and let ***I****(AccessPoint)* be the subset of Δ that ***I*** maps the predicate *AccessPoint* to. Then $<\Delta, \boldsymbol{I}>$ *satisfies AccessPoint(a)* if and only if $\boldsymbol{I}(a) \in \boldsymbol{I}\,(AccessPoint)$. This example illustrates the notion of *satisfaction* for the simplest type of wff. More complicated cases would involve wffs having logical connectives and *quantifiers* over individual variables, but the basic idea is the same. For example, consider a wff of the form $(\forall x)\ F(x)$. A semantic structure $<\Delta, \boldsymbol{I}>$ satisfies the latter if and only if for every element x in Δ, x is a member of $\boldsymbol{I}(F)$.

If a semantic structure S satisfies every wff in a set of wffs W, then S is said to be a *model* for W. The key notion of *entailment* is defined as follows: a set of wffs W *entails* a wff σ if and only if every semantic structure that is a model for W also satisfies σ.

10.3.3 Example: IEEE 802.11 and The Semantics of Association

10.3.3.1 Motivating Considerations and Scenarios

Some of the recent work in cognitive radio references or builds upon the notion of *dynamic spectrum access*, i.e., the ability, both technically and legally, for an unlicensed radio device to access a licensed portion of the spectrum under certain conditions and subject to certain constraints. Access to the 5 GHz band by unlicensed devices, a band used by government radar systems, is an example of such a case. This situation is an attractive case for dynamic spectrum access, because neither the radars nor the unlicensed devices need continuous access to the entire band at all locations in the world.

[1] The reader should be aware that the actual semantics of OWL is more complicated than the account presented here. This is due in part to OWL's role as a *web* language, which means, in part, that it must accommodate XML data types.

The technique known as *dynamic frequency selection* (DFS), an elementary form of dynamic spectrum access, is the current method for unlicensed use of this band. In general terms, the policy for using DFS in the 5 GHz band is as follows. The wireless device must first monitor the radio channel it intends to use for a specified time (60 seconds). If it does not detect a signal (a radar pulse in this case) above a specified threshold after that time, it may use the channel, but it must continue monitoring for radar activity. In the event a radar signal is subsequently detected, the device must conclude its use of the channel, including any network management activity, within a specified amount of time (10 seconds) and must not attempt to access the channel again for at least 30 minutes. There is also a requirement to ensure uniform use of channels across the 5 GHz band.

In terms of community architecture, regulators and affiliated technical committees have codified the description of DFS in formal but not machine-readable terms [23], [29], [11]. Also, some rulemaking bodies have delineated the responsibilities for executing DFS behavior differently depending upon the category of a device. For example, in the United States the Federal Communications Commission (FCC) regulations on this topic are written in terms of Unlicensed National Information Infrastructure (U-NII) *master* and *client* devices. A master is required to have radar-sensing capability. No client is allowed to use a channel without first *associating* with a master and receiving a control signal from the master indicating that the desired channel may be used. Before allowing a channel to be used, a master has the responsibility for determining that there is no radar activity on the channel. Also, a master has the responsibility to monitor the channels of the band for the presence of radar signals and commands any affiliated clients to cease activity when a radar signal is detected. There is an additional consideration for in-service monitoring. Unlike the master, client devices are not required to have radar-detecting capability, but *if* they do have the capability and are performing in-service monitoring then they have the responsibility of sending an alert to the master and ceasing activity on the channel when radar is detected.

Figure 10.1 shows a master device in network #1 detecting the presence of radar and broadcasting a control signal letting all clients know that the channel (e.g., 56) must be vacated. In network #2, the client device has in-service radar-monitoring capability, just like a master. The figure also shows in network #2 another master that responds somewhat differently. It goes above and beyond what is required: when notifying clients of the presence of radar, it also informs them of an alternate available channel, if one exists. All of these devices are in compliance with the FCC rules.

10.3.3.2 Cognitive Spectrum Access: Enhanced DFS

As illustrated in Figure 10.1, a U-NII master can go above and beyond what is required, by providing unsolicited information as to channel availability when informing clients that a particular channel or set of channels must be vacated. Similarly a U-NII client can also provide additional functionality and still be in compliance with the FCC policy.

The two scenarios that follow elaborate on possibilities for enhanced forms of DFS.

10.3.3.2.1 Scenario 1: Master Swapping

In Figure 10.2 a U-NII client (*without* radar-sensing capability) using channel c receives a vacate control signal for c from its current master for that channel. Additionally this master has not provided information concerning the availability of other channels, possibly

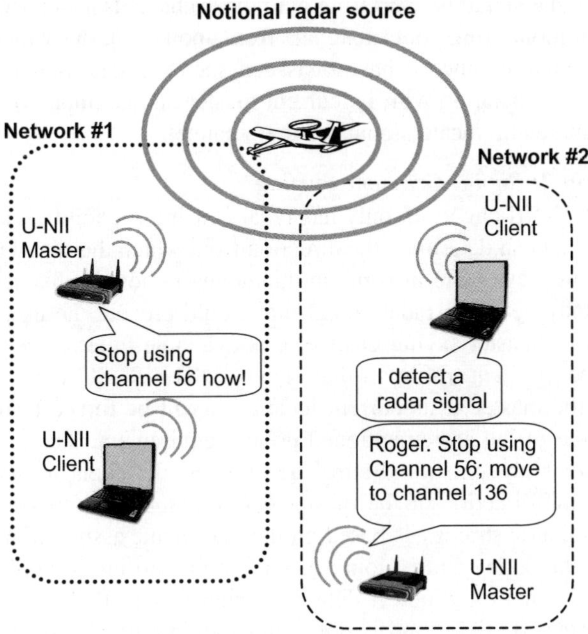

Figure 10.1 DFS examples

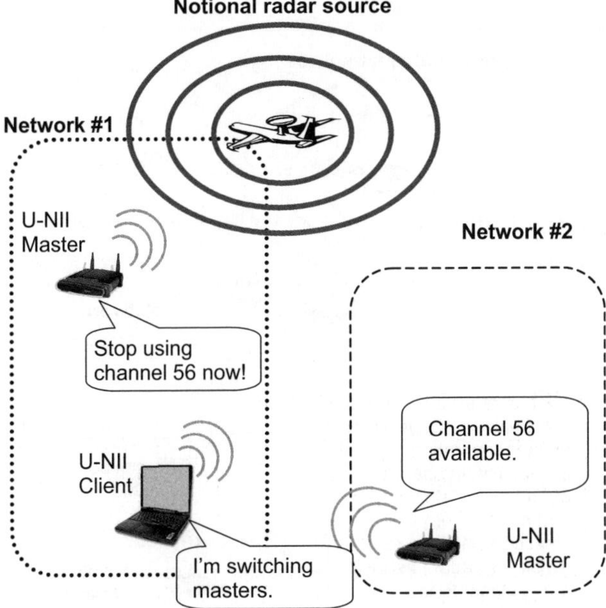

Figure 10.2 Enhanced DFS scenario 1

because the same radar signal is interfering with those channels as well. On the other hand, the client has been monitoring control signals from another master which appears to have availability for channel c, since it has not issued such a vacate message. Consequently, the client drops the association with its current master and attempts to associate with the master that has not sent a vacate signal for that channel.

10.3.3.2.2 Scenario 2: Signal Gradient Surfing

In Figure 10.3 the U-NII client not only has radar-sensing capability it also has direction-finding capability that can determine the direction from which the signal arrived. The latter capability is currently available in some multiantenna radios. In this scenario, the U-NII client simultaneously detects a radar signal that would prevent the use of channel c, and a faint signal from a master saying channel c is available (alternatively the client could detect the presence of client traffic on that channel). The client knows that if it attempts to associate with the master at its current location it will be forced to report the radar to the master by virtue of the aforementioned policy requirement, thus causing a shutdown on the channel. For both selfish and altruistic motives, the client decides to request that its user moves in the direction of the master *before associating* to see if it can lose the radar signal. The client's strategy is based on the reasonable assumption that if the master is compliant with the policy and is doing in-service monitoring it would not be signaling availability on a channel on which it detects a radar signal. If the master has delegated in-service monitoring to clients and is not doing such monitoring itself, then it could be that this client is the first to hear the approaching radar. Presumably a master would have some way of letting clients know that they are responsible for in-service monitoring. So this client would be aware of that requirement. Therefore as it moved closer and closer

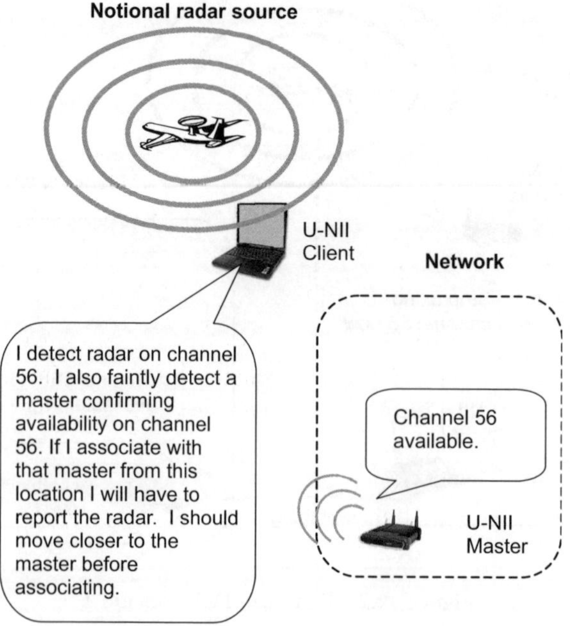

Figure 10.3 Enhanced DFS scenario 2

to the master without losing the radar signal it would reach a point where it would be required to associate with the master and notify it of the radar signal.

10.3.3.3 Commentary on the Scenarios

The basic DFS scenario as well as the enhanced scenarios make use of the notion of *association*, which term is used in the relevant FCC document. Although the connection to IEEE 802.11 does not seem to be explicit in the latter, we may assume that the term is intended to be used in that way. The enhanced DFS scenarios described above appear consistent with the IEEE 802.11 specifications. In particular, the ideas of a station being associated with one access point but monitoring control messages from others and of a station 'deliberately' dropping an association with one access point in order to associate with another, appear to be compliant with both the FCC DFS rules and the IEEE 802.11 specifications.

The issue in the context of this chapter is not whether these enhanced scenarios are good examples of what cognitive radios can do, but rather, how the intended meaning of regulatory and standards specifications can be captured in a way that allows questions concerning *what* constitutes compliance, to be raised and answered in a principled fashion. Using semantic web technologies to encode the semantics of such documents seems to be a natural step in that direction.

10.3.4 Semantics of Association Using OWL-DL Ontologies

It is beyond the scope of this chapter to write a complete set of axioms for *association*. It would not be possible to do that in a thorough fashion without analyzing and formalizing all related notions in the IEEE 802.11 specification. As an example, this section focuses on representing what the specification says about how a station requests access to the medium via an access point. In so doing, we touch upon some substantive differences in the ways in which this information would be represented in OWL versus FOL and also indicate how the use of OWL ontologies contribute to the enterprise of providing semantics.

According to the IEEE 802.11 specification, in order for a station (STA) to use the medium it must first associate with an access point (AP), i.e., STA must transmit a message with certain information to the AP. In particular, the type of the message must be *management* and its subtype must be *association_request*. (The message must also include address identifiers for the STA and AP, but those are ignored here.) We use *Message(x)* to encode 'x is an IEEE 802.11 protocol message,' *msgTypeIs(x, y)* to encode 'the type of message x is y,' and *msgSubTypeIs(x, y)* to encode 'the subtype of message x is y.' The following written axiom encodes this information more compactly:

If

Message(x) & msgTypeIs(x, MANAGEMENT) &

msgSubTypeIs(x, ASSOCIATION_REQUEST)

Then

AssociationRequestMessage(x) (10.4)

where 'MANAGEMENT' and 'ASSOCIATION_REQUEST' are individual constants that denote the aforementioned IEEE 802.11 message parameters. In terms of an OWL ontology, Formula (10.4) states that if an individual is an instance of the class of IEEE 802.11

protocol messages having the given type and subtype then it is an instance of the class denoted by the concept name *AssociationRequestMessage*.

Now let the relational predicate *msgRequestingStation(z, x)* encode 'message z is a request from station x' and let *msgRequestedAccessPoint(z, x)* encode 'message z is a request to access point x.' Given what we have so far, how do we represent the event of a station transmitting such a message to an access point? Here is where FOL and OWL based representations part company.

Using FOL a natural approach would be to use the following three-place predicate *transmitsMsgAtTime(x, z, t)* to encode 'x transmits message z at time t.' In OWL and DL, however, only the equivalent of one- and two-place predicates are allowed. This restriction is not arbitrary; it is one of the features (in combination with others) that helps to make reasoning in those languages decidable and computationally tractable. This is not to say using three-place predicates, in itself, necessarily leads to a system that does not have those properties. Rather, the lesson to be drawn is that sticking to the strictures of OWL-DL is a *sufficient condition* for representing knowledge in a way that guarantees it can be utilized in a *mechanical* fashion.

It is often the case that three-, and higher, place predicates can be eschewed, or rewritten in terms of one- and two-place predicates with equivalent intended semantics. The aforementioned results concerning OWL-DL provide one motivation for doing so. The predicate *transmitsMsgAtTime(x, z, t)* is a case in point. Let us introduce another type of entity, a *message transmission* with corresponding predicate *MessageTransmission(x)*. We think of the latter as something that has a beginning and an end, i.e., has a temporal duration, i.e., we introduce the relational predicates *hasBeginning(x,t)* and *hasEnding(x,t)*. A message transmission must also have a message payload and must be transmitted by a station (which could be an access point): so we need the predicates *hasMessagePayload(x,y)* and *transmittedBy(x,y)*. Similarly for the case of reception, we introduce *message reception* as a kind of temporally extended entity, and introduce the predicate *receivedBy(x,y)*.

To review the discussion and prepare for the next step, consider the following compound expression:

Station(x) & AccessPoint(y) & AssociationRequestMessage(z) &

msgRequestingStation(z, x) & msgRequestedAccessPoint(z, y) &

MessageTransmission(m) & hasEnding(m, t) & hasMessagePayload(m, z) &

transmittedBy(m, x) (10.5)

From an ontological/semantic point of view, Formula (10.5) can be interpreted as picking out certain sets of entities standing in certain relations to each other. One could flesh this out by specifying an appropriate semantic structure. Abstractly this could be viewed as a complex structure consisting of the set of all stations, access points, messages (actually, association request messages) and message transmissions, wherein the station finishes transmitting the association request to the access point *at some time t*. We italicize 'at some time t' to indicate that what Formula (10.5) 'really is about,' i.e., the kinds of entities it denotes are things we have not mentioned yet, namely *events*: occurrences at an instant of time as opposed to things, like transmissions, that occur over a duration. Adding events (instants) to our ontology means that we can now introduce a new predicate, *AssociationRequestTransmissionEvent(x)* for 'x is an association request transmission event' and we can now write the following *rule* (the reason for using 'rule' instead of 'axiom' will be explained below):

> ***If***
>
> *Station(x) & AccessPoint(y) & AssociationRequestMessage (z) &*
>
> *MessageTransmission(m) & Instant(t) & msgRequestingStation (z, x) &*
>
> *msgRequestedAccessPoint (z, y) & hasEnding(m, t) &*
>
> *hasMessagePayload(m, z) & transmittedBy(m, x)*
>
> ***Then***
>
> *AssociationRequestTransmissionEvent(t)* (10.6)

Note that when writing axioms or rules such as Formula (10.6), the convention is to regard all individual variables as universally quantified. In reading Formula (10.6) one would say 'if for all stations x, for all access points,...'

10.3.4.1 Whence an Ontology of Time?

The task of encoding IEEE 802.11 requires positing temporal durations and instants. Does that mean that we need to come up with a semantic account of those notions in order to encode the semantics of IEEE 802.11? It is highly likely that formal use of technical specifications such as 802.11, or policy documents, would require reasoning about temporal entities and their relations: it is difficult to find real-world applications that do not somehow involve temporality.

Because of the ubiquity of temporality it clearly belongs to the family of concepts whose semantics would be defined in what is sometimes termed an 'upper ontology:' an ontology that encodes a semantics for core concepts presupposed by most, if not all, application domains. An example of a proposed ontology of time written in OWL may be found at http://www.w3.org/2001/sw/BestPractices/OEP/Time-Ontology. One of the advantages of using a standard, such as OWL, is the promotion of reuse. This is especially easy to accomplish in the case of a semantic web based standard, due to the requirement of using URIs to identify resources. For example, if the aforementioned ontology has base URI 'http://www.w3.org/2006/owl-time,' which can be represented by the prefix 'otime:' and contains the concept *Instant*, then we can write *otime:Instant* in order to utilize that concept in our IEEE 802.11 ontology. We would also do the same for *hasEnding* which is also defined in owl-time. Additionally we would write an OWL axiom to the effect that *MessageTransmission* is a subclass of the owl-time class *ProperInterval* which represents the set of all entities that exist through some non-zero temporal interval (as opposed to an instant of time).

In the interest of readability we do not use ontology prefixes in our example and discussion.

10.3.4.2 The Need for Rules

In the above example, Formula (10.6) was introduced as a 'rule' and not an axiom. Recall that OWL, because of its DL origins, does not employ individual variables. This restriction makes it impossible to express certain types of relationships among concepts. In Formula (6), for example, it is essential that the station x in *transmittedBy(m, x)* be the same as the requesting station x in *msgRequestingStation (z, x)*. Without using individual variables in such relational predicates there is no general way of constraining the desired relations among the individuals that form the domain of the semantic structure. Therefore, unlike Formula (10.4), which involves only one individual variable x, Formula (10.6) cannot be viewed as an OWL axiom *per se*. This limitation of OWL is well known and is one of the motivators for various *rule language* extensions to OWL [17]. While the need for such extensions is widely accepted in the semantic web community, there is also a need for further research into the logical and computational properties of various ontology and rule language hybrids.

10.3.4.3 Example: DFS Scenario

We close this section with another example of a declarative radio axiom that would be impossible to formalize completely in OWL-DL without a rule language extension. This example will also be relevant to the ensuing discussion.

Recall that the FCC regulations require clients using DFS in the 5 GHz range to vacate within 10 seconds of a radar signal being detected. This performance requirement, however, is implicitly dependent upon the reaction time of the master device. In practice, a master (and its associated clients) can be expected to be in compliance if the master senses for radar within an interval that is short enough to allow clients to vacate within 10 seconds of the appearance of radar. For example, if it takes a client less than 1 second to vacate a channel, then if the master senses every 9 seconds and immediately informs clients when detecting radar, the system should be in compliance. For the purpose of this chapter we assume that this simple DFS implementation suffices. How could we encode the requirement that if a channel is in use, the master must check for radar presence on that channel every 9 seconds?

Let the predicate *Channel(x)* encode 'x is a channel,' let *ChannelInUse(x, t)* encode 'channel x is in use at time t,' let *RadarCheckedAtTime(x, t)* encode 'a radar check of channel x occurs at time t,' let *TimeDifference(t1, t2)* be a function that returns the time difference in seconds between the time values of instants t1 and t2, let *RADAR_CHECK_TIME_THRESHOLD* be an individual constant with the value 9, and finally let *RequireRadar Check(x,t)* encode 'a radar check on channel x is required at time t.' Then consider the following rule:

If

Channel(x) &

Instant(t1) & ChannelInUse(x, t1) &

Instant(t2) & RadarCheckedAtTime(x, t2) &

TimeDifference(t1, t2) $\geq$ *RADAR_CHECK_TIME_THRESHOLD*

Then

RequireRadarCheck(x,t1) (10.7)

Intuitively, Formula (10.7) says that 'if channel *x* is in use at time instant *t1* and a radar check on channel x was conducted at time instant *t2*, where the latter is at least 9 seconds earlier than *t1*, then a radar check on channel *x* is required at time *t1*.'

10.4 Device Architecture for Cognitive Radio and Imperative Semantics

Cognitive radio community architecture looks at the world of wireless from a disinterested third-person point of view. Semantically, the ideal output of an analysis of the world from that perspective would be a logical theory, derived from policy and standards specifications, whose *models* would represent all (and only) the valid ways in which radio systems could interact.

While such a theory would be of great utility in many ways, it would not, in itself, constitute a blueprint for designing and implementing radio *devices* that *act* in accord with the theory. That would be akin to asserting that the laws governing automobile traffic control together with a specification of the highway system entail a complete theory as to how to design and build an automobile that can only move about in a lawful manner. That might *not* be entirely implausible if no automobile existed *before* the specification of the laws and highway system, but, as with radio, the actual situation is reversed.

At the device level, therefore, cognitive radio architecture must take account of the fact that a radio implements a certain set of actions defined in advance by the physics of radio and the current state of radio technology. (Some of those actions involve the ability to observe certain phenomena.) In current parlance, one can think of a radio as an *agent* capable of making certain observations and performing certain basic actions upon command. From a semantic point of view, work on cognitive radio at this level would involve understanding the properties of formal languages used for programming or controlling such an agent.

10.4.1 Example: DFS Scenario Revisited

Consider the rule (Formula (10.7)) that we constructed earlier. The fact that the rule consequent is couched in English in terms of some action being 'required' does not mean that the rule encodes an imperative. Whether or not some action is *required* at a particular time under particular circumstances is a matter of fact that can be true or false. While Formula (10.7) is a reasonable declaration, at the community architecture level, of one way in which a system can implement an aspect of DFS, it cannot, as it stands, be used to control the behavior of a device.

As a step towards a truly imperative representation, consider the following:

If

Channel(x) &

Instant(t1) & ChannelInUse(x, t1) &

Instant(t2) & RadarCheckedAtTime(x, t2) &

TimeDifference(t1, t2) $\geq$ *RADAR_CHECK_TIME_THRESHOLD*

Then

action:RadarCheck(x) (10.8)

This differs from Formula (10.7) only in the consequent. For now let us suppose that *action:RadarCheck(x)* can be interpreted as denoting the act of invoking some software method defined in terms of the underlying traditional radio architecture. If the antecedent of Formula (10.8) is satisfied then the software method corresponding to *RadarCheck* will be invoked with channel x as a parameter.

Removing time as a parameter from *RadarCheck* has unfortunate consequences. The antecedent of Formula (10.8) has declarative semantics, that is, all its models are semantic structures of the sort we discussed earlier. The problem is that any semantic structure that is a model for the antecedent of Formula (10.8) will be a model 'for all time,' so to speak. For example, suppose a model says that a radar check was conducted on channel C at time instant 0 and at time instant 9, and that channel C is in use at the later instant. Then the antecedent of Formula (10.8) is satisfied. In declarative semantics there is no notion of the elements of a semantic structure changing over time, as it were, so if a structure satisfies Formula (10.8) it always satisfies Formula (10.8). Assuming that this model represents the context in which rule Formula (10.8) is evaluated, the action of checking channel C for radar would occur whenever the evaluation occurs, even if, for example, the time of evaluation is at instant 50.

If we had left the time parameter in *RadarCheck* this situation could be avoided in a number of ways. For example, the first part of the action could be to check whether the value of the time parameter is earlier than the current time, and if so, the rest of the action would be aborted. The action would continue only if the time value was the current time. However, it seems far more realistic to suppose that many of the software methods available at the radio device level will not require or accept time as a parameter, especially when the only reason is to determine whether or not the method should continue. Any engineer who routinely designed software based on that methodology would certainly not be on the job very long.

At the semantic level, the example shows where a major part of the divide between declarative and imperative semantics lies. Even though any implemented language based upon declarative semantics must take account of the context in which expressions are evaluated, which includes 'the current model,' the fact is that a declaration is something that can, in principle, be translated into a 'decontexualized' form that does not reference the current context of evaluation/execution while retaining the same meaning. The semantics of imperatives seem to have an irreducible reference to the notion of the *current* context or state in which the expression is evaluated. This distinction is easy to see in ordinary language: police officer x's yelling *Freeze!* at fleeing suspect y at 10 a.m. does not mean the same as the report that *Officer x ordered suspect y to halt at 10 a.m.*

In order to avoid the problem raised above, we recast Formula (10.8) as follows:

If

Channel(x) &

ChannelInUse(x, CURRENT_TIME) &

Instant(t) & LastRadarCheck(x, t, CURRENT_TIME) &

TimeDifference(CURRENT_TIME, t) $\geq$ *RADAR_CHECK_TIME_THRESHOLD*

Then

action:RadarCheck(x) (10.9)

In Formula (10.9) we have introduced a 'special' individual constant *CURRENT_TIME*. Unlike standard individual constants, which are always assigned a single unchanging element of the domain Δ, this constant's intended denotation changes to always denote 'the present moment in time,' so to speak. We have also replaced *RadarCheckedAtTime(x, t)* with *LastRadarCheck(x, t, CURRENT_TIME)* where the later encodes 't is the time that radar was *last* checked on channel x, relative to the current time.' Given this understanding, the components of Formula (10.9) that reference this constant, such as *ChannelInUse(x, CURRENT_TIME)*, will only be satisfied when *CURRENT_TIME* is the time instant of the current runtime context. Interpreted in this way, Formula (10.9) avoids the aforementioned problem because, intuitively, the time of the last radar check on a channel will change appropriately as the current time changes.

A term, such as *CURRENT_TIME*, whose denotation is determined by the context in which a linguistic expression is used, is known as an *indexical*. Traditionally, the study of such terms has not been considered part of formal semantics. However, the point of introducing Formula (10.9) is not to discuss indexicals, but to show how our intuitive understanding of imperative expressions involves notions like *context of evaluation* or *context of execution* that are not part of the standard model theory of declarative language semantics.

10.4.2 Production Rules and Event–Condition–Action Rules

Rather than introduce temporal indexicals such as *CURRENT_TIME*, imperatives can be handled by taking explicit account of the current state of rule evaluation and execution in the entire semantics of the language. In other words, the semantics of rules becomes inseparable from the fact that they are entities that need to be evaluated and executed by some mechanism. Thus, with regard to production rules, one introduces the notion of *working memory*, which can be thought of as a time-varying construct that contains *facts*. At any given time the state of working memory is the context against which a production rule belonging to the system is evaluated. If the antecedent of the rule is satisfied by working memory then the actions specified in the consequent of the rule *may be* executed. We say 'may be' instead of 'are' because whether or not a satisfied rule is executed (allowed to 'fire') usually depends on the particular system and also whether or not other satisfied rules are capable of being fired in the same execution cycle [28]. Typically the actions of a production rule will change the contents of working memory. For example, the rule (Formula (10.9)) above could be written as follows in a production rule language style:

If

Channel(x) &

ChannelInUse(x) &

?item $\leftarrow$ *LastRadarCheck*$(x, t,$ *CURRENT_TIME*$)$ *&*

TimeDifference(CURRENT_TIME, t) $\geq$ *RADAR_CHECK_TIME_THRESHOLD*

Then

action:RadarCheck(x) &

modify: ?item(t $\leftarrow$ *CURRENT_TIME*$)$ (10.10)

This allows the working memory fact that satisfies *LastRadarCheck(x, t, CURRENT_TIME)* to be bound to the variable *?item* at evaluation time. When the actions are performed the radar check will be done as before, but now the element bound to *?item* will be modified to reflect the fact that last radar check was just performed. This will prevent the rule from firing again, until the desired amount of time has passed.

Event–Condition–Action (ECA) rules, which arose in conjunction with work in *active databases* [12], represent a more structured approach to imperatives than production rules, but are overall quite similar. ECA systems formally specify certain types of events recognizable to the system. An ECA rule is typically *triggered* by the occurrence of an event. Once triggered the rule condition is evaluated against the current state of the database (working memory) and if satisfied the specified actions will be executed. To return to our example, consider the following ECA rule:

On :

event:RadarCheckAlarm.isRinging

If

ChannelInUse(x)

Do

action: RadarCheck(x) (10.11)

RadarCheckAlarm represents an object known to the system that can generate events of type *isRinging*. To encode the relevant DFS rule this event would have to occur every 9 seconds assuming, as before, that clients can vacate a channel within 1 second when commanded to do so by the master. When it does, it will trigger Formula (10.11) and the *ChannelInUse(x)* condition will be evaluated against the database. Then *RadarCheck(x)* will be executed for any channel satisfying the condition.

10.4.3 Model-Theoretic Semantics for Imperatives

Historically the semantics of production rules and ECA rules have been specified *operationally*, using notions described in the previous discussion. It is worth noting, however, that model-theoretic accounts of the semantics of similar constructs have been formulated and form the basis for implemented systems [5]. We will give a very brief and rudimentary indication of how this may be done.

Recall that a semantic structure for a declarative language consists of a pair $<\Delta, \boldsymbol{I}>$ where Δ is a domain of individuals and $\boldsymbol{I}$ interpretation that assigns members of Δ to individual constants, subsets of n-tuples from Δ to n-place predicates, etc. The key to extending this semantics is to allow for multiple interpretations, $I_1, \ldots, I_n$ over the elements of the language to exist within a semantic structure. Each $<\Delta, I_i>$ pair in a semantic structure is then a *state*, or, in the parlance of *modal logic*, in which field these ideas first arose, a *world*. In a semantics based purely on modal logic concepts, one would also specify an *accessibility relation* among the various states or worlds. For one state s_2 to be accessible from another state s_1 would 'mean' different things depending upon the intended application. For example, if the states are regarded as databases, then accessibility of states would mean that there is a valid sequence of elementary database updates transforming s_1 into s_2. However, an accessibility relation approach is unnecessary and usually less useful than an approach that provides *transition functions* τ on *actions* in the language. Thus if a is an action and $<\Delta, I_i>$ represents a state in which a can be performed, then τ_a $(<\Delta, I_i>)$ denotes the set of all states that could result from applying a in $<\Delta, I_i>$. Finally, instead of specifying satisfaction of a formula in the language with respect to a simple $<\Delta, \boldsymbol{I}>$ pair, one does that in terms of sets or sequences of states that meet the conditions imposed by the transition functions.

Thus, going back to the DFS example, a transition function could be used to model the sequence of temporal instants – the value of *CURRENT_TIME* in a state would be given by some I_i (*CURRENT_TIME*) in accordance with this transition function – and another could be used to model the *RadarCheck* action in such a way that if, in state $<\Delta, I_i>$, *RadarCheck(x)* were invoked on some channel c, then in any succeeding state in $\tau_{RadarCheck}(<\Delta, I_i>)$ the predicate *LastRadarCheck* would only be true for the time value corresponding to *CURRENT_TIME* in the state $<\Delta, I_i>$. A model for a rule like Formula (10.9) in this kind of semantics would therefore be a sequence of states of this sort whose contents in addition meet the other specific requirements of the rule.

10.5 An Architecture for Cognitive Radio Applications

As we have seen, cognitive radio, as an engineering endeavor, bridges community and device architectures. In doing so, it must bridge linguistic styles of both declarative and imperative semantics. The state of the art in radio device technology, *software defined radio* (SDR), is built upon object-oriented programming, the semantics of which is essentially imperative in nature. However, the semantics of imperative programming languages are not easily represented by the methods we have discussed. Moreover, from a practical point of view, one should not expect radio software engineers – who would probably be best qualified to do the job – to engage in the arduous task of formalizing the semantics of their code using languages that are foreign to their expertise and are not currently part of the tools of their trade, nor likely to be so in the near future. The implication is that from an architectural and software engineering point of view a successful cognitive radio application must somehow integrate two distinct styles of development that have arisen independently, each style having its own history and associated typical platforms or infrastructure.

The situation is shown in Figure 10.4, which also shows components that relate the two platforms and help to unify them into an integrated application. On the one hand, shown on the left side of the figure, there is the SDR platform. On the other hand, shown on the right side, we see components based upon rule and ontology formalisms, as described earlier in this chapter, including appropriate reasoning and execution mechanisms for utilizing both declarative and imperative languages. The *world* and *self* model components will be discussed below. Note, however, that the architecture prescribes two *abstraction layers* as the foundation of the cognitive radio application.

The use of abstraction layers is a familiar methodology in component-based design and its benefits are well known. Typically abstraction layers are based on platform-independent standards or languages. OWL is an example of a standard that could provide a basis for the *ontology and rule abstraction layer* (ORAL) shown in the figure. Another part of the ORAL could be provided by the Rule Interchange Format or RIF, which will be a rule language standard provided by the W3C Semantic Web initiative [32]. The same

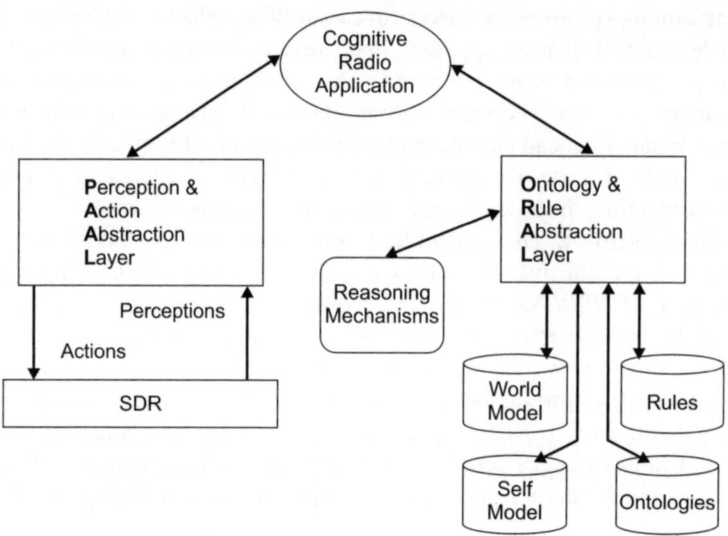

Figure 10.4 Cognitive radio architecture

standards could underlie the *Perception and Action Abstraction Layer* (PAAL) shown in the figure, but it is more likely that the PAAL would be based upon a higher level set of constructs built on top of the former. This point should become clearer in the ensuing discussion.

The *perceptions* and *actions* shown in Figure 10.4 correspond to what others have called 'meters' and 'knobs' with reference to the interfaces provided by an SDR. A cognitive radio must be able to recognize sensory inputs, or patterns thereof, as *being* or *representing* something in its environment. For example sensing a series of pulses with a particular energy over a certain time interval etc. is one thing, but being able to classify that series as a *radar waveform* and creating an instance of that concept to represent its existence as an individual is a higher level capability. Providing a standards-based platform-independent way of achieving such capabilities is the function of the PAAL. Concepts such as *MessageTransmission* and *transmittedBy*, described in our earlier discussion of the semantics of association, are examples of two higher level radio ontology concepts that should be definable in terms of the PAAL. Similarly, in order to allow a cognitive radio to cause the underlying radio infrastructure to perform certain actions ('turn knobs') the PAAL also provides an interface in the opposite direction, i.e., from the cognitive radio application to the SDR platform. Thus, referring to the DFS example we discussed earlier, the PAAL would provide the means for defining a construct like the *RadarCheck(x)* action in terms of SDR operations on SDR defined primitives. Just as the development of standards for the ORAL is a task for the semantic web community, so the development of standards for a PAAL is a task for the communities involved in the future of wireless communication.

The *world model* and *self model* are the two remaining components shown in Figure 10.4. What are they and why are they in the ORAL side of the architecture?

The world and self models are a key part of cognitive *device* architecture. Cognitive radio applications that run on devices will typically require what is widely called 'context awareness' [3], i.e., knowledge of the time-varying circumstances in the environment (world model) as well as knowledge of the capabilities and time-varying internal state of the device (self model). A world model is a time-varying construct that represents the state of the world as perceived by a device. For example, at a certain time a particular signal may be in the world model of a cognitive radio. That 'same signal' may also exist in the world model of another radio located nearby. That is because objects in the world can be perceived and acted upon by multiple agents. A cognitive radio's self model is a time-varying construct that represents the internal state of that radio as well as its basic capabilities, properties and goals (including its representation of its user's goals).

There are a number of reasons for these components being related to the ORAL. These constructs are basically 'instance repositories:' at any given time they contain instances of ontology-defined items. Instances are deposited in these constructs not only through interactions between the radio and the external world, but also because of inferences triggered by the states of these models. So the question of how a cognitive radio application handles the creation, deletion, etc. of instances, is essentially connected to the type of knowledge representation mechanisms the application employs. To the extent that these mechanisms are specifiable through the ORAL standards, the overall application is semantically interoperable, i.e., capable of being rendered in a format that is independent of the particular implementations employed.

To make this discussion more concrete consider the issue of how temporal change is handled in an application. The difference between declarative rule Formula (10.7) and ECA rule Formula (10.11) discussed above, both of which encode information about when to do radar checking, provides an illustration of different ways an application can handle temporal change. Rule Formula (10.7) explicitly quantifies over instances of time which means that the rule assumes an ontology that includes such instances. Rule Formula (10.11) does not explicitly quantify over temporal instances, but only refers to certain specific types of events. Moreover, Formula (10.7) explicitly encodes the 9-second interval requirement, whereas Formula (10.11) is based on the assumption that a particular object, the *RadarCheckAlarm*, will provide the necessary timing. Recall that Formula (10.7) represents a requirement of the DFS policy in a completely declarative way, saying precisely what the policy requires, whereas Formula (10.11) is one way of translating that rule into a device architecture that can provide policy-compliant behavior for a radio. An architecture capable of supporting the execution of Formula (10.11) would not require world and self models with a temporal ontology since it would be enough, as implied above, to keep only instances representing the current state. Of course *something* in the application would need to guarantee that the necessary temporal changes are accounted for correctly in the world and self models. In summary, Formula (10.7) (together with other rules and concepts that have not been discussed) represents at the ORAL level what Formula (10.11) and its execution environment encodes at the application level. There are many ways in which the requirement stated declaratively in Formula (10.7) can be implemented and each of them will require some compatible way of coordinating the dynamic contents of the device's world and self models. For an application to be semantically interoperable the way in which it handles temporal change at the device architecture level (the way in which it maintains world and self models) must be clearly related to temporal constraints and requirements stated at the ORAL level.

10.6 Future of Semantics in Cognitive Radio

Much of the work on cognitive radio heretofore has focused on device architecture. The scope, quality and performance of formal semantics based approaches has been promising [9] [24] [13]. One of the issues confronting all device architecture innovations (not just those based on formal semantics) are the practical limits imposed by current technology (e.g., memory) and cost concerns. For the near future, it is likely that cognitive radio based innovations in device architecture will continue to be focused on government and military applications and platforms. Within that arena there is every reason to expect increased reliance upon semantic methods.

In terms of deployment, cognitive radio technology can also be expected to follow a pattern like that of software defined radio, first being installed in environments such as base stations [6] where computation, memory, networking and power resources are relatively unconstrained compared to the environment of a handheld device.

The main potential growth area for semantic methods in cognitive radio is in community architecture. For reasons we have touched on in this chapter and that are more fully developed in [14], the realization of the full promise of cognitive radio at the device level is directly linked to the incorporation of formal semantic methods in the production of the intellectual artifacts that represent the shared understandings of regulators, standards bodies and implementors. The current perspectives of various wireless standards organizations working groups - such as IEEE 802.21 [22] and the SDR Forum Commercial Working Group [7], both of which rely upon the use of formal semantics for the specification of community architecture - are good indicators of the future applications for formal semantics.

10.7 Conclusion

With the advance of technologies collectively associated with *cognitive radios*, there is a growing need to standardize the data and knowledge structures related to the spectrum environment so as to enable mechanical and automated methods for spectrum access. Such activities are a way to architect the systems and conventions surrounding human use of the radio spectrum so as to make it possible for such radio devices to coexist with themselves and other devices. Without a common understanding of the radio domain and spectrum policy (i.e., the 'rules of the game'), dynamic spectrum access would not be feasible. Additionally, it would almost certainly be necessary for radios to share a common language in which they could communicate special circumstances that would cause the 'rules of the game' (e.g., operational or policy) to be altered or suspended, such as during emergency situations. As discussed in this chapter, the *precision* in expression of knowledge and the ability of that knowledge to be utilized in a *mechanical* fashion as enabled by formal semantics and knowledge representation illustrates the semantic side of cognitive radios.

References

[1] Aitken, J. (2004) Part-of relations in anatomy ontologies: a proposal for RDFS and OWL formalisations. *Proceedings of the Pacific Symposium on Biocomputing*, pp. 166–77, January 6–10, Big Island, Hawaii, USA.

[2] Baader, F. *et al.* (eds.) (2003), *The Description Logic Handbook: Theory, Implementation, and Applications*, Cambridge, Cambridge University Press.
[3] Bellavista *et al.* (2006) Context-aware semantic discovery for next generation mobile systems. *IEEE Communications Magazine, Special Issue on Advances in Service Platform Technologies*, **9**.
[4] Berlemann, S. *et al.* (2005) Policy-based reasoning for spectrum sharing in cognitive radio networks. *Proceedings of the IEEE DySPAN Conference*, November 8–11, Baltimore, MD, USA. IEEE Press.
[5] Bonner, J. and Kifer, M. (1993) Transaction logic programming. *Proceedings of the 10th International Conference on Logic Programming*, June, Budapest, Hungary.
[6] Bose, Vanu (2006) *Software radio in the real world* (invited presentation). *Proceedings of the SDR Forum Technical Conference*, November 14.
[7] Cummings, M. and Subrahmanyam, P.A. (2006) The role of a metalanguage in the context of cognitive radio lifecycle support (invited presentation), *Proceedings of the SDR Forum Technical Conference*, November 16.
[8] Denker, G. (2006) Policy-defined cognitive radios: what are the issues at hand? *Proceedings of the Cognitive Radio Workshop*, April 10, San Francisco, CA, SDR Forum.
[9] Denker, G., Ghanadan, R., Talcott, C. and Kumar, S. (2006) An architecture for policy-based cognitive tactical networking. *Proceedings of MILCOM 2006*, October 23–26, Washington, DC, USA.
[10] Federal Communications Commission (FCC) (2002) *Spectrum policy task force report*. ET Docket No. 02-135, November.
[11] Federal Communications Commission (FCC) (2006) *Report and order*. ET Docket No. 03-122, June 30.
[12] Fraternali, P. and Tanca, L. (1995) A structured approach for the definition of the semantics of active databases. *ACM Transactions on Database Systems*, **20**(4), 414–71.
[13] Ginsberg, A., Poston, J. and Horne, W. (2006) Toward a cognitive radio architecture: integrating knowledge representation with software defined radio technologies. *Proceedings of MILCOM 2006*, October 23–26, Washington, DC, USA.
[14] Ginsberg, A., Horne, W. and Poston, J. (2007) Community-based cognitive radio architecture: policy-compliant innovation via the semantic web. *IEEE Symposium on New Frontiers in Dynamic Spectrum Access Networks (DySPAN 2007)*, April 17.
[15] Gruber, T. (1993) A translation approach to portable ontologies. *Knowledge Acquisition*, **5**(2), 199–220.
[16] Horrocks, I. and Patel-Schneider, P.F. (2004) A proposal for an OWL rules language. *Proceedings of the 13th International World Wide Web Conference (WWW 2004)*, pp. 723–31, May 17–22, New York, USA.
[17] Horrocks, I. (2005) OWL rules: a proposal and prototype implementation. *Journal of Web Semantics*, **3**(1), 23–40.
[18] Institute of Electrical and Electronics Engineers (IEEE) Project 802 (1999a) *Wireless LAN medium access control (MAC) and physical (PHY) layer specifications*, IEEE Std 802.11.
[19] Institute of Electrical and Electronics Engineers (IEEE) Project 802 (1999b), *Amendment 1: high-speed physical layer in the 5 GHz band*, IEEE Std 802.11a-1999.
[20] Institute of Electrical and Electronics Engineers (IEEE) Project 802 (2003), *Amendment 5: spectrum and transmit power management extensions in the 5 GHz band in Europe*, IEEE Std 802.11h-2003.
[21] Institute of Electrical and Electronics Engineers (IEEE) Project 802 (2004), *Amendment 7: 4.9 GHz–5 GHz operation in Japan*, IEEE Std 802.11j-2004.
[22] Institute of Electrical and Electronics Engineers (IEEE) Project 802.21. Draft IEEE standard for local and metropolitan area networks: media independent handover services, http://www.ieee802.org/21/doctree/2006-05_meeting_docs/P802-21-D01-00.pdf. Accessed January 19, 2007.
[23] International Telecommunication Union (ITU) (2003) *Dynamic frequency selection (DFS) in wireless access systems including radio local area networks for the purpose of protecting the radiodeterminantion service in the 5 GHz band*, ITU-R Recommendation M.1652.
[24] Kokar, M., Brady, D. and Baclawski, K. (2006) Roles of ontologies in cognitive radios In B. Fette (ed.), *Cognitive Radio Technology*, pp. 401–33, Elsevier.
[25] MacKenzie, A. and DaSilva, L. (2006) *Game Theory for Wireless Engineers*, Morgan & Claypool.
[26] Mitola, J. (2006) *Cognitive Radio Architecture: The Engineering Foundations of Radio XML*, Wiley-Interscience.
[27] Noam, E. (1998) Spectrum auctions: yesterday's heresy, today's orthodoxy, tomorrow's anachronism. taking the next step to open spectrum access. *Journal of Law & Economics*, **XLI**, 765–90.

[28] Object Management Group (OMG) (2006) *Production rule representation proposal, draft response to OMG RFP br/2003-09-03*, June, www.w3.org/2005/rules/wg/wiki/FrontPage?action=AttachFile&do=get&target=PRR-draft-010606.pdf. Accessed January 19, 2007.

[29] United States of America (USA) (2005) 47 C.F.R. §15.403, 15.407.

[30] World Wide Web Consortium (W3C). Semantic Web, http://www.w3.org/2001/sw. Accessed January 19, 2007.

[31] World Wide Web Consortium (W3C) SWRL: a semantic web rule language combining OWL and RuleM, http://www.w3.org/Submission/SWRL/. Accessed January 19, 2007.

[32] World Wide Web Consortium (W3C). Rule interchange format, http://www.w3.org/2005/rules/. Accessed January 19, 2007.

[33] World Wide Web Consortium (W3C) (2004) *OWL web ontology nanguage overview*, W3C Recommendation, February 10, http://www.w3.org/TR/2004/REC-owl-features-20040210/#s1.1. Accessed January 22, 2007.

11

Security Issues in Cognitive Radio Networks

Chetan N. Mathur and K. P. Subbalakshmi
Stevens Institute of Technology, NJ, USA

11.1 Introduction

One of the primary requirements of cognitive radio networks is their ability to scan the entire spectral band for the presence/absence of primary users. This process is called spectrum sensing and is performed either locally by a secondary user or collectively by a group of secondary users. The available spectrum bands are then analyzed to determine their suitability for communication. Characteristics like signal-to-noise ratio (SNR), link error rate, delays, interference and holding time can be used to determine the most appropriate band. After the spectrum band is selected, secondary transmission in that band takes place. If a secondary user/network detects a primary user transmission, it vacates the corresponding spectrum band and looks for another vacant band. The process of handing off the licensed spectrum band to a primary user is called *spectrum handoff*. We later show that the delay associated with spectrum handoff causes many of the attacks proposed on cognitive networks.

There are two basic types of cognitive radio networks: centralized and distributed. The centralized network is an infrastructure-based network, where the secondary users are managed by secondary base stations which are in turn connected by a wired backbone. In a decentralized architecture, the secondary users communicate with each other in an ad-hoc manner. Spectrum sensing operation in decentralized architecture is usually performed collaboratively. This type of architecture also encompasses coexistence of two or more wireless networks operating in unlicensed bands. An example of this type of network is the coexistence of IEEE 802.11 with IEEE 802.16 [20].

To analyze the security of cognitive networks, we start by introducing some basic security concepts in the context of cognitive networks. Some of the fundamental building

Cognitive Networks: Towards Self-Aware Networks Edited by Qusay H. Mahmoud

blocks of communication security we consider are availability, integrity, identification, authentication, authorization, confidentiality and non- repudiation. We categorize the security issues in cognitive networks into inherent reliability issues and security attacks. Among reliability issues we specifically concentrate on those that are unique to cognitive radio networks like high sensitivity to weak primary signals, unknown primary receiver location, tight synchronization requirement in centralized cognitive networks, lack of common control channel, etc. We then show that each of these vulnerabilities can be utilized by a malicious entity to perform attacks at various layers of communication protocol. Attacks at physical, link, network, transport and application layers are considered.

In the next section, we will provide a brief background on cognitive networks, in Section 11.3 we will look at building blocks of security, specifically as applied to cognitive networks and in Section 11.4 we discuss reliability issues inherent in cognitive radio networks. In Section 11.5 we look at attacks that can occur in each of the layers of the protocol stack as well as cross-layer attacks. In Sections 11.5.1 and 11.5.2 we propose novel attacks on the PHY layer and the link layer, respectively. Network layer attacks are discussed in Section 11.5.3. Most of the network layer attacks are directed against the routing protocol. Transport layer attacks are discussed in Section 11.5.4. Attacks at this layer basically target the TCP protocol and security mechanisms. In addition to the layer-specific attacks, we also consider cross-layer attacks in Section 11.5.6. These are attacks performed in one layer to affect another layer. In Section 11.6, we briefly introduce some recent developments in cognitive network architectures like OCRA [3], Nautilus (http://www.cs.ucsb.edu/htzheng/cognitive/nautilus.html), IEEE 802.22 [10] and DIMSUMnet[6] and discuss their reliability and security issues. The concept behind cognitive networks is to have in-built cognitive (intelligence) capabilities through which the cognitive users detect and prevent intrusions and attacks on the network. In this chapter we show that the cognitive radio networks are still vulnerable to a variety of attacks across various layers of the protocol stack. This is largely because most of the protocols used for cognitive radio networks are borrowed from existing wireless networks. We further suggest that protocols developed for cognitive radio networks need to be improved to make them more intelligent to the surrounding environment, spectrum aware and resilient to intrusions and attacks. In Section 11.7 we propose some future directions to make cognitive radio networks secure and resilient to many of the proposed attacks. Finally, we conclude the chapter in Section 11.8.

11.2 Cognitive Radio Networks

11.2.1 Cognitive Radio Network Functions

11.2.1.1 Spectrum Sensing

One of the primary requirements of cognitive networks is their ability to scan the spectral band and identify vacant channels available for opportunistic transmission. As the primary user network is physically separate from the secondary user network, the secondary users do not get any direct feedback from primary users regarding their transmission. The secondary users have to depend on their own individual or cooperative sensing ability to detect primary user transmissions. Since the primary users can be spread across a huge geographical area, sensing the entire spectral band accurately is a challenging task [7] [11]. The secondary users have to rely on weak primary transmission signals to estimate their presence. Most of the research on spectrum-sensing techniques falls into three categories: transmitter detection, cooperative detection and interference-based detection [2]. The main aim of all these techniques is to avoid interference to primary transmissions. The amount of interference caused by all the secondary users at a point in space is referred to as the interference temperature [39] at that point. When a primary user transmission is taking place, the interference temperature should be below a specified threshold near the primary receivers. However, this is not easy to achieve as the location of the primary receiver is not known to the secondary users. Additionally, when multiple secondary networks overlap, the secondary users scanning the spectrum should not confuse transmissions from secondary users in other secondary networks with primary transmissions.

11.2.1.2 Spectrum Analysis and Decision

Each spectrum band has some unique features owing to its frequency range and the number of users (both primary and secondary) using the band. Spectrum sensing determines a list of spectrum bands that are available; however, the secondary users decide on the most appropriate band from the list of available bands. In addition to the commonly used SNR parameter, some of the characteristics of spectrum bands that can be used to evaluate their effectiveness are interference, path loss, wireless link errors, link layer delay and holding time (expected duration that the secondary user can occupy the band).

11.2.1.3 Spectrum Mobility

Spectrum mobility refers to the agility of cognitive radio networks to dynamically switch between spectrum access. As secondary users are not guaranteed continuous spectrum access in any of the licensed bands and the availability of vacant spectrum bands frequently changes over time, spectrum mobility becomes an important factor when designing cognitive protocols. One of the primary factors affecting spectrum mobility is the delay incurred during spectrum handoff. This delay adversely affects protocols employed at various layers of the communication protocol stack. Another important factor to be considered in spectrum mobility is the time difference between the secondary network detecting a primary transmission and the secondary users vacating the spectral band. Transmissions from secondary users during this period will cause harmful interference to the primary users. The FCC [13] has set upper bounds on the spectrum handoff duration to avoid prolonged interference to primary users.

11.2.2 Cognitive Network Types

11.2.2.1 Centralized Cognitive Networks

In a centralized architecture, the secondary user network is infrastructure oriented. That is, the network is divided into cells; each cell is managed through a secondary base station. These base stations control the medium access and the secondary users as shown in Figure 11.1. The secondary users are synchronized with their base stations and may perform periodic spectrum-sensing operations. The secondary base stations can be interconnected through a wired backbone network.

11.2.2.2 Decentralized Cognitive Networks

In a decentralized architecture, the secondary users are not interconnected by an infrastructure-oriented network. Figure 11.2 represents a decentralized network, where

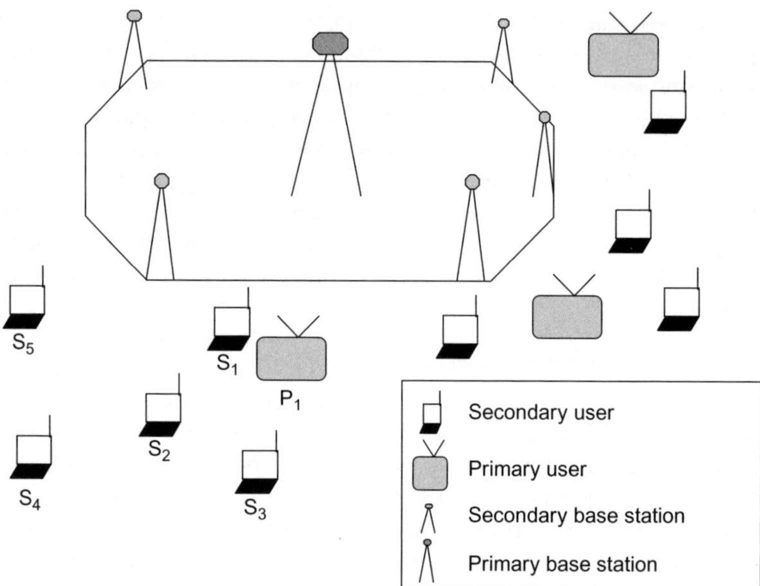

Figure 11.1 Centralized cognitive radio network

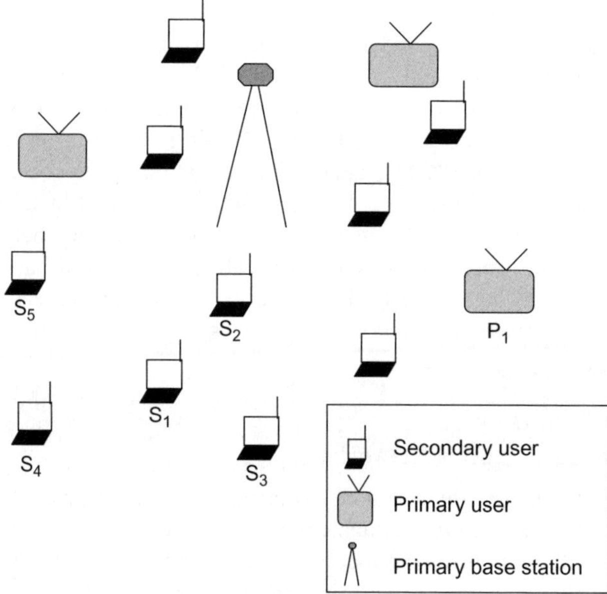

Figure 11.2 Decentralized cognitive radio network

the secondary users communicate with each other in an ad-hoc manner. Two secondary users who are within communication range can exchange information directly, while the secondary users who are not within direct communication range can exchange information over multiple hops.

In distributed cognitive networks, secondary users make decisions on spectrum bands, transmit power, etc. based on either local observations or cooperatively through some utility functions to get near optimum performance for all the secondary users. To illustrate the basics of a collaborative approach, consider the following example (see Figure 11.2), where two secondary users (S_1 and S_2) are operating in a band licensed to a primary base station. S_1 is in the boundary of the transmission range of the primary base station whereas S_2 is closer to the primary base station. Hence S_2 will detect the presence of primary users quickly and easily compared to S_1. Collaborative sensing techniques highlight the fact that if the secondary users share their relative sensing information, then the overall primary user detection for the cognitive network can be improved. However, these protocols do not consider malicious users in the network. Later in this chapter we show that collaborative protocols can be used by malicious users to cause security violations in cognitive networks.

A subclass of decentralized cognitive networks are the spectrum sharing networks, where two wireless networks coexist in an unlicensed band [12]. An example of such a network is the coexistence of IEEE 802.11 and 802.16 [20]. In such networks, a common spectrum coordination channel is established to exchange control information on transmitter and receiver parameters. Primary user identification, spectrum mobility and management functions are not necessary in these class of networks.

11.3 Building Blocks of Communication Security

In this section, we briefly introduce the basic building blocks of communication security. We describe how these building blocks are applied in existing wireless networks like IEEE 802.11 wireless LAN and discuss their importance in cognitive networks.

11.3.1 Availability

One of the fundamental requirements for any type of a network is availability. If the network is down and not usable, the purpose of its existence is defeated. Most of the attacks that we hear about these days like the denial of service (DoS) attacks, jamming attacks, buffer overflow attacks on network queues are all targeted towards rendering the network unavailable either temporarily or permanently [14]. An issue that is closely related to network availability is data availability. This is the availability of data (user information, routing tables, etc.) to the users in the network.

In wireless networks, availability usually refers to the availability of the wireless transmission medium. Several techniques are used to ensure that the wireless communication medium is available for transmission. For example, random back-off mechanism [29] is used to prevent collision (jamming) between multiple users at the medium access control (MAC) sublayer of the IEEE 802.11 link layer.

In the context of cognitive networks, availability refers to the ability of primary and secondary users to access the spectrum. For primary (licensed) users, availability refers to being able to transmit in the licensed band without harmful interference from the secondary users. From the definition of dynamic spectrum access policies [13], spectrum availability for the primary users is guaranteed. For the secondary (unlicensed) users, availability refers to the existence of chunks of spectrum, where the secondary user can transmit without causing harmful interference to the primary users. Although studies have shown that large chunks of licensed spectrum are available for opportunistic use [14], availability of spectrum to the secondary users is not guaranteed. In centralized cognitive networks, availability also refers to the availability of secondary base stations. Security mechanisms should ensure that DoS attacks against secondary base stations are appropriately countered.

11.3.2 Integrity

Data that is in transit in the network needs to be protected from malicious modification, insertion, deletion or replay. Integrity is an assurance that the data received is exactly as sent by an authorized entity. However, some parts of the data are mutable [41], that is, they need to be legitimately modified as they move from one node to the other. For example, the next hop information, timestamp, hop count are all mutable fields. For this reason, most of the techniques used to ensure integrity perform selective field integrity [41] (provide integrity only for the selected non-mutable fields).

Integrity is extremely important in wireless networks because, unlike their wired counterparts, the wireless medium is easily accessible to intruders. It is for this reason that in wireless LANs an additional layer of security is added at the link layer, to make the wireless links as secure as the wired links. The security protocol used in this layer is called the CCMP [9] (counter-mode encryption with CBC-MAC authentication protocol). The CCMP protocol uses the state-of-the-art advanced encryption standard (AES) [15] in cipher block chaining mode [33] to produce a message integrity check (MIC), which is used to verify the integrity of the message by the recipient. These techniques can also be employed in cognitive radio networks.

11.3.3 Identification

Identification is one of the basic security requirements for any communication device. It is a method to associate a user/device with their name or identity. For example, in cellular networks, the mobile devices are provided with an equipment identification called international mobile equipment identifier (IMEI). This identifier is used to uniquely identify the mobile devices in the cellular networks. Similarly, a tamper-proof identification mechanism should be built into the secondary user devices in cognitive networks.

11.3.4 Authentication

Authentication is an assurance that the communicating entity is the one that it claims to be. The primary objective of an authentication scheme is to prevent unauthorized users from gaining access to protected systems. It is a necessary procedure for verifying both an entity's identity and authority. From the service provider's perspective, authentication protects the service provider from unauthorized intrusions into the system. Most of the mechanisms that ensure authentication rely on a centralized certificate authority (CA) that is trusted by all the users in the network. A typical authentication protocol would require the peer entities to get their identities signed (using public key encryption) by the CA and the digitally signed certificates are exchanged and verified by the peers to ensure authenticity. Once the authenticity of peers is established, regular communication is initiated.

In cognitive radio networks, there is an inherent requirement to distinguish between primary and secondary users. Therefore, authentication can be considered as one of the basic requirements for cognitive networks. In centralized cognitive networks, where the primary and secondary base stations are connected to a wired backbone network, it may be easier to have the CA connected to the wired backbone. However, in distributed cognitive networks with a number of secondary users dispersed over a large geographical area, providing the functionalities of a CA can be quite a challenge [42].

11.3.5 Authorization

Different entities in the network have varying levels of authorization. For example, the wireless access point has the authorization to remove a possibly malicious user from access to the network. Other users in the network do not have this privilege. The network access control policy describes the level of authorization for each of the entities. In the context of cognitive networks, we have a unique authorization requirement which we call conditional authorization. It is conditional because the secondary users are authorized to transmit in licensed bands only as long as they do not interfere with primary users' communications in that band. As it is difficult to pinpoint exactly which of the secondary users is responsible for harmful interference to the primaries' transmission, this type of authorization is hard to enforce and even more so in a distributed setting. Hence conditional authorization poses a unique challenge in dynamic spectrum access.

11.3.6 Confidentiality

Confidentiality is closely linked with integrity. While integrity assures that data is not maliciously modified in transit, confidentiality assures that the data is transformed in such a way that it is unintelligible to an unauthorized (possibly malicious) entity. This is achieved by employing ciphers and encrypting the data to be transmitted with a secret key which is shared only with the recipients. The encrypted data is then transmitted and only the recipients with a valid key can decrypt and read the data.

Since the wireless medium is open to intruders, the IEEE 802.11 LAN employs the AES encryption in counter mode in its CCMP protocol [9] to encrypt the data at the link layer as an additional layer of security. The error-prone and noisy nature of wireless medium poses a unique challenge to both data confidentiality and integrity mechanisms. This is because almost all confidentiality and integrity techniques rely on ciphers which are sensitive to channel errors and erasures. This sensitivity property under noisy conditions triggers excessive retransmissions consuming a lot of network bandwidth [28], [31]. This issue is even more pronounced in cognitive networks, where the secondary user access to the network is opportunistic and spectrum availability is not guaranteed.

11.3.7 Non-repudiation

Non repudiation techniques [34] prevent either the sender or receiver from denying a transmitted message. Therefore, when a message is sent, the receiver can prove that the message was in fact sent by the alleged sender. Similarly, when a message is received, the sender can in fact prove that the data received was by the alleged receiver. In cognitive radio network setting, if malicious secondary users violating the protocol are identified, non-repudiation techniques can be used to prove the misbehavior and disassociate/ban the malicious users from the secondary network.

11.4 Inherent Reliability Issues

In this section we point out some of the inherent reliability issues in cognitive radio networks.

11.4.1 High Sensitivity to Primary User Signals

To prevent interference to licensed primary users, secondary users should detect the primary transmissions in the first place. To ensure high probability (e.g. 99%) of non-interference to the primary users, stringent sensitivity requirements are placed on the detectors at the unlicensed secondary devices. There are two prominent ways to detect the primary user's transmissions: (a) energy-based and (b) waveform-based sensing. Energy-based sensing does not require any knowledge of the primary user's transmission signal, it is based on the simple fact that any information-baring signal has finite signal strength (power). However, energy-based sensing techniques are prone to false detections and usually take a longer time when the signal is low power. Waveform-based sensing is applied when information about the waveform and signal patterns of the primary users transmission signal is known. This makes the waveform-based sensing techniques perform better than energy-based sensing in terms of speed and reliability. However, in many instances, primary user transmission signal patterns may not be known to the secondary users. Moreover, the FCC mandates [13], as one of the requirements for cognitive networks, to predict the temperature interference (measured by energy) on nearby primary user's receivers and keep it below a threshold. To ensure compliance with this requirement the sensitivity towards primary user signals in cognitive networks is usually set to be high. This high sensitivity magnifies false detections due to energy-based sensing resulting in inefficient use of spectrum.

11.4.2 Unknown Primary Receiver Location

One issue that has received very little attention so far is the lack of knowledge of the primary receivers' location. In order to minimize the interference to the primary user network, the secondary transmitters need to know the locations of the primary receivers. Most of the interference models that have been studied in the literature [16], propose to minimize interference temperature, but the primary receivers' location is unknown. This may lead to hidden terminal [35] problems in cognitive networks. Some recent works (like [38]) have studied this issue and proposed techniques that can be used to detect primary receivers by detecting the receiver power leakage. However, significant work remains to be done to protect the primary receivers from accidental interference caused by secondary users.

11.4.3 Synchronization Requirement

The secondary users in centralized cognitive radio networks perform fast sensing operations between periods of transmissions (e.g., IEEE 802.22 [10]). These measurements are relayed to the base station which aggregates and determines the presence of primary user transmissions. Therefore, synchronization of secondary users in time is an important requirement to detect the presence of primary users. Even if one secondary user is out of sync with the rest of the secondary users, all other secondary users would in turn detect the energy from the out of sync secondary user and transmit that information to the base station. The base station would then assume that a primary user transmission is going on and may shut down all the secondary user transmissions in that frequency band. This would result in what is known as a missed opportunity [36]; the band is available yet unused by the secondary users. However, synchronization between secondary users is harder to achieve in some spectrum bands like the TV bands that span over a large geographical area.

11.4.4 Lack of Common Control Channel

Unlike other infrastructure oriented wireless networks, cognitive networks lack a predetermined control channel. Therefore, as soon as a secondary user boots up, it needs to search for control signals across the entire spectral band. Additionally, this operation needs to be performed during connection reestablishment and whenever mobile secondary users move out of the coverage area of an existing base station to the coverage area of another base station.

11.4.5 Protocols and Utilities Based on Homo Egualis Model

The utility functions in many coordinated spectrum access protocols are usually based on *Homo egualis* [17] like model. This model is based on an *Homo egualis* society where individuals have an inequality aversion. Such models assume no centralized structure of governance, so the enforcement of norms depends on the voluntary participation of peers. However, in reality this situation rarely exists. Malicious entities tend to violate the *Homo egualis* model for selfish purposes like gaining more bandwidth and resources or to intentionally block others from getting specific resources. More robust models that account for malicious behavior must be employed (more in Section 11.7) to design coordinated access protocols in cognitive networks.

11.5 Attacks on Cognitive Networks

We define an attack on cognitive networks as any activity that results in (a) unacceptable interference to the licensed primary users or (b) missed opportunities for secondary users. An attack is considered strong if it involves a minimal number of adversaries performing minimal operations but causing maximum damage/loss to the primary and or secondary users in the network. In this section we describe attacks on various layers of cognitive networks. Most of the attacks we describe make use of one or more reliability issues we pointed out in the previous section. We consider five layers in the protocol stack, namely, the physical layer, link layer, network layer, transport layer and application layer. In addition to the attacks that are specific to a given layer, we also discuss some cross-layer attacks that can be applied at one layer to affect at another layer.

11.5.1 Physical Layer Attacks

The physical layer is the most basic layer in the protocol stack. It provides the means of transmitting raw signals over the transmission medium. The physical layer determines the bit rate, channel capacity, bandwidth and maximum throughput of the connection. In cognitive networks, the physical layer has the capability to transmit at various frequencies across most of the spectrum band. This makes the physical layer in cognitive networks more complex compared to conventional wireless networks [8]. Therefore, when transmission from one frequency band is switched to another frequency band, the switching process incurs considerable delay in the physical layer of cognitive networks. In this section, we propose some attacks on cognitive networks that specifically target the physical layer.

11.5.1.1 Intentional Jamming Attack

This is one of the most basic types of attack that can be performed by secondary users in cognitive radio networks. The malicious secondary user jams primary and other secondary users by intentionally and continuously transmitting in a licensed band. The attack can be further amplified by using high transmit power, transmitted in several spectral bands. Although simple energy-based detection and triangulation techniques can be used to detect this attack, the time it takes to pinpoint and ban the malicious user impacts severely on the network performance. The attack can be made more dangerous by a mobile malicious secondary user performing the attack in one geographical area and moving to another area before being caught.

11.5.1.2 Primary Receiver Jamming Attack

The lack of knowledge about the location of primary receivers can be used by a malicious entity to intentionally cause harmful interference to a victim primary receiver. The attack is caused when a malicious entity closer to the victim primary receiver participates in a collaborative protocol and requests transmissions from other secondary users to be directed towards the malicious user. Although the interference temperature is kept below a specified threshold at some other point in space, this would still cause continued interference to the primary receiver eventually blocking it out from listening to primary transmissions. Moreover, the naive secondary users causing the interference are oblivious about it.

11.5.1.3 Sensitivity Amplifying Attack

In order to prevent interference to the primary network, some primary user detection techniques have higher sensitivity towards primary transmissions (see Section 11.4). This leads to frequent false detections and missed opportunities for the secondary users. A malicious entity can amplify the sensitivity and hence the number of missed opportunities by replaying the primary transmissions. What makes this attack more lethal is that even an adversary with low transmit power can transmit in spectrum band boundaries and still cause multiple secondary users operating in multiple spectrum bands to incur missed opportunities and render spectrum usage inefficient.

11.5.1.4 Overlapping Secondary User Attack

In both centralized and distributed cognitive networks, multiple secondary networks may coexist over the same geographical region. In such cases, transmissions from malicious entities in one network can cause harm to the primary and secondary users of the other network. This type of attack is hard to prevent because the malicious entities may not be under the direct control of the secondary base station/users of the victim network.

11.5.2 Link Layer Attacks

This is the second layer in the network protocol stack. The main purpose of the link layer is to transfer data from one node to the next node in one hop. The link layer provides the functional means to allow fragmentation of data, error correction and modulation. The medium access control (MAC) layer is one of the important sublayers of the link layer, which controls channel assignment. Fairness is one of the primary requirements for channel assignment protocols. In traditional wireless environments, SNR is considered as one of the main parameters to judge the fairness of a channel allocation scheme. However, in cognitive network environments other parameters such as holding time, interference, path loss, wireless link error rate and delay are just as important as the SNR. For this reason, channel assignment is a more complex operation in cognitive networks. In this section, we propose three novel attacks that can be performed at the link layer in cognitive networks.

11.5.2.1 Biased Utility Attack

A malicious secondary user may selfishly tweak parameters of the utility function to increase its bandwidth. If the secondary users and/or base stations are unable to detect such anomalous behavior, this may result in deprivation of transmission medium for other secondary users. For example, in [39] the authors propose a utility function that is used by the secondary users to determine their bandwidth (in terms of transmission power) with the constraint that the interference temperature due to the secondary transmissions on the primary receivers is below a given threshold. The problem is formulated as a public good game and Nash equilibrium solutions for a global optimum determines the transmit powers of the secondary users. If a malicious user tweaks its utility function to transmit at higher power, it will result in other users getting less bandwidth. Some secondary users may not even get to transmit.

11.5.2.2 Asynchronous Sensing Attack

Instead of synchronizing the sensing activity with other secondary users in the network, a malicious secondary user may transmit asynchronously when other secondary users are performing sensing operations. If the base station or other secondary users consider this as a transmission from a primary user, then this could result in missed opportunities. This attack can be made more efficient by transmitting only during sensing periods [23].

11.5.2.3 False Feedback Attack

For protocols that rely on secondary users exchanging information, false feedback from one or a group of malicious users could make other secondary users take inappropriate actions and violate the goals of the protocol. For example, consider a scenario represented in Figure 11.2. In this decentralized cognitive network, five secondary users, $S_1, \ldots, S_5$, are associated with a secondary base station. Secondary user S_2 is closer to the primary base station. Suppose S_2 was a malicious user. When the primary base station starts using its licensed spectrum band, S_2 senses it but does not reveal that information to other secondary users. Secondary user S_1 on the other hand is just outside the boundary region of the primary base station's transmission range and hence does not sense the presence of the primary user. Similarly, secondary users S_3, S_4 and S_5 also fail to sense the presence of the primary user. Now any transmission from S_1, S_3, S_4 and S_5 may cause harmful interference to the primary receivers within the primary base station's transmission range. A similar attack is possible in centralized cognitive radio network.

11.5.3 Network Layer Attacks

While the data link layer is responsible for node to node (one-hop) packet delivery, the network layer is responsible for end-to-end (source-to-destination) packet delivery. The network layer provides functional means for performing routing, flow control and ensuring quality of service (QoS). Routing refers to selecting paths along the network through which data is transmitted from source to destination. Every node in the network is responsible for maintaining routing information (usually in the form of a table) about its neighboring nodes. When a connection needs to be established, every node determines which of its neighbors should be the next link in the path towards the destination. Some of the routing protocols used in wireless environment are for example dynamic source routing (DSR) and ad-hoc on demand distance vector (AODV) routing [32]. A malicious node in the path can disrupt routing by either broadcasting incorrect routing information to its neighbors or by redirecting the packets in the wrong direction. Several routing attacks have been discovered in wireless ad-hoc networks, most of the attacks can be classified into two categories: routing disruption attacks and resource consumption attacks. Some of the examples of routing attacks are the black hole attack [19] where the malicious node attracts packets from every other node and drops all the packets, the gray hole attack [1] where the malicious node selectively drops the packets, the worm hole attack where the malicious user uses two pairs of nodes with a private connection between the two pairs. The worm hole attack is a dangerous attack since it can prevent route discovery where the source and the destination are more than two hops away. Most of these attacks are prevented by using secure on-demand routing protocols like Ariadne [18] or secure AODV [4], which use cryptographic mechanisms to guarantee integrity of routing information and authenticity of nodes. Although most of the network layer security issues in wireless LANs and wireless ad-hoc networks have been well studied, a similar analysis of network layer security issues in cognitive networks is yet to be performed. In this section we point out some of the unique network layer attacks applied on cognitive networks.

11.5.3.1 Network Endo-Parasite Attack (NEPA)

NEPA [30] assumes at least one compromised or malicious node in the network. The malicious node attempts to increase the interference at heavily loaded high priority channels. Most of the time, the affected links are along the routing path through the malicious nodes towards the wired gateway; hence the attack takes the name of a parasite attack. Under normal channel assignment operation, a node assigns the least loaded channels to its interfaces and transmits the latest information to its domain neighbors. A compromised node launches NEPA by assigning its interfaces the high priority channels. However, it does not inform its neighbors about this change. Since the transmitted information is not verified by neighbors, the network remains unaware of change. It results in hidden usage of heavily loaded channels; hence the attack takes the name endo-parasite, which refers to internal/hidden parasites. The links using these channels experience interference, decrease in available bandwidth and continuous degraded performance.

11.5.3.2 Channel Ecto-Parasite Attack (CEPA)

CEPA [30] is a special case of NEPA with slight modification in the attack strategy. A compromised node launches CEPA by switching all its interfaces to the channel that is being used by the highest priority link. However, the severe nature of this attack makes the detection easy.

11.5.3.3 Low cOst Ripple effect Attack (LORA)

In LORA [30], misleading information about spectrum assignments is transmitted to all the neighbors to push the network into a quasi-stable state. Since the channel assignment of the compromised node is not actually changed, this attack is significantly different from NEPA and CEPA and it is a low-cost (stronger) attack. The attack is relatively more severe then NEPA and CEPA because the effect is propagated to a large portion of the network beyond the neighbors of the compromised node, disrupting the traffic forwarding capability of various nodes for considerable time duration. The attack is launched when the compromised node transmits misleading channel assignment information, forcing the other nodes to adjust their channel assignments. This may generate a series of changes even in the channel assignment of non-neighboring nodes. Note that normally loaded channels are selected instead of heavily loaded ones and an illusion of heavy load on these channels is created, which results in propagation of change upwards in the routing tree. Heavily loaded channels are not selected because such selection will affect the links closer to gateways resulting in quick adjustment to the change and hence no ripple effect will be created.

11.5.4 Transport Layer Attacks

The transport layer provides functional requirements to transfer data between two end hosts. It is primarily responsible for flow control, end-to-end error recovery and congestion control. There are two main protocols that operate in the transport layer, the User Datagram Protocol (UDP) and the Transport Control Protocol (TCP). UDP is connectionless while TCP is connection oriented and guarantees ordered packet delivery. TCP performance is usually measured by a parameter called round trip time (RTT). Errors in the wireless environment cause packet loss which in turn triggers retransmissions. Moreover, frequent spectrum band changes by secondary users due to spectrum handoff at the link layer increases RTT. This phenomenon is utilized in our proposed variant (see Section 11.5.6, of jellyfish attack [1]) to degrade the performance of TCP. Further, different secondary nodes operate in different frequency bands and these bands constantly change. Hence the RTT for a TCP connection in cognitive networks has a high variance.

11.5.4.1 Key Depletion Attack

Unusually high round trip times and frequently occurring retransmissions imply that transport layer sessions in cognitive networks last only for a short duration. This results in large number of sessions being initiated for any given application. Most of the transport layer security protocols like SSL and TLS establish cryptographic keys at the beginning of every transport layer session. A higher number of sessions in cognitive networks and hence the number of key establishments will increase the probability of using the same key twice. Key repetitions can be exploited to break the underlying cipher system [37]. For example, the WEP and TKIP protocols used in IEEE 802.11 link layer are vulnerable to key repetition attacks. The newer and stronger CCMP protocol [9] is designed to exponentially delay key repetitions. However, most of the security protocols used below the network layer are designed taking into consideration the total number of sessions typically occurring in wireless LANs. These protocols need to be reinvestigated in the context of the number of sessions that occur in cognitive networks.

11.5.5 Application Layer Attacks

The application layer is the final layer of the communication protocol stack. It provides applications for users of the communication devices. Some of the basic application layer services include file transfer protocol (FTP), Telnet, email and lately multimedia streaming [27]. Protocols that run at the application layer rest on the services provided by the layers below it. Therefore, any attack on physical, link, network or transport layers impact adversely on the application layer. One of the most important parameters in the application layer is the quality of service (QoS). This is especially important for multimedia streaming applications. Physical and link layer delays due to spectrum handoffs, unnecessary rerouting and stale routing due to network layer attacks and delays due to frequent key exchanges cause degradation of the QoS in the application layer protocols.

11.5.6 Cross-layer Attacks

By cross-layer attacks we mean malicious operations performed at one layer to cause security violations at another layer. In cognitive networks, there is an inherent need for greater interaction between the different layers of the protocol stack. It is for this reason that cross-layer attacks need to be given special attention in cognitive networks. In this section, we discuss the well-known jellyfish attack [1] performed in the network layer to attack the transport layer, and also present a novel routing information jamming attack which is performed in the link layer to attack the network layer.

11.5.6.1 Jellyfish Attack

This attack is performed at the network layer but it affects the performance of the transport layer, specifically the TCP protocol. The goal of this attack is to reduce the throughput of the TCP protocol. There are three variants [1] of this attack: misordering, dropping and delay variance. In the misordering attack, the packets are intentionally and periodically reordered as they pass through the malicious node. This attack takes advantage of TCP's vulnerability to misordered packets [5], which triggers retransmissions and degrades throughput. The second variant is the packet dropping attack, where the malicious node periodically drops a fraction of packets that pass through it. This is similar to gray hole attack (see Section 11.5.3), however, the packets are dropped intelligently by the adversary to coincide with the TCP transmission window. When carefully executed it can cause near zero throughput in the TCP protocol. The third variant of the attack is called delay variance attack, here the packets are randomly delayed as they pass through a malicious node. This causes the TCP timers to be invalid which results in congestion inferences. Although these attacks are primarily proposed for ad-hoc wireless networks, they can be applied to decentralized cognitive networks as well. We propose a fourth variant of the jellyfish attack, where the attacker performs operations in the link layer to attack the transport layer. To perform this attack, the attacker causes the victim cognitive node to switch from one frequency band to another frequency band (using any of the link layer attacks), this causes considerable delay in the network and transport layers. When performed actively observing the TCP traffic timing, the delay caused due to spectrum handoffs can push RTT to round trip timeout (RTO). RTOs result in retransmissions and hence drastic degradation of TCP throughput. Note that all of the attacks above are hard to detect at the network layer but cause DoS attack at the transport layer.

11.5.6.2 Routing Information Jamming Attack

This is a novel cross-layer attack that makes use of the lack of a common control channel in cognitive networks and the spectrum handoff delay to jam the exchange of routing information among neighboring nodes. This results in the network using stale routes and incorrect routing of packets from source to destination. The attack is initiated when a malicious node causes spectrum handoff in the victim node just before routing information is exchanged. When this happens, the victim node stops all ongoing communication, vacates the spectral band, opportunistically selects a new spectrum for transmission, scans the entire spectrum band to identify the neighboring nodes and informs the neighboring nodes of the new frequency. Only after these operations are performed can the victim node receive/transmit the updated routing information from/to its neighbors. Until this period any path that goes through the victim node and its neighbors uses stale routing information. This attack can be extended by performing spectrum handoff attacks on the victim node successively just before routing information exchange. The attack can be made more severe if it is performed along the min-cut between the source and destination nodes.

11.6 Cognitive Network Architectures

In this section, we briefly introduce some recent developments in cognitive network architectures and discuss the attacks presented in Section 11.5 in the context of these architectures.

11.6.1 Nautilus

Nautilus (http://www.cs.ucsb.edu/htzheng/cognitive/nautilus.html) is a distributed, scalable and efficient coordination framework for open spectrum ad-hoc networks. The Nautilus framework addresses the lack of a common control channel faced in distributed cognitive network architectures. Some collaborative spectrum access schemes that do not rely on a centralized entity or a common control channel are proposed. One of the proposed collaborative schemes is based on graph coloring, where a topology-optimized allocation algorithm is used for a fixed topology. For mobile cognitive networks, a distributed spectrum allocation based on local bargaining is proposed, where mobile users negotiate spectrum assignment within local self-organized groups. For resource-constrained cognitive devices, a rule-based spectrum management is proposed, where unlicensed users access spectrum according to some predetermined rules and local spectrum observations. Although these novel techniques make spectrum access in cognitive networks more robust and independent of a centralized authority, other security attacks like spectrum handoff attacks, routing attacks, jellyfish attacks pose significant threat to the Nautilus architecture.

11.6.2 Dimsumnet

DIMSUMnet architecture [6] relies on a spectrum broker who permanently owns and manages the licensed bands (referred to as coordinated access bands (CABs)). The secondary base stations register with radio access network managers (RANMANs). The RANMANs negotiate with spectrum brokers to lease the appropriate portion of the spectrum requested by the secondary base stations. The spectrum broker, who maintains a database of currently available frequency bands, responds to RANMANs with allotted spectrum frequencies and timeslots. After spectrum bands are assigned to the secondary base stations, the secondary users connected to those base stations are informed to switch to the corresponding frequency bands. Since DIMSUMnet is a truly centralized architecture, with spectrum sensing performed by a centralized entity, security mechanisms are easier to implement and adhere to. The sensing attacks [23] that are possible in IEEE 802.22 are harder to implement in DIMSUMnet. However, as the spectrum monitoring function is performed by just one entity the information about spectrum availability may not be as accurate as in the case of distributed sensing performed by IEEE 802.22. Inaccurate spectrum information can be a primary reliability issue in the case of DIMSUMnet.

11.6.3 IEEE 802.22

IEEE 802.22 is a standard for wireless regional area networks (WRAN) [10] that utilize ultra high frequency (UHF) and very high frequency (VHF) television bands between 54 and 862 MHz. The standard mandates a centralized cognitive network architecture, where the secondary IEEE 802.22 base stations manage a unique feature of distributed sensing. To perform distributed sensing, the base station instructs the secondary cognitive user devices (referred to as consumer premise equipments (CPEs) in the standard) to synchronously sense various spectral bands for the presence of primary user activity. These sensing results are periodically collected by the base station, which then performs aggregation on the results to determine the presence/absence of primary users in each of the licensed spectrum bands. Hence, the IEEE 802.22 relies on synchronous sensing. This requirement can be a source of vulnerability as pointed out in [23]. Here, the malicious entity transmits during the sensing period causing all other secondary users to report primary user activity to the base station. This results in inefficient use of spectrum resources. In [23] we provide a digital signature-based solution, which helps the base stations identify the original primary signals from those that are sent by malicious entities.

11.6.4 OCRA (OFDM-based Cognitive Radio Architecture)

The OCRA network [3] is based on orthogonal frequency division multiplexing (OFDM) technology. Here both centralized and distributed cognitive network architectures are considered. To perform spectrum sensing and handoff decisions, OCRA employs a novel OFDM-based spectrum management technique which is based on a physical layer that enables dual mode spectrum sharing [3]. This type of spectrum sharing enables access to existing networks as well as coordination between cognitive users. OCRA proposes to use a novel cross-layer routing technique that jointly considers rerouting and spectrum handoff. To increase the reliability and QOS, multiple transport layer connections over non-contiguous spectrum bands are established.

11.7 Future Directions

In this section we provide some future directions that need to be taken to make secure cognitive radio networks against both accidental and intentional attacks. Most of our proposed solutions are easy to implement (for example, using existing security protocols). However, we also propose solutions (for example, developing analog crypto primitives) that require more work.

11.7.1 Using Existing Security Protocols

Security services provided in cellular, WLAN and wireless ad-hoc networks can be applied to cognitive networks as well. In a centralized wireless network architecture, the backbone network is usually a wired medium. Hence, strong security mechanisms exist that protect this network. It is the last hop between the wireless base stations and the wireless terminals that needs to be protected over the air. As cellular networks are centralized, security solutions in existing cellular networks (3G in particular) could be used as a model to provide security in cognitive networks. In cellular networks, user identity is obtained by using a temporary identity called international mobile user identity. Authentication is achieved by a challenge/response mechanism using a secret key. A challenge/response mechanism is where one entity in the network proves to another entity that it knows a particular secret without revealing it. The UMTS authentication and key agreement (UMTS AKA) is used to achieve authentication. Confidentiality is provided by using the confidentiality algorithm known as f8 and the secret cipher key (CK) that is exchanged as a part of the AKA process. Integrity is provided by using the integrity algorithm, f9 and the integrity key (IK). A block cipher known as KASUMI [21] is the building block of both f8 and f9 algorithms. KASUMI operates on 64 bit blocks and uses a 128-bit secret key. A similar setup could be used in centralized cognitive networks to establish the basic security requirements between the secondary users and the secondary base station.

In decentralized networks, secondary users communicate with each other over one or more hops. Due to the lack of infrastructure, these networks are also referred to as ad-hoc networks. These types of networks usually employ a two-level security mechanism. One level of security is provided at the link layer to protect every hop of communication and the other level of security is employed at the network/transport or application layer to protect the end-to-end communication path. Two most complicated operations in ad-hoc wireless networks are key management and secure routing. Fortunately, there has been a lot of research in this area and several security architectures for IEEE 802.11 multihop wireless ad-hoc networks and mobile ad-hoc networks (MANETs) have been proposed [42]. Decentralized cognitive networks could use security mechanisms employed in ad-hoc wireless networks. Some of the indigenous issues such as lack of a common control channel and use of diverse frequency bands by different secondary users may impose additional constraints on the existing security protocols.

11.7.2 Using Cryptographic Primitives

Most of the attacks performed at the link layer involve a malicious entity masquerading as a primary user. Therefore primary user identification is very important for both centralized and decentralized cognitive networks. We recently proposed a digital signature based primary user identification mechanism [23] that can be used by secondary users to distinguish malicious transmissions from primaries. Further research in the use of cryptographic primitives to solve inherent security issues in cognitive network needs to be performed.

11.7.3 Reactive Security Mechanisms

Reactive security mechanisms that detect malicious activity in cognitive networks need to be developed (see Chapter 12). For example, mechanisms that can detect unusually high spectrum handoffs is useful to prevent jamming and spectrum handoff attacks. Detection mechanisms combined with non-repudiation mechanisms enable secondary users to identify and block malicious users from the network.

11.7.4 Spectrum Aware Approach

There are two ways to handle spectrum mobility and associated delays. One is to make spectrum sensing, analyzing and handoff process fast and transparent to the higher layer protocols. However, spectrum sensing and handoff processes are in their infant stages and it will take a long time for such approaches to materialize. Another approach is a cross-layer methodology to incorporate spectrum mobility as state information in protocols operating in upper layers. Although this approach increases cross-layer dependencies, it will make the entire communication protocol spectrum aware and hence better defend some of the attacks on the upper layer protocols in cognitive networks. For example, routing should consider the operational spectrum band and its frequency characteristics and the transport layer should consider the effect of spectrum handoff on the round trip time and correspondingly adjust the retransmission window.

11.7.5 Robust Security Models

Reliable and robust models need to be developed for collaborative protocols. Instead of assuming a *Homo egualis* (see Section 11.4) type of model, a more secure Byzantine model could be used. The Byzantine model originates from the Byzantine generals' problem [22], which assumes the following scenario. A group of generals of the Byzantine army are camped with their troops around an enemy city. Communicating only by a messenger, the generals must agree upon a common battle plan. However, one or more of them may be traitors who will try to confuse the others. The problem is to find an algorithm to ensure that the loyal generals will reach agreement. Such models have been used to provide fault tolerance in distributed computing and reliability in ad-hoc wireless networks [4]. These models could be used to provide increased security against malicious users in cognitive networks as well.

11.7.6 Develop Analog Crypto Primitives

One of the challenges in incorporating security mechanisms into cognitive networks is that in some frequency bands like the TV band, the primary base stations transmit analog signals (with the exception of HDTVs). Since most of the cryptographic primitives operate in the digital domain, it may not even be possible to incorporate them into analog TV signals. Hence, crypto primitives that work in analog domains need to be developed.

11.7.7 Use Light-weight Security Protocols and Primitives

If secondary users in cognitive networks have mobile equipments with limited processing power and resources, it would be a challenge to provide both cognitive radio capability and security in real time. Light-weight security protocols [24] need to be developed for power/resource constrained environments.

11.8 Conclusions

The main motivation behind cognitive radios [26] has been to increase spectrum utilization by allowing the unlicensed (secondary) users to opportunistically access the frequency band actually owned by the licensed (primary) user. In contrast to other network security architectures, in cognitive radios networks, the users are categorized into two distinct classes: primary users and secondary users. In this chapter, we showed that this categorization gives rise to several security issues that are unique to cognitive radio communications. We also discussed various security aspects such as authentication and authorization of users, confidentiality and integrity of communication as well as identification and non-repudiation of cognitive user devices. Some reliability issues that are inherent in cognitive networks were examined. We then proposed several novel security attacks on different layers of the protocol stack in cognitive networks that make use of one or more of the inherent vulnerabilities. Through these attacks we showed that the fundamental idea behind cognitive networks (to have self-aware networks that offer resilient services and keep the intruders out of it simply by cognition) is not yet fulfilled. We then briefly examined several existing cognitive radio network architectures such as OCRA, Nautilus, IEEE 802.22 and DIMSUMnet with comments on their security. Finally, we suggested some future directions that need to be taken to make the protocols that are employed in cognitive radio networks 'spectrum aware' and hence more resilient to the attacks that we discussed.

Acknowledgements

This work is partially supported by grants from the U.S. Army and the NSF Cyber Trust grant 0627688.

References

[1] Aad, I., Hubaux, J.-P. and Knightly, E.W. (2004) Denial of service resilience in ad hoc networks. *Proceedings of the 10th Annual International Conference on Mobile Computing and Networking (MobiCom '04)*, pp. 202–15, New York, USA.

[2] Akyildiz, I., Lee, W.-Y., Vuran, M.C. and Mohanty, S. (2006) Next generation/dynamic spectrum access/cognitive radio wireless networks: a survey. *Computer Networks*, **50**(13), 2127–59.

[3] Akyildiz, I. and Li, Y. (2006) *OCRA: OFDM-based cognitive radio networks*. Broadband and Wireless Networking Laboratory Technical Report, March 2006.

[4] Awerbuch, B., Holmer, D., Nita-Rotaru, C. and Rubens, H. (2002) *An on-demand secure routing protocol resilient to byzantine failures*. ACM Workshop on Wireless Security (WiSe), September, Atlanta, Georgia. citeseer.ist.psu.edu/article/awerbuch02demand.html.

[5] Blanton, E. and Allman, M. (2002) On making TCP more robust to packet re-ordering. *ACM Computer Communication Review*, -(1).

[6] Buddhikot, M.M., Kolodzy, P., Miller, S., Ryan, K., and Evans, J. (2005) Dim-sumnet: new directions in wireless networking using coordinated dynamic spectrum access. *Proceedings of the 6th IEEE International Symposium on a World of Wireless Mobile and Multimedia Networks (WoWMoM'05)*, pp. 78–85, Washington, DC, USA.

[7] Cabric, D., Mishra, S. and Brodersen, R. (2004) Implementation issues in spectrum sensing for cognitive radios. *Signals, Systems and Computers*.

[8] Cabric, R.B.D. (2005) Physical layer design issues unique to cognitive radio systems. *Proceedings of the IEEE Personal Indoor and Mobile Radio Communications (PIMRC)*, vol. 2, pp. 759–63, September.

[9] Cam-Winget, N., Housley, R., Wagner, D. and Walker, J. (2003) Security flaws in 802.11 data link protocols. *ACM Communications*, **46**(5), 35–9.

[10] Cordeiro, D.B.C., Challapali, K. and Shankar, S. (2005) IEEE 802.22: the first worldwide wireless standard based on cognitive radios. *Proceedings of the IEEE DyS- PAN*, pp. 328–37, November.

[11] Digham, M.A.F. and Simon, M. (2003) On the energy detection of unknown signals over fading channels. *Proceedings of the IEEE ICC*, vol. 5, pp. 3575–9.

[12] Etkin, R., Parekh, A. and Tse, D. (2005) Spectrum sharing for unlicensed bands. *Proceedings of the IEEE DySPAN*, pp. 251–8, November.

[13] FCC (2003), *Notice of proposed rule making and order*, ET docket no 03-222, December.

[14] Ferguson, N. and Schneier, B. (2003) *Practical Cryptography*, John Wiley & Sons, Inc., New York, USA.

[15] FIPS (2001) *Specification for the advanced encryption standard (AES)*. Federal Information Processing Standards Publication 197. http://csrc.nist.gov/publications/_ps/_ps197/_ps-197.pdf.

[16] Ganesan, G. and Li, Y. (2005) Cooperative spectrum sensing in cognitive radio networks. *Proceedings of the IEEE DySPAN*, pp. 137–43, November.

[17] Gintis, H. (2000) *Game Theory Evolving: A Problem-Centered Introduction to Modeling Strategic Behavior*, Princeton University Press.

[18] Hu, Y.-C., Perrig, A. and Johnson, D.B. (2002) Ariadne: a secure on-demand routing protocol for ad hoc networks. *MOBICOM*, pp. 12–23.

[19] Jakobsson, M., Wetzel, S. and Yener, B. (2003) Stealth attacks on ad hoc wireless networks. *Proceedings of the VTC*.

[20] Jing, X. and Raychaudhuri, D. (2005) Spectrum co-existence of IEEE 802.22b and 802.16a networks using CSCC etiquette protocol. *Proceedings of the IEEE DySPAN*, pp. 243–50, November.

[21] Johansson, T (ed.) (2003) A concrete security analysis for 3GPP-MAC. *Fast Software Encryption*, LNCS 2887, Springer, Berlin.

[22] Lamport, L., Shostak, R.E. and Pease, M.C. (1982) The byzantine generals problem. *ACM Transactions on Programming Languages and Systems*, **4**(3), 382–401.

[23] Mathur, C. and Subbalakshmi, K. (2007a) Digital signatures for centralized dsa networks. *Proceedings of the Consumer Communications and Networking Conference (CCNC)*.

[24] Mathur, C. and Subbalakshmi, K. (2007b) Light weight enhancement to RC4 based security for resource constrained wireless devices. *International Journal of Network Security*, **5**(2).

[25] McHenry, M. (2003) *Spectrum white space measurements*. New America Foundation Broadband Forum, June.

[26] Mitola, J. (1995) The software radio architecture. *IEEE Communications Magazine*, May, 26–38.

[27] Mitola, J. (1999) Cognitive radio for flexible mobile multimedia communication. *Proceedings of the IEEE International Workshop on Mobile Multimedia Communications (MoMuC)*, pp. 3–10, November.

[28] Nanjunda, C., Haleem, M. and Chandramouli, R. (2005) Robust encryption for secure image transmission over wireless channels. *Proceedings of the IEEE International Conference on Communications*, vol. 2, May.

[29] Natkaniec, M. and Pach, A.R. (2000) An analysis of the back-o® mechanism used in IEEE 802.11 networks. *Proceedings of the Fifth IEEE Symposium on Computers and Communications*, p. 444, Washington, DC, USA.

[30] Naveed, A. and Kanhere, S.S. (2006) Security vulnerabilities in channel assignment of multi-radio multi-channel wireless mesh networks. *IEEE Globecom*, November.

[31] Reason, J.M. and Messerschmitt, D.G. (2001) The impact of confidentiality on quality of service in heterogeneous voice over IP. *Lecture Notes in Computer Science*, **2216**, 175.

[32] Santivanez, C.A., McDonald, A.B., Stavrakakis, I. and Ramanathan, R. (2002) On the scalability of ad hoc routing protocols. *Proceedings of INFOCOM*.

[33] Schneier, B. (1995) *Applied Cryptography: Protocols, Algorithms, and Source Code in C*, John Wiley & Sons, Inc., New York, USA.

[34] Stallings, W. (1999) *Cryptography and Network Security: Principles and Practice*, Prentice-Hall, Upper Saddle River, NJ, USA.

[35] Tobagi, F. and Kleinrock, L. (1975) Packet switching in radio channels: Part II – the hidden terminal problem in carrier sense multiple-access and the busy-tone solution. *IEEE Transactions on Communications*, **23**, 1417–33.

[36] Visotsky, E., Kurner, S. and Peterson, R. (2005) On collaborative detection of tv transmissions in support of dynamic spectrum sharing. *Proceedings of the IEEE DySPAN*, pp. 338–45, November.

[37] Walker, J. (2000) *IEEE 802.11 wireless LANs unsafe at any key size: an analysis of the WEP encapsulation*. Technical report, Platform Networking Group, Intel Corporation, October. Available: citeseer.ist.psu.edu/558358.html.

[38] Wild, B. and Ramchandran, K. (2005) Detecting primary receivers for cognitive radio applications. *Proceedings of the IEEE DySPAN*, pp. 124–30, November.

[39] Xing, Y., Mathur, C., Haleem, M., Chandramouli, R., and Subbalakshmi, K. (2006a) Priority based dynamic spectrum access with QoS and interference temperature constraints. *Proceedings of the IEEE International Conference on Communications*, vol. 10, pp. 4420–5.

[40] Xing, Y., Mathur, C., Haleem, M., Chandramouli, R. and Subbalakshmi, K. (2006b) Dynamic spectrum access with QoS and interference temperature constraints. *IEEE Transactions on Mobile Computing*, **1**(8).

[41] Zapata, M.G. and Asokan, N. (2002) Securing ad hoc routing protocols. *Proceedings of the 3rd ACM Workshop on Wireless Security*, pp. 1–10, New York, USA.

[42] Zhou, D. (2003), *Security Issues in Ad Hoc Networks*, CRC Press, Inc., Boca Raton, FL, USA.

12

Intrusion Detection in Cognitive Networks

Hervé Debar

France Télécom Division R&D

12.1 Introduction

In this chapter, we are covering the area of intrusion detection. As cognitive networks rely on knowledge acquisition and exchange between network nodes to ensure proper function, it is likely that the knowledge management processes will be the target of attacks. Since the knowledge used to manage the network impacts its configuration, it is necessary to monitor in real time its behavior to ensure that its security properties are preserved. Knowledge exchanged between network components will support cooperative intrusion detection, as suggested in [6]. Furthermore, the knowledge processes can be enriched with diagnosis provided by the intrusion detection system, leveraging the cognitive network processes to provide additional threat response mechanisms.

The chapter is organized as follows: we will first introduce the area of intrusion detection. We will then highlight the threat model and propose a model for applying intrusion detection and prevention concepts to cognitive networks.

12.2 Intrusion Detection

Intrusion detection is the process of detecting unauthorized use of, or an attack upon, a computer or network. The requirements were originally formulated by Anderson [2] and the first intrusion detection system was developed by Denning and Neuman [16] at SRI. For detailed information on intrusion detection, the reader is referred to [5]; we will only introduce thereafter the most important concepts and the tools that are relevant for the remainder of the chapter. We will first present a quick overview of the origins

Cognitive Networks: Towards Self-Aware Networks Edited by Qusay H. Mahmoud

of this area, and then present the detection mechanisms. We will then present security information management. Finally, the trend towards intrusion prevention will be presented with a taxonomy of reaction measures.

12.2.1 Origin and Concepts

Figure 12.1 describes the typical architecture of an intrusion detection system (IDS), according to the Intrusion Detection message exchange Working Group of the IETF [30].

Activity about the monitored system is gathered from a data source and preprocessed into events. The analyzer then decides which of these events are security-relevant – according to the security policy set forth by the security administrator – and generate alerts accordingly. If the IDS has the capability to respond to the detected threat, it may apply the response immediately.

Intrusion detection provides two important functions in protecting information system assets: alerting [12] and response.

The first function is that of a feedback mechanism that informs the security staff of a relevant organization about the effectiveness of other components of the security system. The lack of detected intrusions is an indication that there are no known intrusions, not that the system is completely impenetrable.

The second function is to provide a trigger or gating mechanism that determines when to activate planned responses to an incident. A computer or network without an IDS may allow attackers to leisurely explore its weaknesses. If vulnerabilities exist in networks, a determined attacker will eventually find and exploit them. The same network with an IDS installed is a much more formidable challenge to an attacker. Although the attacker

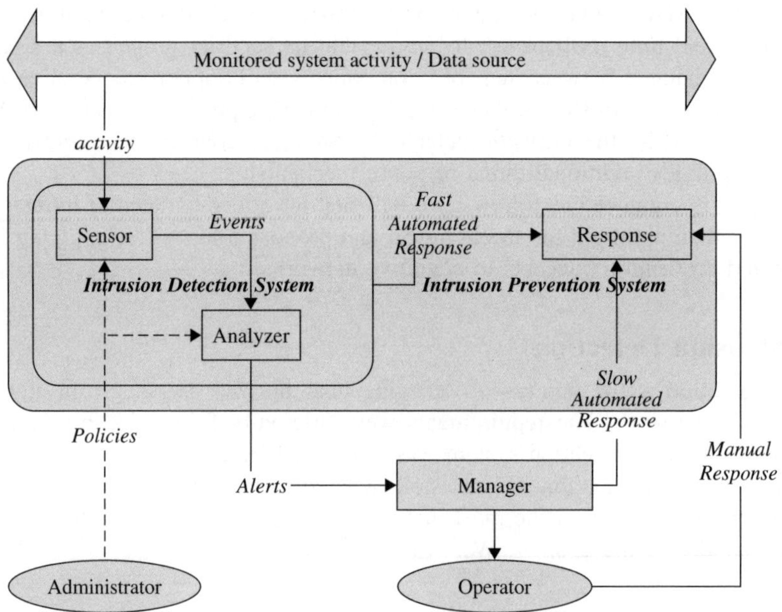

Figure 12.1 Architecture of an intrusion detection or prevention system

may continue to probe the network for weaknesses, the IDS should be able to detect these attempts (even more so if the vulnerability is known), block these attempts and alert security personnel who can take appropriate action.

Note that within the context of cognitive networks, the knowledge management and information exchange may occur at multiple levels. Classic intrusion detection systems link sensor and analyzer functions closely, while the ability of cognitive networks to distribute information gathering may lead to truly distributed detection. Knowledge exchange may also occur at the manager level, thus implementing security information management (SIM) functions in a more natural way. Finally, distribution may also occur at the response level; while distributed response systems have not yet been discussed in the literature, the delegation of response to network management mechanisms holds great promise.

12.2.2 Domain Description

Figure 12.1 highlights the two key aspects of intrusion detection: the data source and the analysis method. These aspects are refined in Figure 12.2 and detailed thereafter.

12.2.2.1 Data Sources

Data sources providing information about the activity of the monitored system are classified into three broad categories: network sources, system sources and application sources.

Network sources consist of packets either grabbed directly from the network by a *sniffer*, or packets traversing an inline network device being analyzed during their presence in the device, or packets traversing an IP stack on the machine. In all cases, the initial format of the information being analyzed is the IP packet (headers and payload), which is then dissected to construct the event stream for analysis. This event stream is of varying complexity, but generally includes at least some defragmentation and reassembly, and verification of conformance to the specifications of the lower levels of the protocol stack. Network packets are an attractive information source because of the initial standard format, and because the operational deployment of network sensors can be realized in passive mode with respect to the monitored system, i.e., with little required configuration changes and little impact on the original behavior. However, this information source is often imprecise and ambiguous, requiring multiple operations before meaningful analysis can be achieved. Also, diagnosis results are often less precise because of the limited amount of interaction and timing constraints imposed on the sensor. Finally, many techniques (for example encryption) may render the flow impossible to analyze.

System sources consist of the various logs available on an information system. The most prevalent log source is the *syslog* service, which offers logging facilities to applications. Many logs kept within *syslog* are very good information sources as the user application provides detailed and security-relevant information; particularly useful in that category is the *auth* log for record keeping of user interactive activity, and the *ssh* log for record keeping of remote access attempts. However, many other logs are not relevant for security; for example, *postgres* database logs kept into *syslog* only indicate commits and rollbacks by default, not access information. Another information source in this category is the security audit system of C2 operating systems, keeping track of interactions between the trusted kernel and the user space. Usage of the C2 audit system is deprecated today, IDS developers building their own system call interception drivers or modules to optimize the information collection mechanisms. Both these information sources deliver fairly accurate information, although kernel-based sources require a lot of formatting work to provide meaningful data. The operational requirements of deploying software on many systems to obtain all the relevant information and of consuming network bandwidth for transport or local CPU resources for analysis currently limit the usage of this information source.

The third category of information sources consists of application logs. The typical example of an application log is the web server (httpd) log file, which stores information about every transaction (request and response) effected by the web server. While the format of a log line is usually extremely concise, it does provide important information for security analysis, because it regroups information about both the request (possibly the attack attempt) and the response (the reaction of the victim). This grouping also reduces the need of preprocessing of the data, even though evasive activity at the application layer still needs to be taken into account. While application logs are a concise and accurate source of information, it must be noted that whatever information not stored in the log is lost, therefore there is no possibility for additional forensics on the data itself; forensics will require access to the original server. Also, many logs do not store useful but implicit information (such as the location and type of the web server in the case of http logs); this information needs to be preserved as part of the log meta data.

12.2.2.2 Detection Algorithms

There are two families of analysis methods: *misuse* detection and *anomaly* detection.

Misuse detection uses the accumulated knowledge on how systems are penetrated to create misuse patterns. The occurrence of such misuse patterns in the event stream triggers the alert. In practice, misuse patterns are often related to known software vulnerabilities. Since alert condition is explicit, misuse detection is easy to document and countermeasures can be proposed. However, it has difficulties detecting unknown threats.

Anomaly detection assumes that an attacker will use the system differently than a normal user. An anomaly detector will construct, either from learning or specifications, a normal or usual behavior of the monitored system. When the detector finds a different usage pattern, it triggers an alert. Since alert condition is implicit, anomaly detection is not limited to known threats, and can also detect deviant usage patterns that are symptomatic of successful intrusions. However, it is also more difficult to associate the alert with a useful diagnosis.

The current state of the art tends to associate these two techniques to get the best detector possible. For example, pattern detection is often preceded by protocol compliance checking to ensure that the protocol exchanges are correct (the pattern detector will be fed with the right information) and that the state of the protocol is appropriate with respect to the attack (e.g. the SMTP exchange is in command mode rather than text mode).

12.2.2.3 Examples of Intrusion Detection Systems

Figure 12.2 shows two examples of network-based intrusion-detection systems: *Snort* [1] [7], and *PAYL* [28] [29]. Snort is a well-known open source intrusion detection system that analyses in real time network traffic to detect known patterns of abnormal use; it thus is an example of misuse detection. Snort uses rules to express these patterns, and will trigger an alert each time a rule matches. Snort also checks for limited protocol compliance at the

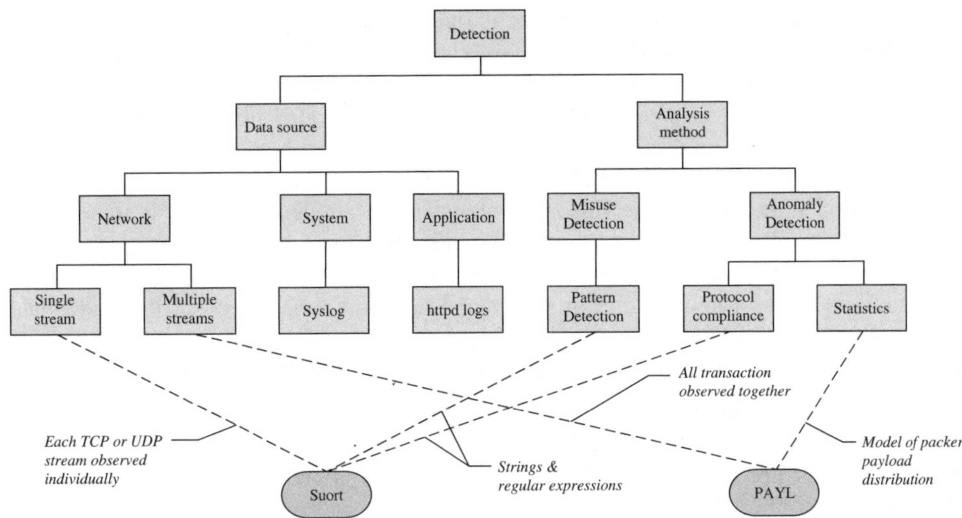

Figure 12.2 Intrusion detection domain characteristics

lower levels of the protocol stack, and will alert if noncompliant packets are found. The Snort rule language is constructed around headers and options. The rule header matches information contained in some of the headers of the packet such as addresses and ports. The rule option narrows the matching condition of the rule, requiring specific flags or specific packet content specified as a string or a regular expression.

PAYL analyzes network traffic using a model of normal traffic, thus following an anomaly detection philosophy. The originality of PAYL is that it models the content of packet payloads, and has very little reliance on packet headers. PAYL very efficiently detects deviations from the normal traffic coming from or going to a particular machine, highlighting usage changes.

These two technologies may provide a useful basis for intrusion detection in cognitive networks.

12.2.3 Security Information Management

While there are a lot of proposed technologies for alert correlation, they share common characteristics, and in particular they address two problems: the volume and the quality of alerts. With respect to alert volume, it is now a well-established fact in all correlation-related papers that the number of alerts needs to be reduced, either because intrusion detection systems are too verbose, or because there are too many information sources. With respect to alert quality, alert correlation also aims at improving the contextual knowledge associated with the alert, to support the alert handler's analysis. Correlation techniques address either one of these two problems or both simultaneously.

12.2.3.1 Classification of Correlation Techniques

12.2.3.1.1 Reduction of Alert Volume

There are four different aspects to the reduction of alert volume. The first dimension is the elimination of alerts as they enter the correlation process. Since we can expect any organization to deploy multiple sensors, it is quite likely that a single root cause can be reported by multiple sensors. The correlation system aims at detecting the fact that this set of alerts has the same root cause and keeps only a single alert identifying this root cause. The detection of duplicates provided by Debar and Wespi [15] offers this property. Of course, some information carried by the eliminated alerts can be added to the remaining one, but in essence the eliminated alerts are excluded from the system.

The second dimension is the fusion of alerts into a single one. Even if these alerts are not exact duplicates related to the same cause, there may be reasons to group them, because they are all parts of the attack process. For example, a teardrop attack requires the attacker to send a few hundred malicious packets to a target, each of them identical and creating the associated alert. The individual alerts have less sense than the fusion of all of them, for example using a simple counter as done in TRM [15], or in CRIM [11].

12.2.3.1.2 Combination of Volume Reduction and Pertinence Improvement

The third dimension of volume reduction is the aggregation of alerts. In many cases, a single alert is not sufficient to warrant an operator's attention. However, if multiple alerts sharing similar characteristics occur, the operator may want to focus his attention on some aspects of the alerts. The most obvious interesting characteristics to aggregate on are the type of attack, the target of the attack or the source of the attack, because in each case, there are actions that can be taken by an operator to reduce the threat. Such early correlation mechanisms, e.g. [15], share similar characteristics. With respect to the two previous approaches, aggregation includes not only reduction but also some sort of pertinence improvement, since it highlights a particular aspect of the general threat that may be of interest to the operators.

The fourth dimension of volume reduction is the synthesis of alerts. The hypothesis behind this dimension of volume reduction is that alerts do not occur in isolation, but are part of an attack scenario that we should reconstruct. The most obvious examples of attack scenarios are worms with multiple propagation vectors, such as Codered or Nimda. To infect a machine, these worms will make several attempts to connect to the target machine, with slightly different requests. If we can assemble in some ordered way the alerts representing these connection attempts, we can have a more synthetic image of the threat and deal with it more accurately. Examples of such worm detection systems have been developed, and this category of correlation mechanisms has been widely studied by for example Cuppens [10, 11], Porras *et al.* [23], Ning *et al.* [20], Nig and Xu [21] and Dain and and Cunningham [12]. Again, while this mechanism is used for alert reduction, the identification of an attack scenario also participates in the improvement of alert information.

12.2.3.1.3 Pertinence Improvement

With respect to pertinence, and in addition to the pertinence improvement also brought by the two last reduction dimensions, a lot of effort has been devoted in the technical intrusion detection community to implement systems capable of correlating alert information with vulnerability information. The goal of this type of correlation is to rank the intrusion detection alerts by order of importance, to make sure that alerts exploiting a vulnerability that is effective on the monitored information system are handled with a high priority by the operators.

In the same line of thought, there has been some research carried out on identifying attackers' intention and application response, to verify whether the attack could have succeeded or not. This is particularly material in a web environment, where analyzing web server return codes yields useful information with respect to the success or failure of the attacker [26].

12.2.3.2 Issues with Alert Correlation

While having pertinent alerts is important for intrusion detection, there are also interesting challenges that need to be addressed in this area.

12.2.3.2.1 Correlation in Sensors

First of all, many correlation algorithms currently implemented in the consoles should be implemented in the sensors. While implementing at the manager's level makes sense in a research environment because there are less development constraints than in sensors, there are numerous cases where only the correlated alert should be sent to the manager.

The most obvious case of this factor is the correlation between vulnerability information and alerts, because this vulnerability information is local in essence. Multiple machines may share the same IP address, e.g., in organizations using network address translation for branch locations. Also, the association between a physical machine (e.g., a MAC address), a machine name and a machine address may require access to local DHCP server logs.

The same is true for aggregation algorithms. If we are aggregating alerts, it generally means that the only information that should go to the management console is related to the aggregate, e.g., a deviation from the norm. These aggregation algorithms should include the possibility to incrementally aggregate information from multiple sources, so that the deviation from normal behavior is detected as early as possible. This early detection should also facilitate including diagnosis information in the alert.

12.2.3.2.2 Relationship Between Pertinence Correlation and Intrusion Prevention

Alert pertinence correlation is less important when it comes to intrusion prevention. If one can reliably identify an incoming threat, the obvious objective is to block that threat at the perimeter, and not wait for the server response. This leaves us with the simple result that for pertinence correlation to be useful there needs to be work on alerts that have passed the perimeter. The most obvious consequence of this result is that we should be looking at pertinence correlation not in general but within some perimeter. This is in fact a positive result, since we cannot expect to have a global knowledge of all machines connected to the Internet, but only of machines that are within our perimeter and registered within our inventory.

This also implies that we should be looking at pertinence correlation also from the point of view of internal intrusion detection sensors, where vulnerabilities may encompass a much larger definition than the general acceptance and limitation of the term as a software program flaw, and should include the notion of configuration vulnerabilities, model of information flows, hidden channels and probably many others. This may explain why the most straightforward style of pertinence correlation has not received a lot of attention from the research community, since the basic detection technology we would need is not available today.

12.2.4 Reaction

Since cognitive networks are adaptive by nature, the ability to react to threats is essential. Beyond intrusion detection, we will propose here a simple model highlighting the various

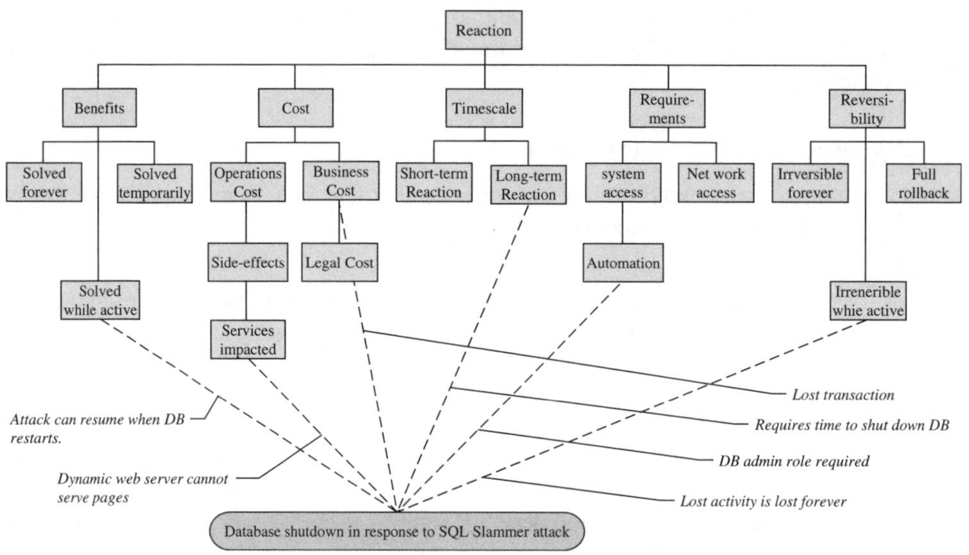

Figure 12.3 Taxonomy of reaction features

facets that compose reaction to alerts. This is coherent with the current trend towards intrusion prevention, where devices not only detect threats but also stop them. The state of the art on countermeasures is much less mature than the intrusion detection area, but we nevertheless propose a taxonomy of reaction features in Figure 12.3.

The *reversibility* feature of the reaction models the fact that the effects of the application of a particular countermeasure can be undone. Some countermeasures are not reversible, such as dropping packets. The lost packet cannot be resurrected. Others include commit and rollback capabilities similar to those of a database; as long as the commit has not been effected, the results of the changes can be undone. The *requirements* feature of the reaction models the level of privileges that is needed to set up the countermeasure, including system and network access, and in the case of system whether the reaction can be automated or requires manual intervention. The *timescale* feature models the time that is required to deploy the countermeasure. Packet drops in firewalls can occur immediately, while shutting down a service or system properly, or propagating a change in network route, will take more time. The timescale feature can be used to order the application of countermeasures, and ensure that coverage remains complete over time. The *cost* feature models the impact of the countermeasure on the protected system. This impact can have technical aspects (e.g., degraded performance) but also business and legal aspects. The *benefits* feature models the gain obtained while applying the countermeasure, and in particular the fact that the countermeasure solves the problem to some degree.

Figure 12.3 includes one use case of a countermeasure that shuts down a database server in response to a SQL Slammer worm attack. The countermeasure is irreversible as long as the database is down, since it cannot answer transactions. It requires database administrator privileges and is a long-term action – many attacks can reach the database server before the action is completed. If the database hosts dynamic content for a web server, the business impact is all lost transactions, and the web server cannot operate properly. Finally, while the database remains down, the attack cannot succeed. Many other responses can be envisioned, such as network blocks at firewalls, or patching the database server so that it is not vulnerable to the attack. These other countermeasures will have other characteristics according to the feature set, and a comprehensive evaluation of all these countermeasures will lead to a better management of countermeasures.

In the case of cognitive networks, this scenario is enriched by the fact that the database server shutdown information may be propagated to other nodes in the network, reducing the volume of traffic that will go to the terminated server and timeout. It may also, as the same information may be available on other nodes, re-route the requests to the remaining servers. In essence, cognitive networks open additional opportunities for threat response at the network, application and information levels.

12.3 Threat Model

Since cognitive networks are constantly evolving to ensure trust and performance, technologies developed in the intrusion detection area can be of help to solve some of these problems. Conversely, the inherently changing aspects of cognitive networks create new challenges for such technologies.

When looking at a security problem, it is necessary to assess the threat before prevention, detection and reaction mechanisms can be proposed. One possible approach to determining threats against cognitive networks is to recognize that the field of autonomic networking is also looking at the perspective of having systems and networks that include self-protecting and self-healing functionalities. A general description of the security aspects of autonomic networking can be found in [7]; thereafter, we will attach ourselves to the impact, target and attacker aspects of the threats.

12.3.1 Model and Constraints

We consider that a cognitive network consists of two main parts: content and transport. Content is related to the services that the user obtains from the network, such as exchanging mail or having voice conversations with other users. Transport is related to the way this content is transferred from one point of the network to another. Note that in many cases, administrative information will also be transported by the network; thus, for network operators, the management content must ensure that the user content properties are preserved.

We do not believe that threats against user content in cognitive networks will be any different from threats in the current Internet. Whenever services are offered, attackers will attempt to shut them down, use them for free, or modify their content. Worms and viruses as well as software flaws will participate in the picture, and most of the mechanisms that are in place today for protecting individual nodes from attack will remain in place. The equation change introduced by cognitive networks is the collaborative ability for threat detection and response, i.e., the fact that network nodes do not cope with the threat individually. However, the wealth of information provided by these systems is also a greater opportunity for attackers.

In Figure 12.4 we adapt the model described in [7] of a generic network node to focus it on the management functions. According to [7], a network node has two functions, a management function that controls its behavior and a service function that provides

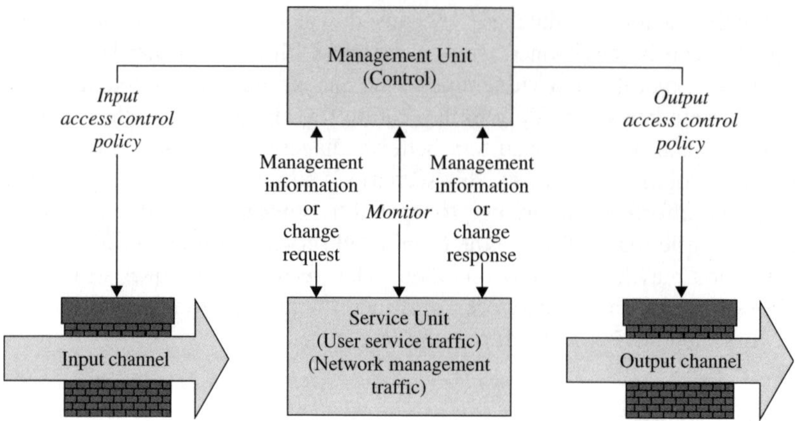

Figure 12.4 System model

actual service to users. Service can either be a network service (routing packets) or an application service (serving web pages, file blocks, etc.). With respect to security, the management unit controls the inputs and outputs of the service unit, and monitors its function. Input and output access control technologies such as firewalls are available for any networked node today, and monitoring standards such as RMON, SNMP or intrusion detection systems enable a close watch on the appropriate behavior of the service unit. Since the management traffic will follow the same path, we will now focus our model on the required management security functions.

Access control for management decides who can request what from the management unit. From a network standpoint, this models domain ownership responsibility; a network operator will naturally enable other nodes in its domain to interact fully with the other nodes in the same domain, while requests from other domains will be rejected. Higher up in the protocol stack, access control mechanisms may also analyze the type of request and not only its source. For example, read access may be allowed for certain outside entities, while write access will not. Finally, the object of the request may also play a role in the decision of granting access to the information, some of it being considered private and the other public. We thus reach the classic access control triple (subject, action, object) to construct our security policy.

Since we are likely to have a large number of network nodes and services, each hosting a large amount of management information and being able to act upon a large number of operational variables, our security policy is likely to contain a huge number of such triples. Fortunately, many abstractions to this model exist, for example the normalized RBAC (role-based access control) mode [18] and the organization-based access control model (OrBAC) [2]. In the case of cognitive networks, we advocate the usage of the latter, since it provides beyond abstraction of subjects to roles an abstraction of actions and objects, as well as the notions of *organization* and *context*. The concept of organization enables sharing of security policies across domains while keeping the attribution intact (i.e. the domain in which the policy is valid). The concept of context enables dynamic rules according to the current state of the network node, state that can be influenced by the management change requests from other nodes. The reader is referred to [16] for an example of how these context changes can be handled and change the way users are authorized to access services.

12.3.2 Threat Impact

The impact of a threat is classically divided in three areas: availability, integrity and confidentiality.

12.3.2.1 Availability

One of the current threats today is related to the availability of networks, systems and services. Targeted denial of services can effectively prevent users receiving services, resulting in loss of revenue for the victim. Denial of service attack on some part of the communication infrastructure (for example the DNS servers) can also be used as accessories for other types of attacks (e.g. confidentiality or integrity).

In this respect, cognitive networks offer an attractive means for both the attacker and the network operator. For the attacker, as the configuration of the network can be often and dynamically altered, introducing false information and using the natural reconfiguration capability of the network is an attractive alternative solution for attacks, which is not easy to carry out today. It is effectively difficult for a malicious attacker to create its own routing elements and flood others to divert traffic from the most appropriate route. Such situations occur, but it is generally a consequence of human error rather than of a concerted malicious effort.

For the network operator, the same mechanisms that help the attacker also provide more options for eliminating the malicious traffic. By detecting or even forecasting congestion in network links and taking the appropriate routing decisions, the network operator can reduce the demand on the network links and share the load more evenly, avoiding the risk of catastrophic failure on one particular point that would cascade in the whole network. Even better, if cognitive networks are able to exchange information at peering points, cooperating network operators reduce the overall load brought onto the network and trace the origin of the attack further back than is currently possible. These mechanisms not only ensure that the impact of the attack is lessened, but also that the source is found more rapidly.

Traceback mechanisms currently rely on routers that will either forward propagate (e.g., itrace) or backward propagate (e.g., capabilities [4]) information related to the origin of the traffic. Given the likelihood of finding the appropriate attack path with the default itrace mechanism, Mankin *et al.* [19] propose an intention-driven itrace, where the probability of tracing the right network stream is increased by the declaration that the trace information will be used. The cognitive model for itrace needs to be reworked to ensure that information provided by the upper stream routers out of the domain of the victim is both timely and accurate. While this creates additional burden on the infrastructure to generate and protect the appropriate measurements, the gain obtained by exchanging information with peers to obtain traces closer to the origin of the network outweighs this burden.

12.3.2.2 Integrity

Note that in the previous example, the request and the trace information must both be authentic and timely. If the attacker can generate misleading trace requests, it can reduce the availability of the trace-generating network, and possibly prevent it from generating other traces requested by legitimate entities, if the trace resource is constrained somehow (performance or storage constraints being almost certain to occur, given the experience we have in network intrusion detection). Legitimate trace requests could therefore not be served. By generating malicious traffic with spoofed sources, the attacker can also induce countermeasures, such as traffic shaping, which will reduce the capability of legitimate network entities to communicate. Both the originator and the recipient of the trace messages must carefully assess the information they provide to the other entity, and in particular the likelihood of spoofing, to reduce this risk.

These control messages must also be protected in transit. Since management information can be seen during its transit through other nodes, the management chain must ensure that the management information or change request remains unchanged during transit. In the context of the traceback mechanism, change in this information can deflect the model of upstream traffic to hide the real source of malicious traffic. Worse, this enables a requesting node to obtain measurement information related to user activity on the other nodes, resulting in information leakage and potential privacy breach.

The integrity and authenticity issue extends to all the management information carried in the cognitive network. Again, the RBAC and OrBAC generalized access control models enable the definition of context-sensitive management information flows, ensuring that messages requesting management information or management action are properly authenticated and signed, and that some trust negotiation has taken place between the subject node (or its domain owner) and the object node (or its domain owner), so that credentials are provided and can be evaluated on each side of the communication.

12.3.2.3 Confidentiality

Management information also possesses confidentiality properties. In the traceback case, part of the measurements taken may be confidential, as their disclosure may reveal sensitive topological information such as routes, router names or active interfaces. This information may then be used to mount targeted attacks against the equipment.

Furthermore, many intrusion detection systems will harbor confidential information. Anomaly detection systems include knowledge about the behavior of the monitored system that are extremely sensitive because they provide the attacker with information about the configuration of the system and the appropriate activity thresholds to avoid detection. Misuse systems are often optimized to only alert on vulnerabilities that are relevant for the information system they monitor, thus again both providing contextual information about the information system and ways to escape detection (using for example specific encodings).

In a cognitive network, it is quite natural that this information will be exchanged between network nodes, to seed or synchronize anomaly profiles, or to detect attacks closer to the source. Leakage of this monitoring information can have catastrophic consequences. The current technique of Google hacking [24] enables information gathering and penetration testing of a web site using only Google-indexed information. In a cognitive network, where each node provides information to the others, confidentiality associated to authentication is a prime requirement for ensuring that sensitive information is not disclosed during transport and is delivered to the appropriate recipient.

In such settings, we expect that trust-building techniques and trust negotiation will be extremely important in reaching consensus on trusted nodes. Concepts such as trust, and techniques such as reputation building and negotiating security policies, not only cipher suites as is done in current network protocols such as TLS or SSL, seem a required avenue for further research.

12.4 Integrated Dynamic Security Approach

As shown in [25], the OODA (Observe, Orient, Decide, Act) loop approach described by J. Boyd [6] provides an interesting support for modeling a system where states change according to information provided by external sources. In Figure 12.5, we present a model for the security of cognitive networks based on the definition of operational planes and OODA loops operating on each of these planes. In this model, we will blend the concepts from intrusion detection and cognitive networks to offer a comprehensive picture of the mechanisms of intrusion detection that can operate in a cognitive network and the interactions with other components.

The model presented in [25] includes information provided by the environment. In our case, three categories of management information are modeled. Generically called *policies*, the first category is information that flows from the top (network operator) down to the network nodes, and describes the operational objectives of the cognitive network. Generically called *alerts*, the second category of information flows from the bottom to the top and signals issues encountered during the operation of the network.

12.4.1 Description of the Operational Planes and their Interactions

In Figure 12.4, we describe a generic architecture of a network node as being composed of two parts: a service unit that carries out the business of the network node and a management unit that controls and configures the service unit. We will naturally find these two units as the two lower planes of the model presented in Figure 12.5. The third plane, the policy plane, represents interactions between the network and its operators, and would typically be executed on a management station.

Each plane exchanges information with the lower and higher levels. For the policy plane, the higher level is built around high-level concepts, including the security policy and business objectives that are given to the network operator by its organization associated with legal and technical constraints. For the network plane, the lower level is composed of the set of observations that are processed by the service unit, for example packets processed by a router or a firewall, or web requests processed by a web server. These observations can be combined together or used separately by multiple mechanisms operating in parallel or sequence, if the input access control, monitoring and output access control systems takes advantage of these multiple categories of information.

The top-down information flow is described as follows. The policy plane processes a policy often expressed in textual form, and provides a formal security policy to the management plane. This formal policy is analyzed by the correlation plane to add contextual information and segment the policy according to the enforcement component capabilities

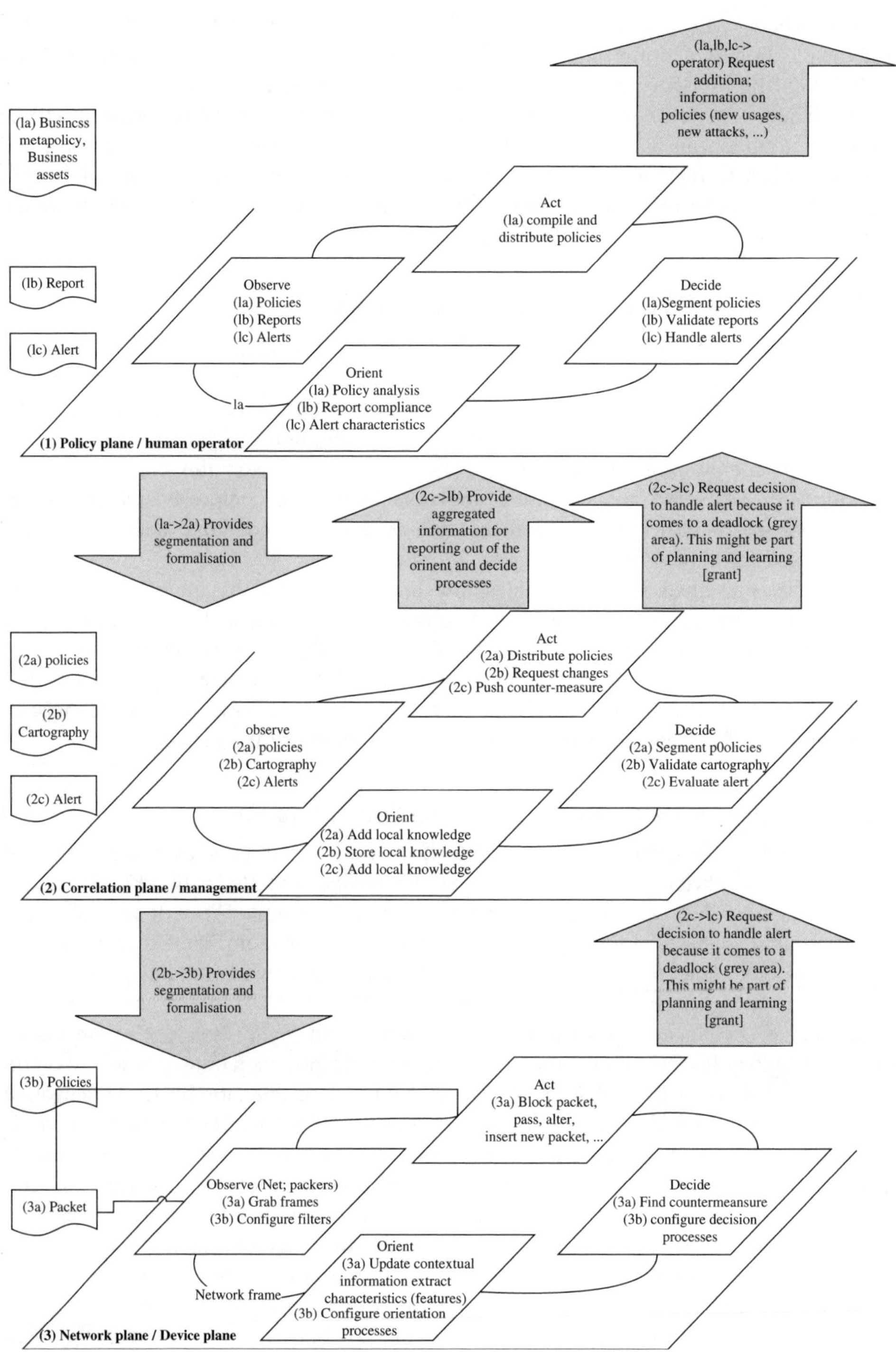

Figure 12.5 Cognitive networks planes and OODA loops

and requirements (input access control, monitoring and output access control). Finally, the device plane receives policies as configuration files that implement at the concrete level in the three functions of the desired policy. Note that this top-down policy flow must configure the components in each node to provide input (e.g. logs or traces) into the bottom-up flow.

The bottom-up information flow is described as follows. The device plane receives requests for service (packets, web requests,). If one of the requests is found to require action, an alert will be sent to the management plane. There are multiple conditions for sending alerts, which will be presented later. The management plane upon receiving an alert enters a handling phase where it will seek to assess the risk presented by the incoming request, possibly in the light of past alerts or contextual information about the actual content of the information system. The management plane forwards to the policy plane alerts that cannot be resolved, and reports on the overall behavior of the monitored environment, to help evaluate the deployment and effectiveness of policies.

12.4.2 Operations within the Network Plane

Operations on the network plane are mapped onto a particular network node. They are located on a single hardware platform, possibly shared by multiple virtual nodes. In the case of colocated virtual nodes, each of the virtual nodes operates its own network plane.

12.4.2.1 Operations on Policies

The network plane applies the policies that have been built as configuration files higher up. The Observe phase receives the policies and verifies that they originate from a legitimate entity; this phase is responsible for input access control with respect to these policies. The Orient phase segments the policies to configure the various components of the network node. The Decide phase validates the received policies, and the Act phase applies the actual change in configurations.

These policies configure the data acquisition in the Observe phase (e.g., filters for packets), the feature extraction phase in the Orient phase, the Decision taken when particular combinations of features are found, and the application of the Decision (response to the request and output access control) in the Act phase. All these functions are currently implemented in various intrusion detection or prevention systems, although they tend to blend together Orient and Decide, and limit the capabilities of the Act to generating alerts or dropping network connections. For example, Snort can be configured with BPF filters to focus the packet stream it receives and its rule engine both orients and decides, a decision being reached individually for each signature specified. The base Snort is limited in its Act phase to sending alerts, but extensions such as Snort-inline can terminate connections, and extensions such as SnortSam can reconfigure firewalls.

Each of these phases can generate alerts when an invalid policy is received. This is identical to the rule compilation phase in intrusion detection systems. Since tempo is of particular importance in OODA loops, the loop operating on policies should go through the four phases as rapidly as possible. However, the speed of execution of the loop is counterbalanced by the speed at which the requests can be phased out if service degradation is considered important. Consider a system where the policy would decide that under certain conditions traffic must switch from cleartext to encrypted (e.g switch from HTTP to HTTPS for web requests). Switching services mean that requests currently being served may abort, sometimes disgracefully, if communications are stopped in the middle of data transfer. Depending on the desired effect and the severity of the attack, the transition may preserve already established transactions or not. Also, this capability of preserving transactions may not exist in all network nodes; for examples, the iptables firewall requires a flush of context (hence a loss of transaction information) when changing filtering rules.

12.4.2.2 Operations on Packets

The Observe phase receives the requests (e.g., packets) and acquires them according to its filter set. Network nodes localized on the same hardware platform could operate in parallel, the filter set being used to differentiate the requests they receive. The filters can also operate some of the input access control, for example by filtering packets only from predefined addresses.

The Orient phase examines the request and extracts particular features. Examples of feature extraction mechanisms include the string matching and regular expression matching of Snort, but also computation of derived features such as volume or packet payload character distribution as proposed by PAYL. The feature extraction phase can require some storage capability, for example to handle fragmented network traffic. This storage capability can impact the availability of the network node, if the volume of the storage is not in line with the volume of requests processed by the network node. The feature extraction process is a requirement for the monitoring function.

The Decide phase examines the features extracted from the incoming request (or requests in the case of accumulative features) and forms a decision according to the policy. The easiest decision is to satisfy the request (e.g., route the packet, respond to the web request), but other decisions should be taken in the case of policy breach, according to both the configuration and the capabilities of the network node, as shown earlier in the differences between Snort, Snort-inline and Snortsam. This synthesizes both the access control and the monitor functions of the network node.

The Act phase realizes the service of the request. Again, the tempo of the OODA loop is important at this stage, and it is driven by the speed at which the requests arrive into the node. For example, for a network node that is connected to a 10 GB link, the time budget to process the smallest packet is 32 microseconds, according to the Ethernet specifications. This time budget gives a lower bound on the time available to the execution of the OODA loop and to the capacity of the network node to handle the most severe operational conditions.

12.4.3 Operations Within the Management Plane

While the description of the management plane may sound centralized, this is not the case. The functions can be distributed on multiple network nodes, provided that the appropriate communication protocols enforce integrity, confidentiality and availability of the communication.

12.4.3.1 Operations on Policies

The management plane receives a formal policy description. The Orient phase segments the policy according to the aspects that are applicable to the network node and the region of the network the node is responsible for managing. Typical segmentation mechanisms include network address spaces and services available on the various nodes. The Decide phase segments the security policy according to the capabilities and the localization of the tools that will enforce it. Finally, the Act phase translates each policy bloc into the configuration language of the particular tool, for example iptables scripts, Snort rulesets, or Apache configurations.

12.4.3.2 Operations on Cartographic Information

Cartographic information defines the relationship between network nodes, and describes the context in which the security policy is applied and the relationship between the various components of the network. As such, cartographic information is used to configure all the phases of the OODA loop on the management plane. Additional functions such as being able to collect cartographic information dynamically (e.g., by requesting it from other network nodes) can also be envisioned, to ensure that each node obtains the information it needs to handle the alerts.

12.4.3.3 Operations on Alerts

The objective of the OODA alert loop at the management plane is to assess the impact of the events that have been observed at the lower level more finally, and to take advantage of a more global view of the activities (both normal and abnormal) occurring on the cognitive network to highlight anomalies that are not perceptible at the lower level.

The Observe phase receives the alerts and validate them. The Orient phase adds contextual information to the alert, related to past history or configuration information that is not available at the lower level. It can also compute features that require a more comprehensive view of the activities occurring on other network nodes, or that are too costly to be computed at the network plane (e.g., because they are too computation intensive). The Decide phase selects the appropriate additional actions that can be taken for the alert, in addition or in substitution to the actions that have already been taken at the network plane. Finally, the Act phase applies the decided actions, and informs both the network plane and the policy plane.

12.4.4 Operations Within the Policy Plane

12.4.4.1 Operations on Business Policies

The objective of the policy plane is to formalize the business policy into a formal security policy. This process is typically iterative, and will impose several executions of the OODA loop before the policy is sufficiently formal and can be pushed down to the lower level. The Observe phase acquires the policy statements, e.g., as a Word file. The Orient phase analyzes the text to extract keywords according to the vocabulary of the security policy, trying to match the content of sentences to concrete subjects, actions and objects, and possibly to abstract entities such as roles, activities and views. In the OrBAC world, the Decide phase organizes the various policy statements, constructs the relationships as hierarchies of roles, activities and views, and mappings between subjects and roles, actions and activities, and views and objects. It also adds information about the organization and the various contexts in which the business operates. The Decide phase also segments the policy to verify that all the elements of the policy can be enforced on the cognitive network, and that all nodes of the network are covered by the policy (i.e., that no node of the network is left out of the configuration process).

The policy should also specify the interaction of the network with its peers. This aspect of security policies including policy and trust negotiation is very much a research subject and the current state of the art does not propose practical solutions to this problem today.

12.4.4.2 Operations on Reports

Reports are used by the policy plane to review the application of the security policy by the network and the compliance of operations with the business objectives of the organization, highlight the most prevalent threats, and possibly revise the policy according to the changes in the threat landscape. In this respect, the operations made on the report by the different phases of the OODA loop are similar to the ones made on the business policies, even though the input information is different.

12.4.4.3 Operations on Alerts

Infrequently, a severe condition will be diagnosed on the network, which cannot be solved automatically by the device or management planes. In this case, the Observe phase receives and validates the alert, the Orient phase verifies that alert features do not match the currently deployed policy, and the Decide/Act phases provides assistance to the operators in resolving the problem.

12.5 Discussion

From a security perspective, the key issue to answer when looking at new technology is whether it will be more secure than the existing ones. Unfortunately, new technologies bring with them new threats, and cognitive networks in this respect are not different. As such, it is impossible to factually demonstrate, at this stage, that cognitive networks will be more secure than IP networks. In particular, the mix between wired and wireless network is likely to leave open the door to saturation attacks, as network equipment realizing the bridge between the two worlds will need to be open to many devices. Furthermore, from a security perspective, the inherent dynamicity introduced in cognitive networks means that there cannot be a fixed static security policy, but that the security policy will need to take into account the same parameters that influence the configuration of the network and its behavior. Dynamic security policies are a domain that has not been well studied in the literature, and the solution to this problem will have an impact on cognitive networks. However, we expect that intrusion detection may benefit from cognitive networks in at least two dimensions: detection and response.

12.5.1 Paradigm Changes in Detection

By definition, cognitive networks monitor their activity to respond to environmental changes. This is also the basis of intrusion detection. By introducing additional variables whose monitoring is seamlessly integrated into the architecture of the network, one enables a tight coupling with intrusion detection models. This capability ensures both that the monitored variables effectively represent the activity of the cognitive network (a fact that is generally assumed in classic intrusion detection systems, but is rarely shown true in practice) and that the information is available in a timely manner to the detection module, hence shortening the response time.

Furthermore, the distributed nature of the monitoring in cognitive systems effectively decouples the information acquisition from the information analysis tasks in intrusion detection. This means that more complex variables can be monitored, which represent security information not at a single point of the network but over multiple nodes. Thus, distributed attacks, be they denial of service, distributed hosting of phishing sites or spam sending, can be naturally detected without the need for costly data transmission, thus potentially solving a key challenge in today's security mechanisms.

Further, if cognitive networks effectively span multiple responsibility realms, one could imagine detection taking into account not only a network operator's own security posture, but also its neighbors and peers capabilities and actions.

12.5.2 Paradigm Change in Response

The second paradigm change introduced by cognitive networks is related to threat response. At a minimum, since a cognitive network node exposes a configuration interface, this provides the security information management system with a large number of options for responding to the threat. Furthermore, the cognitive network architecture may enable practical traceback mechanisms, which have consistently been impossible to deploy in the classic Internet.

Beyond additional response due to the large number of network nodes, additional response capabilities are introduced by the intelligent migration of services or information from node to node. Response mechanisms available today are likely to have the side effect of disabling service to users regardless of whether they participate in the attack or not. In a nutshell, most response mechanisms shift properties from availability to confidentiality or integrity, but there are very few of them that preserve all three properties equally. By offering the capability to negotiate configurations among their nodes, cognitive networks provide a framework for migrating services and information away from the threat while preserving their availability.

12.6 Conclusion

In this chapter, we have presented an introduction to intrusion detection systems, and a model for their adaptation to cognitive networks. Intrusion detection and prevention technologies can provide support for the self-protecting and self-healing capabilities that will be required of a sustainable cognitive network infrastructure. This model, based on the Observe-Orient-Decide-Act loop, describes both the information that needs to flow between the network nodes and the operations they must support to maintain the properties specified by the security policy. The OODA loop also provides information about the performances that these cognitive network nodes must offer in order to support these functions.

However, there is a number of issues that remain unsolved and must be studied for an effective application of intrusion detection technologies to cognitive networks. It is quite clear that misuse detection alone will not solve the security issues we face, because of the lack of comprehensive knowledge about vulnerabilities and attack processes. However, anomaly detection algorithms may observe changes in network conditions that impact them and vice versa, thus introducing instability between the model and the network. These unstable conditions may be difficult to model and control, and should be addressed in all planes of the model. Furthermore, the model does not address the position of the detection and reaction nodes with respect to one another, and the constraints in the request data flow that result from these respective positions.

References

[1] Abbes, T., Bouhoula, A. and Rusinowitch, M. (2004) On the fly pattern matching for intrusion detection with Snort. *Annals of Telecommunications*, **59**, 9–10.

[2] Abou El Kalam, A., Baida, R.E., Balbiani, P. *et al.* (2003) Organization based access control. *Proceedings of the 4th IEEE International Workshop on Policies for Distributed Systems and Networks (Policy'03)*, June.

[3] Anderson, J.P. (1980) *Computer Security Threat Monitoring and Surveillance*, James P. Anderson Company, Fort Washington, Pennsylvania.

[4] Anderson, T., Roscoe, T. and Wetherall, D. (2003) Preventing Internet denial-of-service with capabilities. *Proceedings of Hotnets-II*, November, Cambridge, MA, USA.

[5] Bace, R.G. (2000) *Intrusion Detection*, Macmillan Technical Publishing.

[6] Boyd, J. (1986) *A Discourse on Winning and Losing: Patterns of Conflict*.

[7] Chess, D., Palmer, C. and White, S. (2003) Security in an autonomic computing environment. *IBM Systems Journal*, **42**(1).

[8] Clark, D., Partridge, C., Ramming, C. and Wroclawski, J. (2003) A knowledge plane for the Internet. *Proceedings of the 2003 Conference on Applications, Technologies, Architectures, and Protocols for Computer Communications (SIGCOMM '03)*, pp. 3–10, August 25–29, Karlsruhe, Germany.

[9] Coit, C.J., Staniford, S. and McAlerney, J. (2001) *Towards faster string matching for intrusion detection or exceeding the speed of Snort*. DARPA Information Survivability Conference and Exposition, June.

[10] Cuppens, F. (2001) Managing alerts in multi-intrusion detection environments. *Proceedings of the 17th Annual Computer Security Applications Conference (ACSAC'01)*.

[11] Cuppens, F. and Miège, A. (2002) Alert correlation in a cooperative intrusion detection framework. *Proceedings of the IEEE Symposium on Security and Privacy*, May, Oakland, CA, USA.

[12] Dain, O. and Cunningham, R. (2003) Fusing a heterogeneous alert stream into scenarios. *Proceedings of the 2001 ACM Workshop on Data Mining for Security Applications*, pp, 1–13, November.

[13] Debar, H. and Wespi, A. (2001) Aggregation and correlation of intrusion-detection alerts. *Proceedings of the 4th Symposium on Recent Advances in Intrusion Detection (RAID 2001)*, September, Davis, CA, USA, pp. 85–103.

[14] Debar, H., Curry, D. and Feinstein, B. (2007) *The Intrusion Detection Message Exchange Format*, RFC 4765.

[15] Debar, H., Dacier, M. and Wespi, A. (2000) A revised taxonomy for intrusion-detection systems. *Annals of Telecommunications*, **55**(7–8).

[16] Debar, H., Thomas, Y., Boulahia-Cuppens, N. and Cuppens, F. (2006) Using contextual security policies for threat response. *Proceedings of the 3rd GI International Conference on Detection of Intrusions and Malware, and Vulnerability Assessment (DIMVA)*, July, Germany.

[17] Denning, D.E. (1987) An intrusion-detection model. *IEEE Transactions on Software Engineering*, **13**(2), 222–32.

[18] Ferraiolo, D.F. and Kuhn, D.R. (1992) *Role-based access control. Proceedings of the 15th National Computer Security Conference*.

[19] Mankin, A., Massey, D., Chien-Lung, W., Wu, S.F. and Lixia Zhang (2001) On design and evaluation of 'intention-driven' ICMP traceback. *Proceedings of the 10th International Conference on Computer Communications and Networks*, pp. 159–65.

[20] Ning, P., Cui, Y. and Reeves, D.S. (2002) Constructing attack scenarios through correlation of intrusion alerts. *Proceedings of the 9th ACM Conference on Computer and Communication Security*, pp. 245–54, Washington, DC, November.

[21] Ning, P. and Xu, D. (2003) Learning attack stratagies from intrusion alerts. *Proceedings of the 10th ACM Conference on Computer and Communications Security* (CCS '03), pp. 200–9, Washington, DC, October.

[22] Northcutt, S. and Novak, J. (2003) *Network Intrusion Detection*, QUE.

[23] Porras, P.A., Fong, M.W. and Valdes, A. (2002) A mission-impact-based approach to INFOSEC alarm correlation. *Proceedings of the 5th International Symposium on Recent Advances in Intrusion Detection (RAID 2002)*, p. 95–114, September.

[24] Skoudis, E. (2004) *Google Hacking for Penetration Testers*, Syngress Publishing.

[25] Thomas, R.W., DaSilva, L.A. and MacKenzie, A.B. (2005) Cognitive networks. *Proceedings of the 1st IEEE International Symposium on New Frontiers in Dynamic Spectrum Access Networks (DySPAN 2005)*, pp. 352–60, November.

[26] Vigna, G., Robertson, W., Kher, V. and Kemmerer, R.A. (2003) A stateful intrusion detection system for world-wide web servers. *Proceedings of the Annual Computer Security Applications Conference (ACSAC 2003)*, pp. 34–43, December, Las Vegas, USA.

[27] Viinikka, J. and Debar, H. (2004) Monitoring IDS background noise using EWMA control charts and alert information. *Proceedings of the 7th International Symposium on Recent Advances in Intrusion Detection (RAID 2004)*, September, Sophia Antipolis, France.

[28] Wang, K. and Stolfo, S.J. (2004) Anomalous payload-based network intrusion detection. *Proceedings of the 7th International Symposium on Recent Advances in Intrusion Detection (RAID 2004)*, September, Sophia Antipolis, France.

[29] Wang, K., Cretu, G. and Stolfo, S.J. (2005) Anomalous payload-based worm detection and signature generation. *Proceedings of the 8th International Symposium on Recent Advances in Intrusion Detection (RAID 2005)*, Seattle, Washington, USA.

[30] Wood, M. and Erlinger, M. (2007) *Intrusion Detection Message Exchange Requirements*, RFC 4766.

13

Erasure Tolerant Coding for Cognitive Radios

Harikeshwar Kushwaha, Yiping Xing, R. Chandramouli and K.P. Subbalakshmi

Stevens Institute of Technology, NJ, USA

13.1 Introduction

In recent years, the proliferation of spectrum-based services and devices for uses such as cellular communication, public safety, wireless LAN and TV broadcast has forced society to become highly dependent on radio spectrum. This dependency and explosive growth in demand for radio resources is propelled by a host of factors: the economy has moved towards the communication-intensive service sector; the workforce is increasingly mobile; and consumers have been quick to embrace the convenience and increased effciency of the multitude of wireless devices available today.

In the United States, the Federal Communication Commission (FCC) regulates access to spectrum. These regulations have led to reservation of spectrum chunks for specific purposes; for example, 824–849 MHz and 1.85–1.99 GHz frequency bands are reserved for licensed cellular and personal communications services (PCS) and require a valid FCC license, whereas 902–928 MHz, 2.40–2.50 GHz, 5.15–5.35 GHz and 5.725–5.825 GHz frequency ranges are reserved as free-for-all unlicensed bands [2]. This strict long-term spectrum allocation as shown in Figure 13.1 is space and time invariant and any changes to it happen under strict FCC control.

Considering the increase in demand for freely available, i.e., unlicensed radio spectrum, it is clear that the necessary radio spectrum will not be available in the future, due to the limited nature of radio resources in the current unlicensed frequency bands. Since many services require protection against interference, most of the radio spectrum is allocated to

Cognitive Networks: Towards Self-Aware Networks Edited by Qusay H. Mahmoud

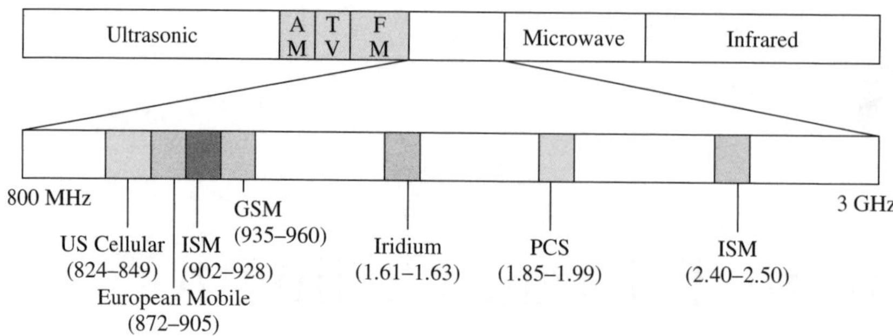

Figure 13.1 FCC's static spectrum allocation

traditional licensed radio service. Such licensing of spectrum has led to several extreme problems uncovered by recent spectrum utilization measurements [14].

As discussed in [1], a large portion of the allocated spectrum is highly underutilized. According to the FCC [6], temporal and geographical variations in the utilization of the assigned spectrum range from 15% to 85%. Although the fixed spectrum assignment policy generally served well in the past, there has been a dramatic increase in the access to the limited spectrum for mobile services in recent years. This increase is straining the effectiveness of the traditional spectrum policies.

The limited available spectrum and the inefficiency in the spectrum usage necessitate a new communication paradigm to exploit the existing wireless spectrum opportunistically. Recent FCC Proceedings [5] propose the notion of secondary spectrum access to improve the spectrum utilization. This allows dynamic access to the unused parts of the spectrum owned by the primary license holder called the primary user (PU) to become available temporarily for the secondary (non-primary) user (SU). This dynamic access of spectrum by secondary users, which is facilitated by the use of cognitive radios, is one of the promising ideas that can mitigate spectrum scarcity, potentially without major changes to incumbents. Fortunately, the advances in software-defined radio (SDR) [15] [25] have enabled the development of flexible and powerful radio interfaces for supporting spectral agility.

The cognitive radio devices communicate by using only the identified opportunities, without interfering with the operation of primary users. The main functions of the cognitive radios are summarized as follows:

- Detection of unused or available spectral bands.
- Selection of the best available channels to meet the secondary user requirements.
- Maintenance of the seamless communication requirements during transition to a different spectral band.

The main objective of the cognitive radio is to obtain the best available spectrum from the temporarily unused spectrum referred to as the spectrum hole. Since most of the spectrum is already licensed to primary users, the most important challenge is to share the licensed spectrum without interfering with the transmission of primary users. This sharing of spectrum is considered secondary usage of spectrum. Conceptually, primary users still own the spectral resources and have primary access rights, however, the secondary users could use these spectral resources by guaranteeing the interference preservation of the primary users.

As proposed in [26], there are two fundamental issues regarding the secondary usage of spectrum:

1. Detection of PU usage by secondary users.
2. Maintenance of the SU communication in case a PU appeared.

The detection of PUs can be accomplished by either of two ways: *negotiated spectrum sharing* or *opportunistic spectrum sharing*.

In negotiated spectrum sharing, whenever a PU decides to make use of its spectral band, it explicitly announces its arrival and the SU gets a chance to relocate to a different spectral band before the PU arrives. It ensures a completely interference-free communication owing to *a priori* declaration of all spectral claims. However, this would require the change of legacy systems in order to enable secondary usage.

On the other hand, in opportunistic spectrum sharing, the PU usage is automatically monitored by SUs with the help of cognitive radios. This monitoring could be done by the SUs themselves or by some centralized device which broadcasts the PU activity to the SUs. The main advantage of this approach is that no changes have to be made to legacy systems as the PU is unaware of the secondary usage of its spectrum.

In this scenario, SUs have to transmit at a relatively low power level in order to protect PU transmission from interference when they arrive. As a result, the PU arrival will most likely corrupt the payload data of SU transmission on the infrared spectrum as PUs might reclaim currently used spectrum of SUs. Although PUs might capture their spectrum randomly, link maintenance becomes an issue for SUs.

The PU arrivals on the SU link forces them to restructure their communication link and reduces the SU system performance. In order to overcome the problems caused by PU arrival on the SU link, the following techniques have been proposed in [26].

- The SU link should be structured in such a way that the probability of PU appearance in the currently used spectral resources by the SU is minimized.
- The SU's payload data should contain some error-correcting mechanism so as to compensate for the loss incurred due to PU arrival on the SU link.

The first technique is implemented by dividing the bandwidth into a large number of subchannels as shown in Figure 13.2. An SU selects a set of subchannels in such a way that they lie in different PU bands. This minimizes the loss of subchannels upon arrival of a particular PU as no PU can cause the complete breakdown of the SU link. This is also called the spectrum pooling concept. Since the arrival of a PU acts like an erasure for payload data on the SU link, it causes the SU to lose all the packets that are being transmitted over the subchannel which was under that particular PU's band.

In counteracting the effect of PU interference over the SU link, the second technique of error correction plays an important role in the link maintenance mechanism. There are two approaches to it: channel coding approach and source coding approach. Channel coding is used to compensate for the loss due to PU appearance and source coding is used to recover the content up to a certain quality depending upon the number of packets received.

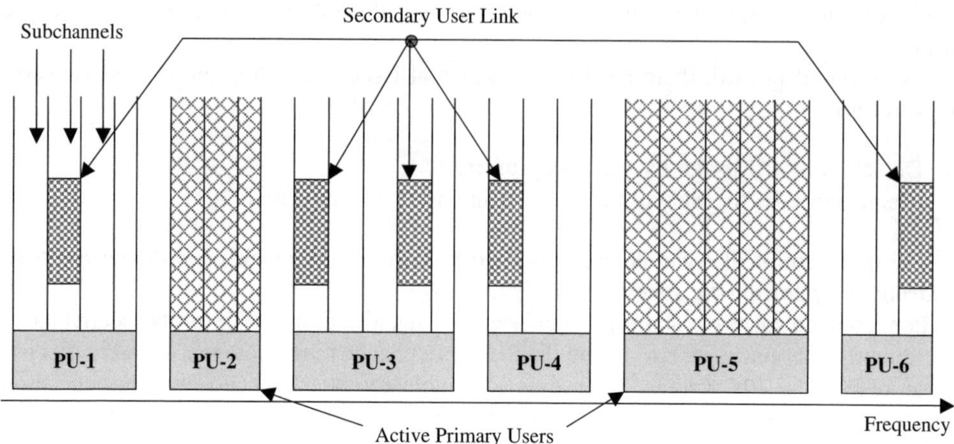

Figure 13.2 Spectrum pooling concept

This chapter is organized as follows. Section 13.2 discusses the spectrum pooling concept. An overview of erasure channels is presented in Section 13.3. Section 13.4 deals with traditional erasure codes and Section 13.5 describes digital fountain codes. Multiple description codes are discussed in Section 13.6. We provide some applications of these codes in Section 13.7. Finally, Section 13.8 concludes the chapter.

13.2 Spectrum Pooling Concept

The notion of spectrum pooling was first introduced in [17]. In this resource-sharing strategy called spectrum pooling the primary user would get the highest priority. Once a primary user appears in a frequency band all secondary users transmitting in this band would have to leave immediately, giving priority to the primary user.

A cognitive radio-based spectrum pooling concept has been developed in [24]. A COgnitive Radio approach for Usage of Virtual Unlicensed Spectrum (CORVUS), a vision of a cognitive radio-based approach that uses allocated spectrum in a opportunistic manner to create virtual unlicensed bands, i.e. bands that are shared with primary users on a non-interfering basis, has been proposed in [3]. The principles of the CORVUS system are explained below.

13.2.1 CORVUS System

The basic assumptions of the CORVUS system are as follows

- There is plenty of spectrum available for sharing by secondary users.
- Secondary users are capable of using cognitive radio techniques to avoid interfering with primary users if present.

In this system, the SUs have to keep monitoring the presence of PUs at regular intervals and as soon as a PU is found using its spectrum band, the SU must vacate that particular band and try to relocate to some other band.

The principal idea of the spectrum pooling concept in CORVUS is depicted in Figure 13.2. This system covers a certain bandwidth W. Within this spectral range, several PUs legally own different parts of the spectrum resulting in a theoretical occupancy of the whole spectrum. However, as different PUs do not always use all their spectrum at a certain time and location this temporarily unused spectrum is available for secondary usage. SUs within this model use these temporarily available spectral resources to meet their own communication needs. In order to accomplish this the whole bandwidth is divided into N subchannels, each with a bandwidth $w = W/N$ to form a spectrum pool.

The shaded frequency bands indicate that the PU is currently active and consequently this frequency band cannot be used by any secondary user. Out of the remaining PU bands, a set of subchannels is selected to construct a secondary user link (SUL) in order to satisfy the communication needs of a secondary user. An SUL is a set of subchannels that varies depending on the PU activity on the used subchannels. As soon as a corresponding PU wants to make use of its spectrum, all SUs have to immediately vacate the corresponding subchannels giving precedence to the primary user.

13.2.2 Reliable Communication among Secondary Users

In order to achieve reliable communication among secondary users against the loss of used subchannels due to the PU reclaim, two means have been proposed in [26] to reduce the PU influence on the SUL.

13.2.2.1 Subchannel Selection

In order to decrease the influence of the appearance of PUs on used spectral resources, an intelligent selection of subchannels forming an SUL is required. Instead of selecting a contiguous set of subchannels to form an SUL, subchannels should be scattered over multiple PU frequency bands. Ideally, an SUL should consist of only one subchannel per PU frequency band. This technique ensures a low effect of the PU appearance on an SUL. Since only one subchannel is used from any PU frequency band, the payload data on only one subchannel will be lost and only that subchannel need to be vacated in case a PU appears.

13.2.2.2 New Subchannel Acquisition

In order to maintain the required quality of service (QoS), the SUL needs to be updated to compensate for the loss of spectral resources due to a PU appearance. Whenever a subchannel is lost, a reconfiguration of the SUL becomes necessary. The procedure for reconfiguration of an SUL is illustrated in Figure 13.3.

13.3 Overview of Erasure Channels

The secondary user link can be modeled as an erasure channel. In erasure channels packet loss occurs at the receiver due to various reasons, e.g., if the bit error correcting code fails, then the erroneous packet may not be passed on to the higher layers in the protocol stack. Therefore, the application layer sees this as an erasure. In erasure channels, only two possibilities are considered: a packet is either received correctly or is lost.

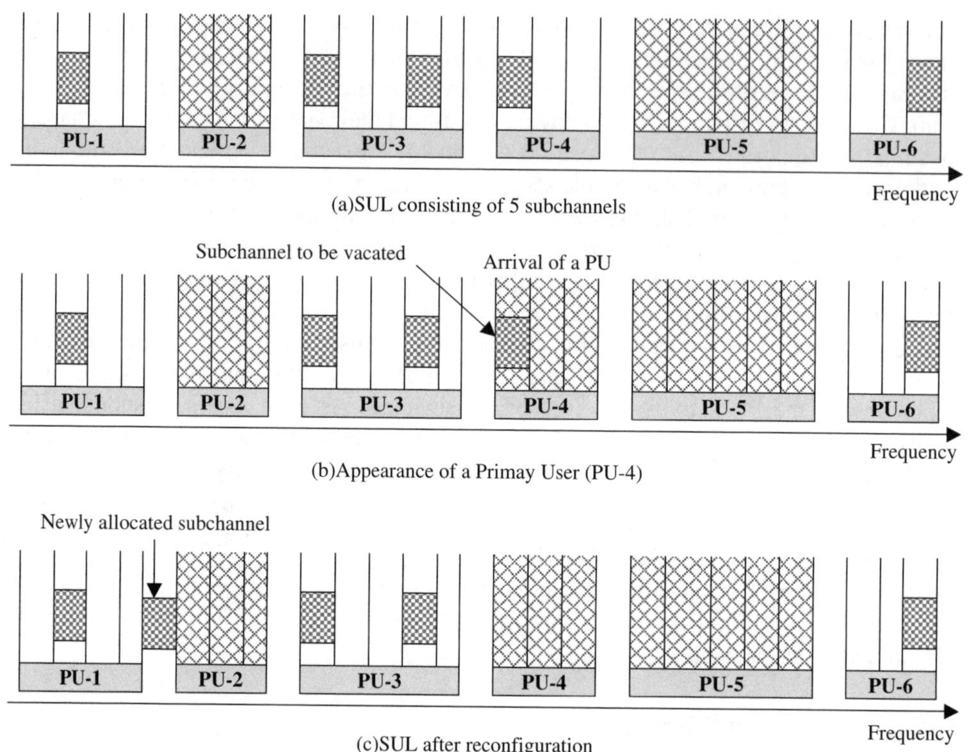

Figure 13.3 SUL recongfiuration

In a wireless network, packet loss generally occurs due to the following reasons:

- In a secondary usage scenario, packets transmitted on a subchannel of a SUL are lost upon reclaim of that subchannel by a primary user.
- Packets may get discarded on the way to their destination for various reasons such as buffer overflows and congestion control at intermediate nodes.
- Packets may get corrupted due to noise and interference.

In this chapter, we mainly consider erasures due to the first reason, i.e., subchannel reclaim by a primary user. An M-ary erasure channel is shown in Figure 13.4, where all input packets $\{1, 2, \ldots, m\}$ have a probability $1 - p$ of being received correctly and a probability p of being lost (i.e., erased). If we consider the capacity of the perfect channel (i.e., $p = 0$) to be C then the capacity of the erasure channel is $(1-p)C$ irrespective of the erasure model.

The traditional methods for communicating over such channels employ a feedback channel from the receiver to the sender that is used to control the retransmission of the erased packets. These retransmission protocols have the advantage that they work regardless of the erasure probability p but they perform poorly in the following circumstances:

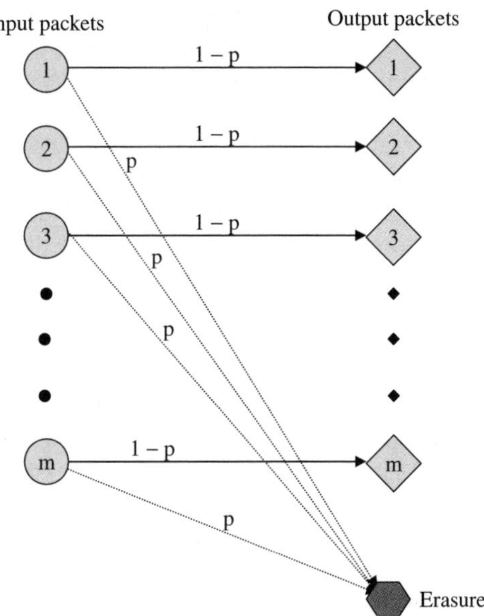

Figure 13.4 M-ary erasure channel model

- If the channel is very bad (i.e. if p is high), then retransmission protocols will introduce too much delay, thereby causing overall performance to suffer.
- In multicast or broadcast scenario, if all receivers request for retransmission then the total number of retransmissions at the sender will be very high and will decrease the overall efficiency.
- If the sender and receiver are geographically too far then the effect of propagation delay due to retransmissions may be significant.

In order to deal with the above problems of feedback-based protocols, new transmission solutions using erasure-correcting codes, that require no feedback, have been proposed in [12, 13]. So if some bits in a packet are lost or if bit errors could not be corrected using a forward error correcting (FEC) code then it may be recovered using erasure-correcting codes. It is required that these codes be capable of correcting as many erasures as possible and at the same time have fast encoding and decoding algorithms.

13.4 Traditional Erasure Codes

Traditional erasure codes are typically block codes with a fixed rate, i.e. K input packets are used to generate N output packets with $N - K$ redundant packets and code rate K/N. For example, an (N, K) Reed–Solomon code [11, 16] has the ideal property that if any K of the N transmitted packets are received then the original K packets can be recovered perfectly. But this code has the following limitations:

1. In a high loss channel, for a fixed value of N, the receiver may not receive K out of N packets. In order to recover original K packets, a retransmission is necessary causing duplicate reception of the previously received packets and thus decreasing the channel efficiency.
2. A more serious limitation is with respect to its encoding and decoding complexity. The reason is that, standard algorithms for encoding and decoding Reed–Solomon codes require quadratic time, which are too slow for even moderate values of K. One way to deal with this problem is to encode over small blocks of data. This approach reduces the total computation, but can significantly increase the number of packets that must be obtained before complete decoding.

Another code is a class of LDPC codes called Tornado codes. Tornado codes are a class of erasure-correcting codes that have linear encoding and decoding times in N. Software-based implementations of Tornado codes are about 100 times faster on small lengths and about 10,000 times faster on larger lengths than software-based Reed–Solomon erasure codes while having only slightly worse overhead, i.e., slightly more than K packets need to be received to successfully recover original K packets. In Reed–Solomon code, every redundant packet (constraint) depends on every input packet and this increases encoding and decoding complexity. On the other hand, in Tornado codes, each constraint depends only on a few input packets, on average, only a constant number of operations are required to generate each redundant packet (or constraint). This makes Tornado codes linear and have fast encoding and decoding times.

Even if Tornado codes have fast encoding and decoding complexity, they still suffer from a major drawback. Since it is a fixed rate code, the value of N must be fixed for a given value of K ahead of time. And if the channel erasure probability p is large then the receiver may receive more than K packets, which is required in order to recover the original K packets. Unfortunately, there is no way to extend this code on the fly to generate more output packets as the demand arises.

Subsequent works [11, 20] in this field has led to codes that overcome the above problems by developing rateless codes, also called digital fountain codes. These codes are discussed in the next section.

13.5 Digital Fountain Codes

In this section, we discuss a new class of erasure-correcting codes, called digital fountain codes [4, 13], that can be used to provide protection against erasures caused by a primary user appearance over a secondary user link. These codes are capable of providing protection from the effects of packet loss irrespective of the loss model of the secondary user link. By recovering lost data packets without requesting retransmission from the sender, these codes provide reliability in various network applications [16] such as multicast, parallel downloading, video streaming etc. And, like a water fountain producing an endless supply of water drops, any of which can be used to fill a glass, these fountain codes can generate an unlimited number of encoded output packets, any of which can be used to recover the original input packets. They have the following characteristics:

- The ability to generate a potentially limitless amount of encoded data from any original set of source data and provide reliable message delivery over extremes of low to high network losses.

- The ability to recover the original data from a subset of successfully received encoded data regardless of which specific encoded data has been received.
- Exceptionally fast encoding and decoding algorithms, operating at nearly symmetric speeds that grow only linearly with the amount of source data to be processed and independently of the actual amount of network loss.

In the following sections, we discuss Luby transform codes (LT codes) and raptor codes.

13.5.1 LT Codes

Luby transform (LT) codes [11, 12, 22] are the first practical realization of a rateless code. These codes were introduced by Luby in 1998 for the purpose of scalable and fault-tolerant distribution of data over computer networks. It works on the following fundamental principle: the encoder can be thought of as a digital fountain that produces a continuous supply of water drops called encoded packets or output packets. Let's say the original file or message data consists of K source packets. The length of each input packet can be arbitrary, from 1 bit to $l > 1$ bit. Now, in order to completely decode the received stream, we need to hold a bucket under the digital fountain and collect drops until the number of drops in the bucket is slightly larger than K. This means that decoding does not depend on which packets were received but only on the number of received packets.

LT codes are rateless in the sense that the number of output packets that can be generated from the original data is potentially limitless and their number can be determined on the fly. The general idea behind the rateless codes is the following: given the data block (file) divided into K equal length packets of length l bits, generate infinite number of encoded packets of the same length l, such that any set containing a little more than K randomly selected encoded packets is sufficient for file reconstruction. This means that the receiver is needed just to collect slightly more than K packets, irrespective of its ordering, to successfully recover the original file. LT codes also have very small encoding and decoding complexities. If the original data consists of K input packets then each output packets can be generated independently of all other output packets, on average with $O(\ln(K/\delta))$ operations, and the K original input packets can be recovered from any N output packets with probability $1 - \delta$, where N is $O(\sqrt{K} \ln^2(K/\delta))$ more than K, for $\delta > 0$. Luby describes these codes as universal due to their being near optimal for every erasure channel and being very efficient as the data length grows.

13.5.1.1 LT Encoder

The task of the encoder is to generate an infinite stream of encoded packets given the file to be transmitted. We assume that the original file is segmented into K packets of length l bits each. Let the source packets be $(x_1, x_2, \ldots, x_K)$, then the encoded packets $(y_1, y_2, \ldots)$ can be generated in the following manner:

1. A random degree d is selected according to a degree distribution $\rho(d)$ on $\{1, 2, \ldots, K\}$.
2. d out of K packets are uniformly selected at random from the original source packets $(x_1, x_2, \ldots, x_K)$.

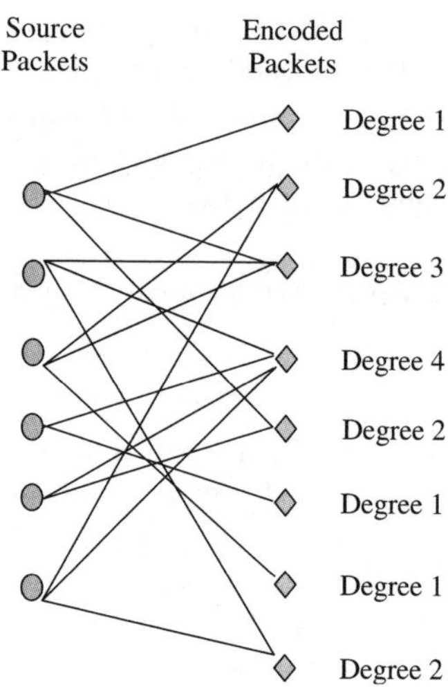

Figure 13.5 A bipartite graph representing LT encoding

3. The d selected packets are added using bitwise XOR operation to produce an encoded packet.
4. Steps 1–3 are repeated to produce new encoded packets.

This process is represented with a bipartite graph as shown in Figure 13.5, connecting the source packets to the encoded packets. When transmitting packets using a traditional erasure code both the sender and receiver are aware of the exact description of the encoding methods used. This is not the case with rateless codes, since the code is being generated concurrently with the transmission. Therefore, in order to be able to recover the source data from the encoded packets, it is necessary to transmit a description of the code structure together with the encoded packets. In other words, the receiver needs to know the exact graph, i.e., which k source packets (but not their values) have been used to generate a particular output packet. It could be accomplished by any of the following manner [11]:

- Sender and receiver can use identical pseudo-random number generators, seeded by a clock if it is synchronized.
- By a pseudo-random process which can determine the degree and the connections given a key, say κ. Then the sender generates κ to compute the output point and transmits the key in the header of the packet. If packet size is much larger than the key size (32 bits or so) then it adds only a small overhead.

13.5.1.2 LT Decoder

After having received a slightly larger than K encoded packets and the bipartite graph, the decoder tries to recover the source packets by a message passing algorithm. The LT decoding process can be described as follows:

1. From the set of received encoded packets, search for a packet y_i that is connected to only one source packet x_j. If there is no such encoded packet then this decoding algorithm cannot proceed further and fails to recover all the source packets.
2. Copy the value associated with the selected degree-one node to its only neighboring source packet, i.e., set $x_j = y_i$. Remove this degree-one output packet (y_i).
3. Add (modulo 2) x_j to all the encoded packets that are connected to this source node (x_j).
4. Remove all edges connected to the source packet (x_j).
5. Repeat steps 1–4 until all x_j have been determined.

This is a simple decoder, but it turns out that the degree distribution function, $\rho(d)$, is a critical part of the LT encoder and decoder design. The decoding process will fail at step 1 if there is no output packet of degree one. Therefore, for good decoding performance, an appropriate selection of degree distribution is required.

13.5.1.3 Degree Distributions

There are two fundamental tradeoffs in the design of degree distribution:

- At the end of each iteration of the decoding process, the set of output packets that have degree one called ripple should be small to avoid redundancy in encoded packets. On the other hand, if the size of the ripple is too small and drops to zero at any iteration then the decoding process halts at that point.
- The degree of encoded packets should be small to reduce the decoding complexity as it depends on the number of edges in the bipartite graph. On the other hand, if most of the encoded packets have small degrees then there is a high probability that some of the source packet may not be connected to any of the output packets, thereby they cannot be recovered.

In the following sections we discuss *ideal soliton distribution* and *robust soliton distribution* as proposed by Luby in [11].

13.5.1.4 Ideal Soliton Distribution

A good degree distribution should have the property that the ripple size never gets too small or too large. In order to satisfy the above criteria an ideal soliton distribution has been proposed as shown in Equation (13.1):

$$\rho(d) = \begin{cases} 1/K & \text{if } d = 1 \\ 1/d(d-1) & \text{for all } d = 2, 3, \ldots, K \end{cases} \tag{13.1}$$

This distribution displays ideal behavior in terms of the expected number of encoded packets required to recover the data. In this distribution we need at least $K \ln(K/\delta)$ edges in our bipartite graph. The decoding operation uses all the edges at least once. The overall running time for the decoding is $O(K \ln(K)$. For this distribution the expected degree is $\ln(K)$.

This distribution behaves well. However, it turns out that this is not a good distribution to use in reality. In practice a slight deviation from its expected behavior may create a situation where there is no output packet with degree one, and this will cause the decoding process to halt. In order to avoid this problem, the robust soliton distribution has been discussed.

13.5.1.5 Robust Soliton Distribution

While the ideal soliton distribution is optimal in some ways, it performs rather poorly in practice. However, it can be modified slightly to yield the robust soliton distribution denoted by $\omega(d)$ and defined in the following manner. It has two parameters, σ and δ. It is designed to ensure that the expected number of degree-one output packets is about

$$S = \sigma ln(K/\delta)\sqrt{K} \tag{13.2}$$

rather than one throughout the decoding process. The parameter δ is a bound on the probability that the decoding fails to run to completion. The parameter σ adjusts the size of the ripple (S). Now define

$$\tau(k) = \begin{cases} (S/K)\dfrac{1}{d} & \text{for} \quad d = 1, 2, \ldots, (K/S) - 1 \\ (S/K)ln\left(\dfrac{S}{\delta}\right) & \text{for} \quad d = K/R \\ 0 & \text{for} \quad d > K/R \end{cases} \tag{13.3}$$

Add the ideal soliton distribution $\rho(d)$ to $\tau(d)$ and normalize to obtain $\omega(d)$ as follows:

$$\omega(d) = \frac{\rho(d) + \tau(d)}{\beta} \tag{13.4}$$

where insert orange highlighted text, see hard copy [is the normalization constant chosen to ensure that $\omega(d)$ is a probability density function. This constant β also determines the number of encoded packets, N, required to recover the original K source packets with probability $1 - \delta$ where $N = \beta K$. These distributions are shown in Figure 13.6.

Luby's analysis [11] explains how the small-d end of τ has the role of ensuring that the decoding process gets started, the spike in τ at $d = K/S$ ensures that every input packet is likely to be connected to an output packet at least once. Luby's key result is that receiving $K + 2\ln(S/\delta)$ output packets ensures that all input packets can be recovered with probability at least $1 - \delta$. The only disadvantage is that the decoding complexity grows as $O(K\ln(K))$, but it turns out that such a growth in complexity is necessary to achieve capacity. However, slightly suboptimal codes with decoding complexity $O(K)$ called raptor codes have been developed by Shokrollahi [20].

13.5.2 Raptor Codes

Raptor codes [20] extend the idea of LT codes one important step further. LT codes suffer in that an averge degree of $O(\ln(K))$ is needed to cover every source packet with high probability. As a result, the decoding complexity also increases as K becomes large. Raptor codes are designed to achieve linear time encoding and decoding complexity. These codes accomplish this task by first precoding the source packets by an appropriate fixed erasure code, say some LDPC code, before passing it to an LT encoder as shown in Figure 13.7.

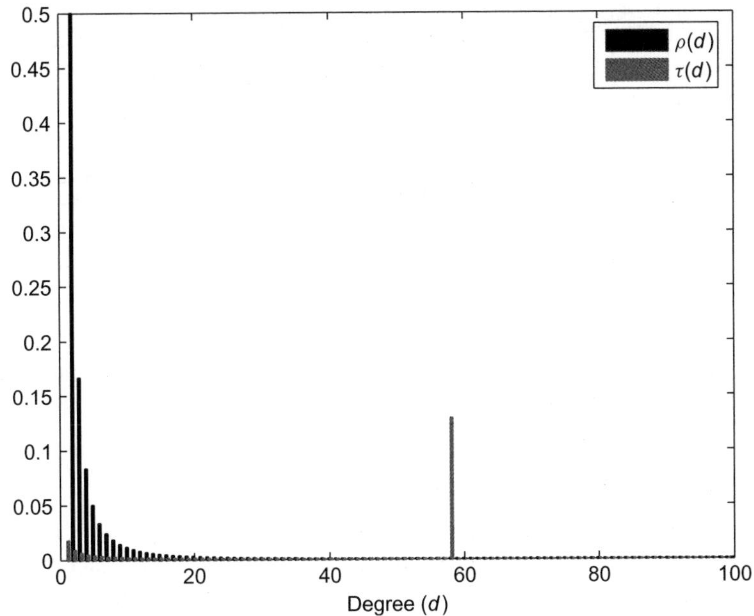

Figure 13.6 The distributions $\rho(d)$ and $\tau(d)$ for $K = 10000$, $\sigma = 0 : 15$ and $\delta = 0 : 1$

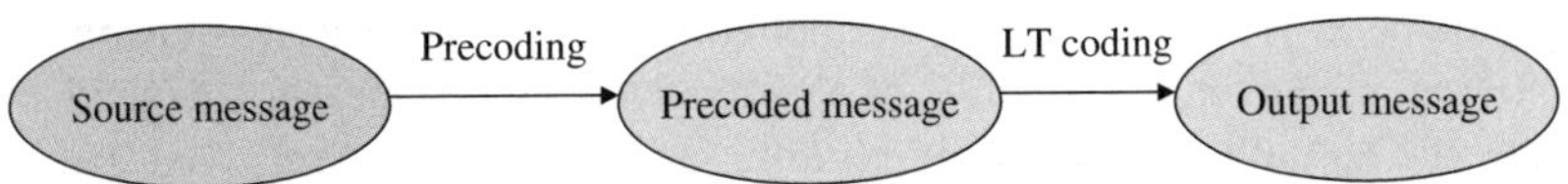

Figure 13.7 Raptor codes

Let the original message be M, and the pre-encoded version of the message be M' then this M' becomes the input to the LT encoder. Now, we don't need to recover every packet of M' in order to recover M, but just a constant fraction $1 - \varepsilon$ of the number of packets in M'. The main idea of the precoding can be explained as follows: LT codes have a complexity of the order of $\ln(K)$ per packet, because the average degree per output packet is $\ln(K)$. Since raptor codes use a precode, all input packets of the LT encoder need not be covered during encoding with an LT encoder. So raptor codes use LT codes with very small average degree which is generally constant and that makes the encoding and decoding linear in the number of packets.

For example, let the original message contain K packets, and after precoding message M' has K' packets. And suppose a fraction ε of K' remains uncovered during encoding with LT codes. Then if $K' \geq K/(1 - \varepsilon)$ then once a slightly more than K' of the output packets have been received, we can recover the original message M as shown in Figure 13.8. In this figure, $K = 9$ source packets are precoded to generate $K' = 13$ input packets for the LT code, which then produces output packets. Once $N = 15$ output packets have been received, we can see that three of the input packets have not been covered so far by this low degree LT encoder but still 10 of the input packets can be recovered and then the original message can be obtained from the precode decoder.

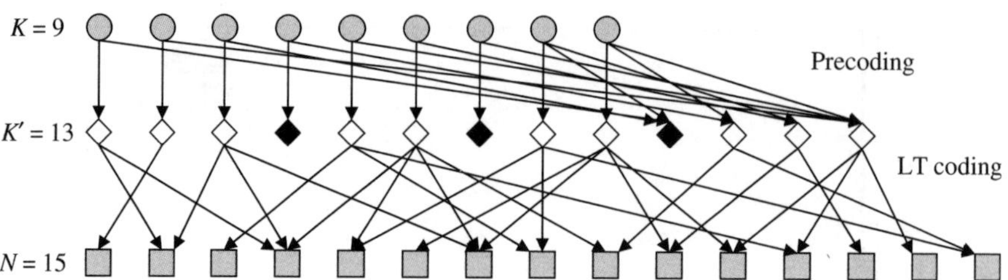

Figure 13.8 Raptor code schematic showing the uncovered input packets by a low degree LT code

Raptor codes currently offer the best approximation to a digital fountain code. Some analysis of LT codes and raptor codes are presented in [8], [19].

13.6 Multiple Description Codes

Multiple description coding (MDC) is a recent source coding approach that is gaining popularity as a viable coding mechanism for delivery of multimedia data over networks, especially because networks typically have multiple paths between the transmitter and the receiver. In such a scheme, several representations of the source, called descriptions, are generated such that the quality of the received signal is proportional to the number of descriptions received and not which descriptions were actually received [7], [18]. Such a coding scheme is well suited to packet networks and fading wireless channels. Unlike layered coding, however, the quality of the received video does not depend on which descriptions are actually received only on how many of them are received and hence we can consider MDC as a generalization of the layered coding.

MDC is a form of data partitioning, thus comparable to layered coding as it is used in MPEG-2 and MPEG-4. Yet, in contrast to MDC, layered coding mechanisms generate a base layer and n enhancement layers. The base layer is necessary for the media stream to be decoded, enhancement layers are applied to improve stream quality. However, the first enhancement layer depends on the base layer and each enhancement layer $n + 1$ depends on its subordinate layer n, thus can only be applied if n was already applied. Hence, media streams using the layered approach are interrupted whenever the base layer is missing and, as a consequence, the data of the respective enhancement layers is rendered useless. The same applies for missing enhancement layers. In general, this implies that in lossy networks the quality of a media stream is not proportional to the amount of correct received data. A general example of a MDC system with two descriptions is given in Figure 13.9.

In a secondary usage scenario, this type of coding could be useful in recovering the original multimedia to a certain quality. If a large number of subchannels is captured by PUs then the channel coding schemes discussed in the previous sections may not be able recover enough number of packets or it may take an unacceptably longer time to recover sufficient number of packets for the LT decoding process to start. But if the original media is converted into descriptions with MDC then we can still construct the image up to a certain quality depending upon the number of descriptions received. Some MDC video coding algorithms have been discussed in [18], [21].

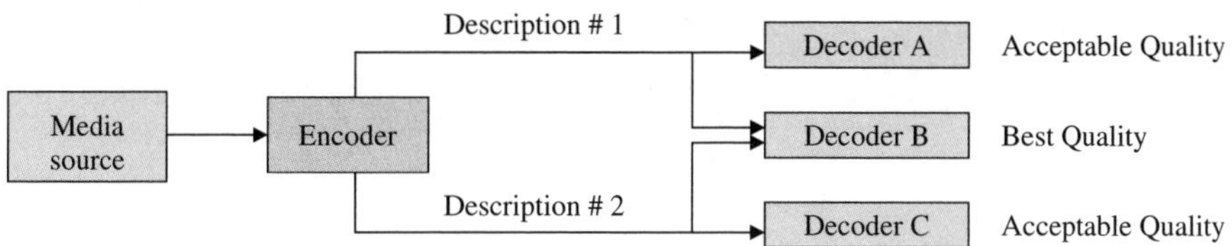

Figure 13.9 An example of MDC system with two descriptions

13.7 Applications

In this section, we discuss the secondary spectrum access with LT codes based on [9]. In the secondary usage scenario, an SU selects a set of subchannels from the PU bands as discussed in Section 13.2. The SU is required to vacate the subchannel as soon as the corresponding PU becomes active on that subchannel. This forces the secondary user to loose packets on that subchannel. To compensate for the loss caused by the PU appearance, the source packets are encoded with LT codes. Let the secondary user have a message M of K packets to transmit. And let the LT decoder need at least N packets in order to recover original K packets with probability $1 - \delta'$. Then in order to compensate for the loss due to PU appearance, we add some more redundancy, say X, which depends on PU arrival probability p. If PU arrival is quite frequent then we need to use a high value of X, if PU arrives occasionally then a small value of X is sufficient.

As discussed in [26], let one packet be transmitted per subchannel, then the total number of subchannels used by SU is $Q = N + X$, and this transmission will be successful only if at most X of the subchannels are lost due to PU appearance. Therefore, the probability of successful transmission for the secondary users is given by

$$P_{success} = \sum_{i=0}^{X} \binom{N+X}{i} p^i (1-p)^{N+X-i} \tag{13.5}$$

Let T_{frame} be the time required for transmission of each packet and W be the bandwidth per subchannel. Then the spectral efficiency of the secondary user link (η) can be computed by

$$\eta = \frac{(1-\delta) N P_{success}}{S \times W \times T_{frame}} \tag{13.6}$$

The LT decoding error probability δ', evaluated via simulation for robust soliton distribution with parameters $\sigma = 0:1$ and $\delta = 0:5$ is shown in Figure 13.10 for different values of K and redundancy factor $\chi = X/K$.

As seen in Figure 13.10, the LT decoding error is very high for small values of redundancy, and after some value of χ, it becomes very small and almost constant. So if a large number of subchannels are captured by PUs, then the total received overhead will be small and will result in a high LT decoding error probability. But if source packets are converted into descriptions with multiple description coding then even with a small overhead, the original data can be recovered up to a certain quality. Various applications of fountain codes in video streaming, broadcast, multicast and parallel downloading are presented in [11, 16, 23].

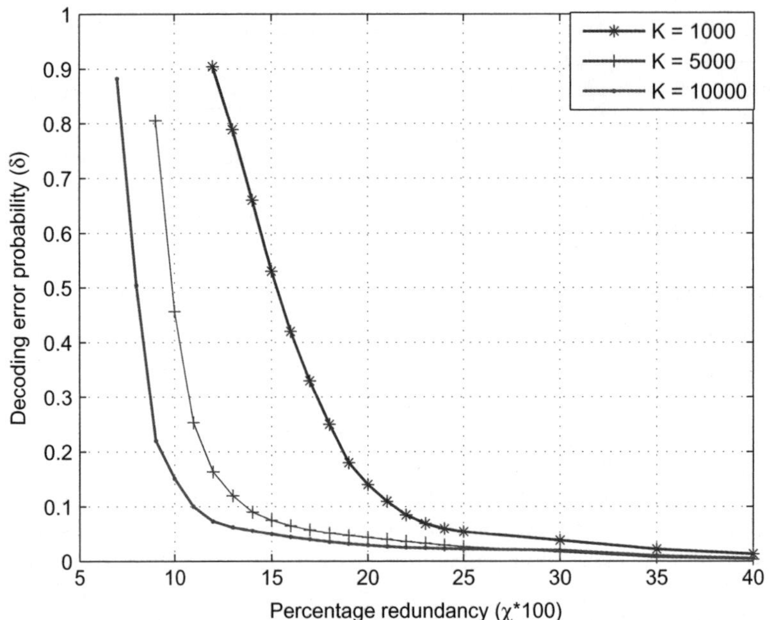

Figure 13.10 LT decoding error probability

13.8 Conclusion

This chapter gives an overview of some coding aspects of dynamic spectrum access with cognitive radios. The spectrum pooling concept is presented for secondary usage of spectrum. The loss on a secondary user link is modeled as an erasure and erasure tolerant coding has been proposed to compensate for it.

Traditional erasure codes such as Reed–Solomon codes and tornado codes have been discussed as fixed erasure codes. Then digital fountain codes such as LT codes and raptor codes have been presented as rateless codes. Multiple description codes have been discussed as a source coding approach for robust communication in secondary usage unreliable scenario. Their applications in secondary spectrum access with cognitive radios have been discussed. Some simulation results on LT codes decoding error probability are also presented.

References

[1] Akyildiz, I.F., Lee, W., Vuran, M.C., Mohanty, S. (2006) NeXt generation/dynamic spectrum access/ cognitive radio wireless networks: a survey. *Computer Networks*, **50**, 2127–59.

[2] Buddhikot, M.M., Kolody, P., Miller, S., Ryan, K. and Evans, J. (2005) DIMSUMNet: new directions in wireless networking using coordinated dynamic spectrum access. *Proceedings of the IEEE WoWMoM*, p. 7885, June.

[3] Cabric, D., Mishra, S.M., Willkomm, D., Broderson, R.W. and Wolisz, A. (2005) A cognitive radio approach for usage of virtual unlicensed spectrum. *Proceedings of the 14th IST Mobile Wireless Communications Summit*, June, Dresden, Germany.

[4] DF RaptorTM Technology (2005) *Advanced FEC technology for streaming media and data distribution applications*. A white paper from www.DigitalFountain.com.

[5] FCC (2002) Spectrum Policy Task Force, ET Docket no. 02-135, November.
[6] FCC (2003) ET Docket No 03-222, Notice of proposed rule making and order, December.
[7] Goyal, V.K. (2001) Multiple description coding: compression meets the network. *IEEE Signal Processing Magazine*, **18**(5), 74.
[8] Karp, R., Luby, M. and Shokrollahi, A. (2004) Finite length analysis of LT codes. *Proceedings of the ISIT*, p. 37.
[9] Kushwaha, H. and Chandramouli, R. (2007) Secondary spectrum access with LT codes for delay-constrained applications. *Proceedings of the IEEE Consumer Communications and Networking Conference*, January.
[10] Luby, M. (2002) LT codes. *Proceedings of the 43rd Annual IEEE Symposium on Foundations of Computer Science*, pp. 271–82, November.
[11] Luby, M., Watson, M., Gasiba, T., Stockhammer, T. and Xu, W. (2006) Raptor codes for reliable download delivery in wireless broadcast systems. *Proceedings of the Consumer and Communications Networking Conference (CCNC 2006)*, Las Vegas, NV.
[12] Mackay, D.J.C. (2003) *Information Theory, Inference, and Learning Algorithms*, Cambridge University Press, Cambridge, UK.
[13] MacKay, D.J.C. (2005) Fountain codes. *IEEE Communications*, **152**(6), 1062–8.
[14] McHenry, M. (2002) *Dupont Circle Spectrum Utilization During Peak Hours*, The Shared Spectrum Company.
[15] Mitola, J. (1995) The software radio architecture. *IEEE Communications*, **33**(5), 26–38.
[16] Mitzeumacher, M. (2004) Digital fountains: a survey and look forward. *Proceedings of the IEEE Information Theory Workshop*, October, San Antonio, Texas.
[17] Motola, J. (1999) Cognitive radio for flexible mobile multimedia communications. *Proceedings of the IEEE International Workshop on Mobile Multimedia Communications*, November.
[18] Puri, R. and Ramchandran, K. (1999) Multiple description source coding using forward error correction codes. *Proceedings of the IEEE* Asilomar Conference on Signals, Systems, and Computers, October, Asilomar, CA.
[19] Shokrollahi, A. (2004) Raptor codes. *Proceedings of the ISIT*, p. 36.
[20] Shokrollahi, A. (2006) Raptor codes. *IEEE Transactions on Information Theory*, **52**(6).
[21] Somasundaram, S. and Subbalakshmi, K.P. (2003) 3-D multiple description video coding for packet switched networks. *Proceedings of the IEEE International Conference on Multimedia and Expo*, July, Baltimore, Maryland.
[22] Vukobratovic, D. and Despotovic, M. (2005) On the packet lengths of rateless codes. *Proceedings of the EUROCON 2005 – The International Conference on Computer as a Tool*, November, Belgrade, Serbia & Montenegro.
[23] Wagner, J., Chakareski, J. and Frossard, P. (2006) Streaming of scalable video from multiple servers using rateless codes. *Proceedings of the IEEE International Conference on Multimedia and Expo*, July.
[24] Weiss, T. and Jondral, F. (2004) Spectrum pooling: an innovative strategy for the enhancement of spectrum efficiency. *IEEE Commununications*, **42**, S8S14.
[25] *White paper on regulatory aspects of software dened radio* (2000) SDR forum document number SDRF-00-R-0050-v0.0.
[26] Willkomm, D., Gross, J. and Wolisz, A. (2005) Reliable link maintenance in cognitive radio systems. *Proceedings of the IEEE Symposium on New Frontiers in Dynamic Spectrum Access Networks (DySPAN 2005)*, November, Baltimore, MD.

Index

Cognitive Networks: Towards Self-Aware Networks Edited by Qusay H. Mahmoud